Surface Engineered Materials and Applications

Volume I

Surface Engineered Materials and Applications
Volume I

Edited by **Reece Hughes**

𝒞LANRYE
INTERNATIONAL

New Jersey

Published by Clanrye International,
55 Van Reypen Street,
Jersey City, NJ 07306, USA
www.clanryeinternational.com

Surface Engineered Materials and Applications
Volume I
Edited by Reece Hughes

International Standard Book Number: 978-1-63240-475-6 (Hardback)

Printed in the United States of America.

Contents

Preface IX

Chapter 1 **Formation Process and Properties of Phytic Acid Conversion
Coatings on Magnesium** 1
Jian-Rui Liu, Yi-Na Guo, Wei-Dong Huang

Chapter 2 **Influence of Coolant in Machinability of Titanium Alloy (Ti-6Al-4V)** 8
Nambi Muthukrishnan, Paulo Davim

Chapter 3 **Finite Element Analysis of the Material's Area Affected during a
Micro Thermal Analysis Applied to Homogeneous Materials** 14
Yoann Joliff, Lénaïk Belec, Jean-François Chailan

Chapter 4 **The Optical Parameters of $Zn_xCd_{(1-x)}Te$ Chalcogenide Thin Films** 22
Umeshkumar P. Khairnar, Sulakshana S. Behere, Panjabrao H. Pawar

Chapter 5 **Effect of Nitriding on Wear Behavior of Graphite Reinforced
Aluminum Alloy Composites** 27
Bhujang Mutt Girish, Bhujang Mutt Satish,
Hanyalu Ramegowda Vitala

Chapter 6 **Analysis of Mobile Phone Reliability Based on Active
Disassembly Using Smart Materials** 34
Zhifeng Liu, Liuxian Zhao, Jun Zhong, Xinyu Li, Huanbo Cheng

Chapter 7 **Modeling of Adsorption of Bi(III) from Nitrate Medium by
Impregnated Resin D2EHPA/XAD-1180** 42
Nasr-Eddine Belkhouche, Nacera Benyahia

Chapter 8 **Effect of Self-Assembled Monolayers on the Performance of
Organic Photovoltaic Cells** 47
Hanène Bedis

Chapter 9 **Effect of Annealing on Structural, Morphological, Electrical and
Optical Studies of Nickel Oxide Thin Films** 56
Vikas Patil, Shailesh Pawar, Manik Chougule, Prasad Godse,
Ratnakar Sakhare, Shashwati Sen, Pradeep Joshi

Chapter 10 **Development of the Biopolymeric Optical Planar Waveguide with Nanopattern**
Seung H. Yoon, Won T. Jeong, Kyung C. Kim, Kyung J. Kim, Min C. Oh, Sang M. Lee   63

Chapter 11 **AZ91 Magnesium Alloys: Anodizing of Using Environmental Friendly Electrolytes**
N. A. El Mahallawy, M. A. Shoeib, M. H. Abouelenain   69

Chapter 12 **H$_2$ and CH$_4$ Sorption on Cu-BTC Metal Organic Frameworks at Pressures up to 15 MPa and Temperatures between 273 and 318 K**
Yves Gensterblum   81

Chapter 13 **High Pressure Water-Jet Technology for the Surface Treatment of Al-Si Alloys and Repercussion on Tribological Properties**
Md. Aminul Islam, Zoheir Farhat, Jonathon Bonnell   88

Chapter 14 **Microstructure, Corrosion, and Fatigue Properties of Alumina-Titania Nanostructured Coatings**
Ahmed Ibrahim, Abdel Salam Hamdy   97

Chapter 15 **Electrochemical Characterization of Plasma Sprayed Alumina Coatings**
Magdi F. Morks, Ivan Cole, Penny Corrigan, Akira Kobayashi   103

Chapter 16 **Effect of Conduction Pre-heating in Au-Al Thermosonic Wire Bonding**
Gurbinder Singh, Othman Mamat   108

Chapter 17 **Finite Element Analysis of Elastic-Plastic Contact Mechanic Considering the Effect of Contact Geometry and Material Properties**
Abodol Rasoul Sohouli, Ali Maozemi Goudarzi, Reza Akbari Alashti   112

Chapter 18 **Oxidation Behaviour of a Newly Developed Superalloy**   117
I. V. S. Yashwanth, I. Gurrappa, H. Murakami

Chapter 19 **Surface and Bulk Defects in Cr-Mn Iron Alloy Cast in Metal and Sand Moulds: Characterization by Positron Annihilation Techniques**
Parthasarathy Sampathkumaran, Subramanyam Seetharamu, Chikkakuntappa Ranganathaiah, Jaya Madhu Raj, Pradeep Kumar Pujari, Priya Maheshwari, Debashish Dutta, Kishore   123

Chapter 20 **Improvement in Tribological Properties of Surface Layer of an Al Alloy by Friction Stir Processing**
Soheyl Soleymani, Amir Abdollah-zadeh, Sima Ahmad Alidokht   131

Chapter 21   The Behaviour of Superalloys in Marine Gas Turbine
Engine Conditions   137
I. Gurrappa, I. V. S. Yashwanth, A. K. Gogia

Chapter 22   Interfacial Actions and Adherence of an Interpenetrating Polymer
Network Thin Film on Aluminum Substrate   143
Weiwei Cui, Dongyan Tang, Jie Liu, Fan Yang

Chapter 23   Importance of Surface Preparation for Corrosion Protection
of Automobiles   149
Narayan Chandra Debnath

Chapter 24   Epitaxial Ge Growth on Si(111) Covered with Ultrathin SiO$_2$ Films   161
Alexander A. Shklyaev, Konstantin N. Romanyuk,
Alexander V. Latyshev

Chapter 25   Dynamic Impact Absorption Behaviour of Glass Coated with
Carbon Nanotubes   171
Prashant Jindal, Meenakshi Goyal, Navin Kumar

Chapter 26   Effect of Carbon Content on Ti Inclusion Precipitated in
Tire Cord Steel   176
Yuedong Jiang, Jialiu Lei, Jing Zhang, Rui Xiong, Feng Zou,
Zhengliang Xue

Chapter 27   Characterization of Pectin Nanocoatings at Polystyrene and
Titanium Surfaces   180
Katarzyna Gurzawska, Kai Dirscherl, Yu Yihua, Inge Byg,
Bodil Jørgensen, Rikke Svava, Martin W. Nielsen,
Niklas R. Jørgensen, Klaus Gotfredsen

Permissions

List of Contributors

Preface

The application of surface engineering is vital to the success of almost every product. A sub-discipline of materials science, Surface Engineering essentially deals with the surface of all solid matter. It is the surface phase of any solid which interacts with the surrounding environment. Due to this interaction, the surface phase can degrade over time due to various factors such as corrosion, wear and tear and numerous others. This would definitely affect the processes in which the surface is involved. Thus, there's a need to make the surface robust to the environment. Surface engineering involves altering the properties of the Surface Phase in order to reduce its degradation, by making it immune to environmental factors. Basically, there are 3 ways of modifying surface properties: by thermal/mechanical means, altering surface chemistry and by coating a new material to the surface.

The structures which are created in Surface Engineering are not found naturally. Artificial methods such as plasma treatment, surface functionalization and activation and application of thin film coatings are used to treat surfaces. Thus, Surface Engineering has applications in other disciplines such as mechanical, electrical and chemical engineering as well.

Surface Engineering finds its applications in a diverse array of fields. Wear-resistant and corrosion-resistant properties at the required substrate surfaces have been developed using these principles. In construction, road surfacing depends to a large extent on the fundamentals of Surface Engineering. Biomedical, textile, petroleum, automotive and aerospace, all these branches apply techniques of Surface Engineering.

I would like to thank all the contributing authors and the publishing house.

Editor

Formation Process and Properties of Phytic Acid Conversion Coatings on Magnesium

Jian-Rui Liu, Yi-Na Guo, Wei-Dong Huang

State Key Laboratory of Solidification Processing, Northwestern Polytechnical University, Xi'an, P. R. China

ABSTRACT

A chromium-free conversion coating treatment for magnesium by phytic acid solution was studied. The formation process of phytic acid conversion coating was studied through measuring the open circuit potential (OCP) and weight change of the pure magnesium in the different conversion treatment time. The morphologies and compositions of the coatings were determined by SEM and EDS respectively. The conversion coating has the multideck structure with net-like morphology which is similar to the chromate conversion coating, and is mainly composed of Mg, P, O and C. The contents of C and P and the size of the cracks in different layers decrease from the external layer to the inner layer. The hydroxyl groups and phosphate carboxyl groups in the coating which have the similar properties to organic paintcoat are beneficial to the combination of substrate and organic paintcoat. The formation mechanism and thickness variation of the conversion coatings are also discussed.

Keywords: Magnesium, Chemical Conversion Coatings, Phytic Acid, Formation Mechanism

1. Introduction

As the lightest structural alloys, magnesium alloys are attractive for automotive and aerospace applications to achieve significant reduction in energy conservation and green-house gas emission because of its good combination of mechanical properties and castability. However, the poor corrosion resistance of magnesium alloys which results from the properties of metal magnesium is one of the critical factors limiting its wide applications, especially in an environment containing corrosive ingredients [1-3]. Surface treatments can improve the corrosion resistance of magnesium and its alloys, which have been extensive researched in recently years [4]. In the surface treatment methods, Chemical conversion treatment is an effective and simple method, such as chromate [5], permanganate [6], phosphate [7,8], phosphate/permanganate [9,10], etc. Chromate conversion treatment as a conventional chemical conversion method, it has been widely used in industry for long time. But the conversion solution containing toxic hexavalent chromium carcinogen is harmful to the environment, which has been restricted and forbidden in many countries. The pollution level of the treatment solutions containing phosphate or permanganate is less than that of chromate, although the metal ions and PO_4^{3-} existing in the solutions have certain

harmfulness to environment. Therefore it is urgently needed to develop new environment-friendly surface treatments for magnesium and its alloys.

Phytic acid ($C_6H_{18}O_{24}P_6$, inositol hexaphosphate ester), an innocuity macromolecule natural compound with 24 oxygen atoms, 12 hydroxyl groups and 6 phosphate carboxyl groups, is extensively applied in the area of surface protection of metals due to its particular physical and chemical properties which has powerful capability of chelating with many metals ions [11,12]. The researches [13-16] indicate that the chemical conversion coatings based on phytic acid solution have better corrosion resistance for magnesium alloys, which can substitute for those harmful methods. It can be seen from these research results that the properties of conversion coatings are closely related to their microstructure and compositions, which greatly depend on the conversion treating parameters such as treating time, temperature and concentration of the solution.

In order to further know the formation mechanism of phytic acid conversion coating and the influence of processing parameters on the microstructure and corrosion resistance of the chemical conversion coating, in this study, the formation process of the phytic acid conversion coating was investigated by electrochemical and weighting method. The detailed micro morphology evolution of the conversion coating during the treatment procedure was also examined. Furthermore, the microstructure and the corrosion procedure of the conversion coating were studied by scanning electron microscopy (SEM) and electron energy spectrum (EDS). The experimental results could provide theoretical references for the analysis and study of phytic acid conversion coating formed on magnesium alloys to accelerate the application of phytic acid in the surface protection of magnesium alloys in the future.

2. Experimental

2.1. Conversion Treating Procedure

The magnesium used in this study was commercial magnesium ingot, and its main compositions are (in wt%): Mg > 99.95, Al 0.009, Mn 0.0055, Fe 0.0027, Cu 0.0022, Ni 0.0004, Si < 0.01. Phytic acid and NaCl used here were analytical reagent, and the water was de-ionized water.

The Magnesium ingots were cut into cubic specimens with the dimension of $10 \times 10 \times 10$ mm^3. Each cube was embedded into epoxy resin with only one side of exposure as the working area. Then the working area was polished with grade 600 and grade 1000 carborundum papers, cleaned with de-ionized water and acetone, dried in hot air. After that, the specimens were immersed in phytic acid solution to obtain phytic acid conversion coating on their surface. Subsequently, the samples were taken out, rinsed by floating de-ionized water, and dried at room temperature. The weight of the sample was measured by an electronic balance with an accuracy of 0.1 mg during the conversion treatment process.

2.2. Characterization

The surface morphology of the conversion coating was observed on a scanning electron microscopy (SEM, Model Discoverer S570, made by Hitachi Company of Japan). Meanwhile the compositions of the coating were characterized using the energy dispersive spectrum (EDS, Model Discoverer S360, made by Cambridge Company of UK). Corrosion properties of the conversion coatings in neutral 3.5 wt% NaCl solutions were measured using the immersion experiments.

The Open Circuit-potential curve used to study the formation mechanism of the conversion coatings on magnesium were measured using CHI600B electrochemical workstation at $20 \pm 1°C$. Measurements were conducted in the phytic acid solution using a traditional three-electrode system, a saturated calomel electrode (SCE) was used as reference electrode, a platinum (Pt) was used as auxiliary electrode, and the sample was used as working electrode.

3. Results and Discussion

3.1. Formation Process of Phytic Acid Conversion Coating

The formation process of phytic acid conversion coating was studied by detecting the open circuit potential (OCP) and weight change of the pure magnesium in the different conversion treatment time. **Figure 1** shows the open circuit potential-time curve and weight-time curve of pure magnesium immersed in 0.5 wt% phytic acid solution under pH = 3 at 25°C. It can be seen from **Figure 1(a)** that the OCP of the specimen increased during the whole conversion treatment procedure from approximately –2.20 V to about –2.05 V, which proved that a phytic acid conversion coating formed on the surface of magnesium. In the first 1600 s, the OCP of magnesium presented a relatively swift increase and accompanied by the intense formation of hydrogen, which indicated the phytic acid conversion coating was forming quickly during this period. As can be seen, the OCP of the magnesium

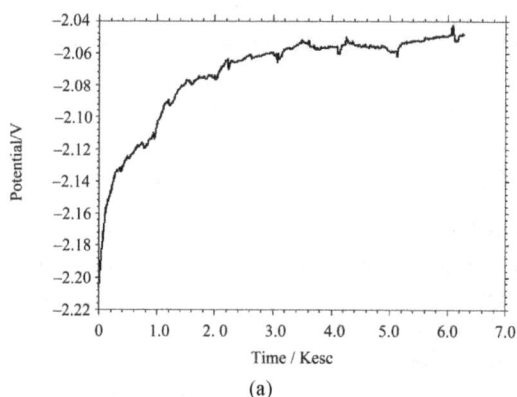

(a)

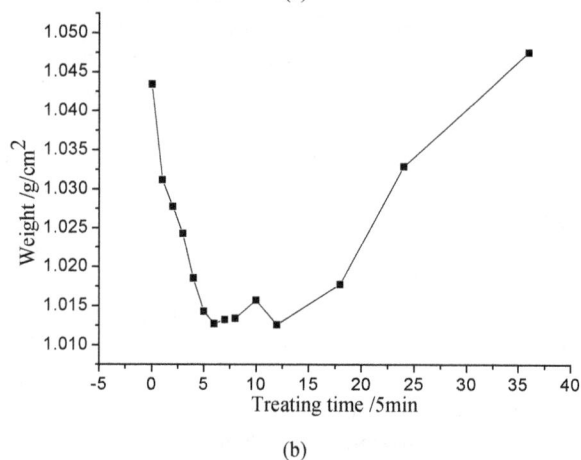

(b)

Figure 1. Phytic acid conversion coating on magnesium surface during the conversion treatment processing. (a) Open circuit potential-time curve; (b) Weight gain to time curve.

after 1600 s became relative stabilization. It suggested that after 1600 s, the transmission of ions and the release of hydrogen were restrained as the conversion coating had basically formed, which result in the formation speed slowing up of the coating. The potential fluctuation of OCP during this period was due to the formation and exfoliation of the coating. It can be seen from **Figure 1(b)** that the weight of specimen decreased markedly before the first 1500 s, then tended to remain constant from 1500 s to 4500 s, after that, it began to recover. The decrease of specimen's weight at first can be attributed to the reasons. Firstly, Mg dissolves in the acidic solution forming Mg^{2+} ions. Secondly, in the beginning, the deposition rate of magnesium phytate compounds is not as quick as the ionization rate of magnesium. The magnesium phytate compounds begin deposition on the substrate surface to form conversion coating when the concentration of magnesium phytate compounds in solution achieves a certain level. The slight weight change between 1500 s and 4500 s suggests that the whole reaction system reaches dynamic equilibrium, a relatively thick conversion coating forms during this period. After that, the transmission of magnesium ions becomes very difficult because of the coating. As a result, the effect of the deposition of magnesium phytate compounds on specimen's weight change turns to be more obvious.

Figure 2 shows the SEM micrographs of the coatings formed in different times. It displays the evolution of the phytic acid conversion coating during treatment process. Seeing from **Figure 2(a)**, there were some white flocs

formed and distributed non-uniformly on the surface of specimen. The chemical components of the flocs were consisted of C, P, O and Mg through EDS analysis, which indicated that the white flocs were magnesium phytate compounds. As treatment time prolonged, the white flocs increased and connected with each other to form a relative uniformity films of magnesium phytate on the specimen surface. As can be seen from **Figures 2(b,c)**, the thickness of the layer was so thin at this moment that the polishing trace could be seen.

Seeing from **Figure 2(d)**, when the treatment time reached 5 min, an integral coating with some meshy cracks on it was formed on the specimen, which was similar with other kinds of conversion coatings [17,18]. The grinding trace couldn't be seen there, which suggested that the phytic acid conversion coating became comparatively thick and flat. Additionally, it was notable that magnesium phytate compounds still existed on the surface of specimen. After 5 min, it can be seen that the crack quantity of the conversion coating decreased with the increase of treatment time (**Figures 2(e-h)**). The conversion coating remained the "dry-mud" morphology all the time. The quantity and magnitude of the cracks changed with the variation of the processing parameters. The cracks could become very small if the specimen was treated under the suitable processing parameters.

3.2. The Mechanism Analysis of Phytic Acid Conversion Coating Forming

Phytic acid is a large molecular polyatomic acid com-

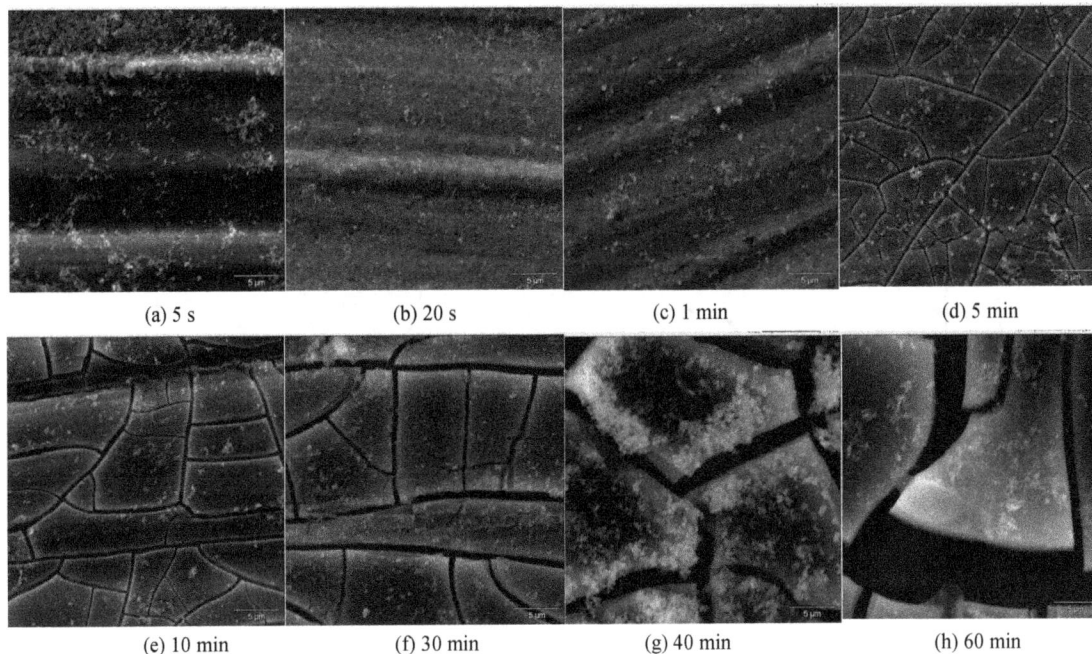

(a) 5 s (b) 20 s (c) 1 min (d) 5 min (e) 10 min (f) 30 min (g) 40 min (h) 60 min

Figure 2. The surface SEM morphologies of the specimen treated in the bath containing 0.5 wt% phytic acid under pH = 3 at 25°C for different time.

pound. It may present multilevel ionization in water as follows [16]

$$RH_{12} + H_2O = RH_{11}^- + H_3O^+ \qquad (1)$$

$$RH_{11}^- + H_2O = RH_{10}^{2-} + H_3O^+ \qquad (2)$$

$$\cdots\cdots$$

$$RH^{11-} + H_2O = R^{12-} + H_3O^+ \qquad (3)$$

Here R refers to $C_6H_6O_6(PO_3)_6$. The ionization degree of phytic acid depends on the pH values of the solution. At the same time of phytic acid ionization, the element Mg on substrate surface occur anodic reaction. Mg dissolves in the solution forming Mg^{2+}.

$$Mg - 2e \rightarrow Mg^{2+} \qquad (4)$$

The H_3O^+ in solution happen cathode reaction with the electrons of Mg releasing to form hygrogen escaping.

$$2H_3O^+ + 2e \rightarrow 2H_2O + H_2 \uparrow \qquad (5)$$

The Mg^{2+} ions react with phytate groups to form magnesium phytate compounds.

$$RH_{10}^{2-} + Mg^{2+} \rightarrow MgRH_{10} \qquad (6)$$

When the concentration of insoluble magnesium phytate compounds reached a certain level, it gradually deposited on the surface of magnesium substrate forming the discrete flocs. With the deposit increase, the discrete flocs combined each other through chemiadsorption effect to form the continuous conversion coatings. The conversion coatings are interlaced with each other, because one of the six phosphate carboxyl groups in phytic acid locates in *a* place and the others locate in *e* place, among them four of which are in the same level. This kind of coating is relatively more compact than the natural oxide film. So it can effectively insulate the contact of

magnesium substrate and environment media, and improve the corrosion resistance of magnesium and its alloys.

3.3. The Morphology and Composition of Phytic Acid Conversion Coating

Infrared absorption spectroscopy method was conducted to further investigate the chemical composition of the conversion coating, and the result is shown in **Figure 3**. Seeing from **Figure 3**, the strong and wide absorption peak at 3392.38 cm^{-1} was attributable to hydroxyl group, the band at 1646.81 cm^{-1} was probably due to the carboxyl group, and the band at 1052.33 cm^{-1} was assigned to the phosphate radical or hydrogen phosphate radical. The result confirmed further that the conversion coating was formed mainly by the deposition of the magnesium phytate compounds. The phytic acid conversion coatings have the similar properties with organic paintcoat because of the hydroxyl groups and phosphate carboxyl groups, which are beneficial to the combination of substrate and organic coating.

Figure 4 exhibits the transverse section and the line scan of the phytic acid conversion coating formed in solution containing 0.5% phytic acid at 25°C and pH = 3 for 30min. It can be seen from **Figure 4(a)** that the conversion coating was compact and well connected with magnesium substrate. Seeing from **Figures 4(b-e)**, the content of Mg reached its minimum level at the interface between substrate and conversion coating. The changes of the content of C, P and O were similar to each other, all of them increased at the interface between substrate and conversion coating, and then reached the peaks. The slight increase of the content of C, P and O at the interface can be explained that the compounds of the conversion coating were embed into the aperture between the

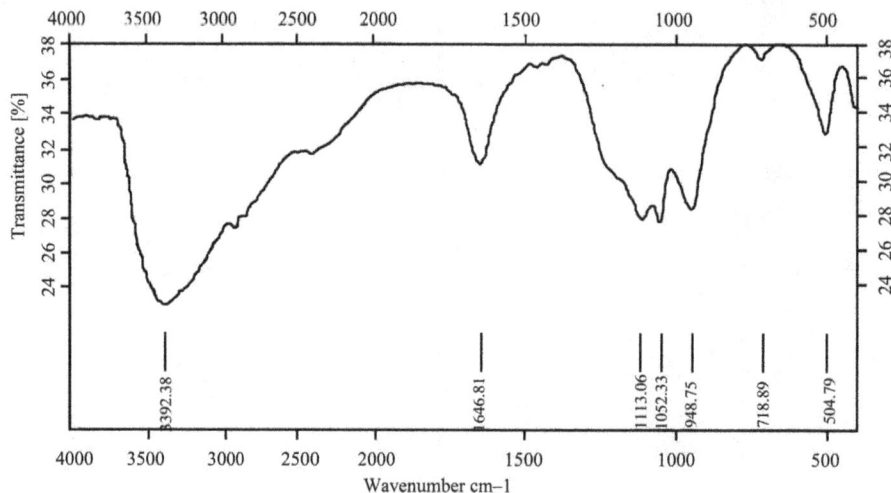

Figure 3. Infrared absorption spectroscopy of the phytic acid and magnesium formed compound.

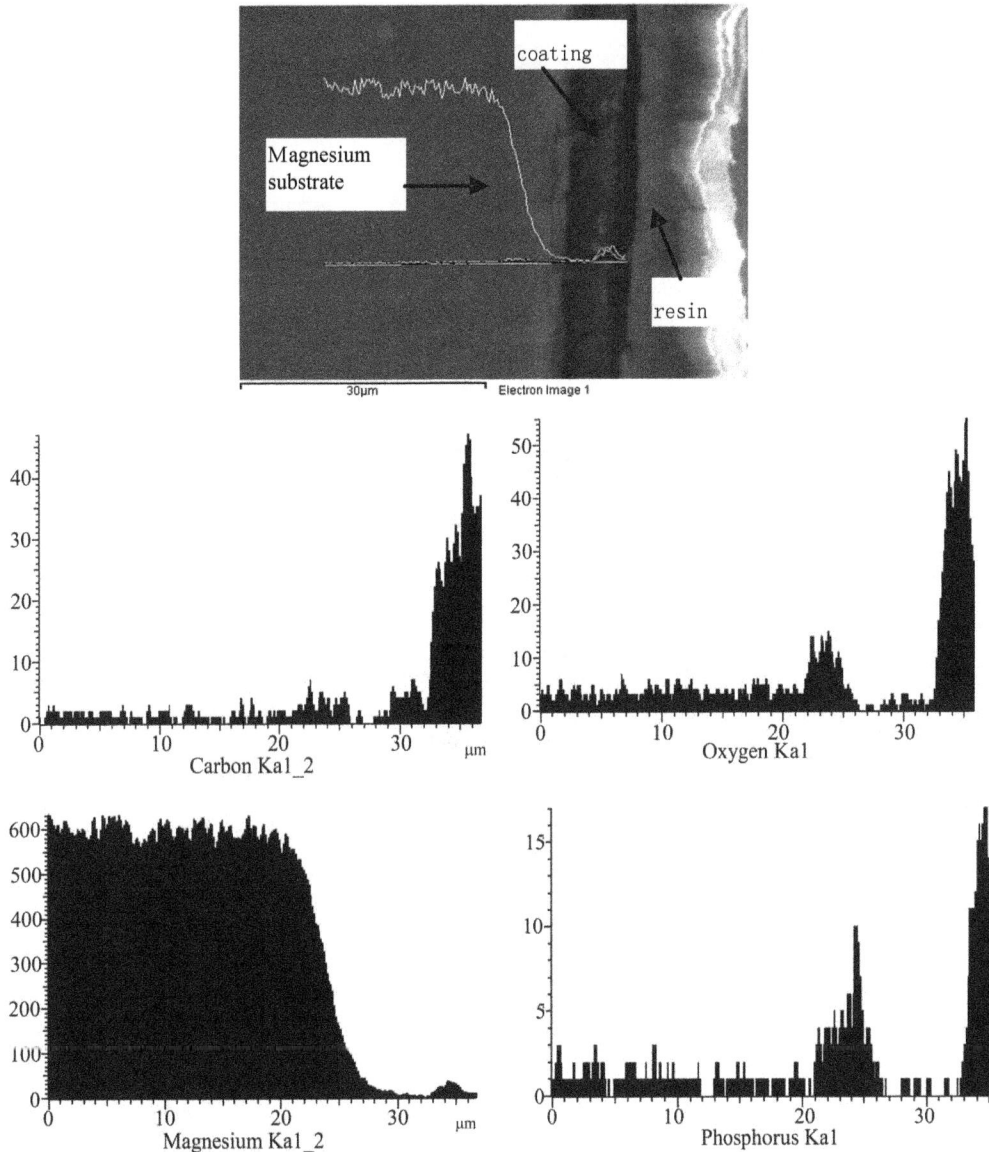

Figure 4. Distribution of X-ray intensity of each element detected by electron microprobe analysis (EMA) in the magnesium substrate and the conversion coating formed in solution containing 0.5wt.%phytic acid at 25°C under pH = 3

substrate and the conversion coating when the sample was grinded. The thickness of the conversion coating varied from 1.0 μm to 15 μm based on the processing parameters.

The above sample with the phytic acid conversion coating had been immersed in the 3.5 wt% NaCl solution for 120 h, then taken out, washed with distilled water, dried at room temperature and finally analyzed by SEM and EDS again. The results are shown in **Figure 5** and **Table 1**. It can be seen from **Figure 5** that the conversion coating after immersion became more disorganized, and showed multideck structure. In addition, the cracks existed anywhere and the size of them became smaller from the external layer to the inner layer. In **Figure 5**, points 1, 2 and 3 denoted the different locations from the inner layer to the external layer. The analysis results of element content for the three points by EDS were listed in **Table 1**. Seeing from **Table 1**, the different layers of corroded conversion coating were also mainly consisted of C, P, O and Mg. But the content of the same element was distinct in different locations. The content of Mg and O increased gradually from the external layer to the inner layer, among them the mass percentage of O changed slightly. On the contrary, the content of C and P gently decreased from the external layer to the inner layer. The reason of this phenomenon is that once a conversion

Table 1. The The content of C, O, P and Mg in the different points by EDS.

Element	Element/wt%			Atom/%		
	1	2	3	1	2	3
C	5.74	13.64	19.36	9.23	20.8	30.67
O	43.33	41.99	30.94	52.30	48.19	36.79
Mg	40.74	29.22	12.58	32.37	22.07	9.85
P	7.04	13.89	35.66	4.39	8.24	21.91

Figure 5. Surface morphology of the conversion coating corroded in 3.5 wt% NaCl solution for 120 h.

coating was formed on the surface of magnesium, it would become difficult that the phytic acid molecules or anions in the solution diffused to the inside of the layer and reacted with the substrate. It can be inferred that the cracks of the conversion coating did not stretch to the substrate of magnesium according to the component of the point in the inner layer (point 1). This indicated that the conversion coating could prevent magnesium from being corroded farther.

The multideck structure of the conversion coating was closely related to the forming of the cracks. The phytic acid molecules or anions in the solution could pass the conversion layer easily to react and form multideck structure because of the cracks in the conversion coating.

There were two possible reasons for the cracks of the conversion coating. Firstly, the amount of hydrogen would form because of the reaction between magnesium and phytic acid during the conversion treatment. The hydrogen transiting the surface layers resulted in the formation of cracks. Secondly, in the process of drying, the evaporation of water in the conversion coating gradually caused the stress to make the cracks.

The experimental results indicated that as the time of the conversion coating being immersed in the 3.5 wt% NaCl solution increased, the external layer of conversion coating gradually fell off and the inner layer would ex-

pose. However, the inner layer was also effectively protective, because the size and quantity of cracks on it were relatively small.

4. Conclusions

Phytic acid conversion coating has the multideck structure with netlike morphology which is similar to the chromate conversion coating. The elements of every layer are mainly composed of Mg, P, O and C. The contents of C and P and the size of the cracks in the layer decrease from the external layer to the inner layer. The hydroxyl groups and phosphate carboxyl groups in the coating which have the similar properties with organic paintcoat are beneficial to the combination of substrate and organic paintcoat. The thickness of the conversion coating varies from 1.0 μm to 15 μm according to the processing parameters.

5. Acknowledgements

This work was supported by the Research Fund of the State Key Laboratory of Solidification Processing (NWPU) (Grant Nos. 29-T-2009).

REFERENCES

[1] Z. H. Chen, H. G. Yan, J. H. Chen, Y. J. Quan, H. M. Wang and D. Chen, "Magnesium Alloys," Chemical Industry Press, Beijing, 2004.

[2] G. Baril and N. Pehere, "The Corrosion of Pure Magnesium in Aerated and Deaerated Sodium Sulphate Solutions," *Corrosion Science*, Vol. 43, No. 3, March 2001, pp. 471-484.

[3] F. C. Liu, W. Liang, X. R. Li, X. G. Zhao, Y. Zhang and H. X. Wang, "Improvement of Corrosion Resistance of Pure Magnesium via Vacuum Pack Treatment," *Journal of Alloys and Compounds*, Vol. 461, No. 1-2, August 2008, pp. 399-403.

[4] J. E. Gray and B. Luan, "Protective Coatings on Magnesium and its Alloys—A Critical Review," *Journal of Alloys and Compounds*, Vol. 336, No. 1-2, April 2002, pp. 88-113.

[5] H. Zhang, G. C. Yao, S. L. Wang, Y. H. Liu and H. J. Luo, "A Chrome-Free Conversion Coating for Magnesium—Lithium Alloy by a Phosphate-Permanganate Solution," *Surface and Coatings Technology*, Vol. 202, No. 9, February 2008, pp. 1825-1830.

[6] H. Umehara, M. Takaya and S. Terauchi, "Chrome-Free Surface Treatments for Magnesium Alloy," *Surface and Coatings Technology*, Vol. 169-170, 2003, pp. 666-669.

[7] W. Q. Zhou, D. Y. Shan and E. H. Han, "Structure and Formation Mechanism of Phosphate Conversion Coating on Die-Cast AZ91D Magnesium Alloy," *Corrosion Science*, Vol. 50, No. 2, February 2008, pp. 329-337.

[8] G. Y. Li, J. S. Lian, L. Y. Niu, Z. H. Zhang and Q. Jiang, "Growth of Zinc Phosphate Coatings on AZ91D Magnesium Alloy," *Surface and Coatings Technology*, Vol. 201, No. 3-4, October 2006, pp. 1814-1820.

[9] M. Zhao, S. S. Wu, J. R. Luo, Y. Fukuda and H. Nakae, "A Chromium-Free Conversion Coating of Magnesium Alloy by a Phosphate-Permanganate Solution," *Surface and Coatings Technology*, Vol. 200, No. 18-19, May 2006, pp. 5407-5412.

[10] Z. C. Kwo and S. S. Teng, "Conversion-Coating Treatment for Magnesium Alloys by a Permanganate-Phosphate Solution," *Materials Chemistry and Physics*, Vol. 80, No. 1, April 2003, pp. 191-200.

[11] Y. Kuniji and M. Yoshio, "Metal Surface Coating Agent," US Patent 4 341 558, 1982.

[12] H. F. Yang, Y. Yang, Y. H. Yang, H. Liu, Z. R. Zhang, G. L. Shen and R. Q. Yu, "Formation of Inositol Hexaphosphate Monolayers at the Copper Surface from a Na-Salt of Phytic Acid Solution Studied by *in Situ* Surface Enhanced Raman Scattering Spectroscopy, Raman Mapping and Polarization Measurement," *Analytica Chimica Acta*, Vol. 548, No. 1-2, August 2005, pp. 159-165.

[13] J. R. Liu, Y. N. Guo and W. D. Huang, "Study on the Corrosion Resistance of Phytic Acid Conversion Coating for Magnesium Alloys," *Surface and Coatings Technology*, Vol. 201, No. 3-4, October, 2006, pp. 1536-1541.

[14] X. F. Cui, Y. Li, Q. F. Li, G. Jin, M. H. Ding and F. H. Wang, "Influence of Phytic Acid Concentration on Performance of Phytic Acid Conversion Coatings on the AZ91D Magnesium Alloy," *Materials Chemistry and Physics*, Vol. 111, No. 2-3, October 2008, pp. 503-507.

[15] X. F. Cui, Q. F. Li, Y. Li, F. H. Wang, G. Jin and M. H. Ding, "Microstructure and Corrosion Resistance of Phytic Acid Conversion Coatings for Magnesium Alloy," *Applied Surface Science*, Vol. 255, No. 5, 2008, pp. 2098-2103.

[16] C. H. Liang, R. F. Zheng, N. B. Huang and L. S. Xu, "Conversion Coating Treatment for AZ31 Magnesium Alloys by a Phytic Acid Bath," *Journal of Applied Electrochemistry*, Vol. 39, No. 10, October 2009, pp. 1857-1862.

[17] K. H. Yang, M. D. Ger, W. H. Hwu, Y. Sung, Y. C. Liu, "Study of Vanadium-Based Chemical Conversion Coating on the Corrosion Resistance of Magnesium Alloy," *Materials Chemistry and Physics*, Vol. 101, No. 2-3, February 2007, pp. 480-485.

[18] M. Zhao, S. E. Wu, P. An and J. R. Luo, "Influence of Surface Pretreatment on the Chromium-Free Conversion Coating of Magnesium Alloy," *Materials Chemistry and Physics*, Vol. 103, No. 2-3, June 2007, pp. 475-483.

Influence of Coolant in Machinability of Titanium Alloy (Ti-6Al-4V)

Nambi Muthukrishnan[1], Paulo Davim[2]

[1]Department of Mechanical Engineering, Sri Venkateswara College of Engineering, Sriperumbudur, India;
[2]Department of Mechanical Engineering, Campus Santiago, University of Aveiro, Averio, Portugal.

ABSTRACT

Application of titanium alloy has increased many fields since the past 50 years. The major drawback encountered during machining was difficult to cut and the formation of BUE (Built up Edge). This paper presents the tool wear study of TTI 15 ceramic insert (80% Aluminum oxide and 20% Titanium carbide) on machining Ti-6Al-4V at moderate speed with and without the application of water soluble servo cut S coolant. Titanium alloy is highly refractory metal and machining titanium is challenging to the manufacturers. Experiments were carried out on medium duty lathe. Application of coolant tends to reduce tool wear and minimize adhesion of the work material on the cutting tool during machining and also improves the surface finish. Result provides some useful information

Keywords: *Titanium Alloy, Machining, TTI 15 Ceramic Insert, Coolant, Surface, Roughness, Tool Wear*

1. Introduction

Machinability of a material can be defined and measured as an indication of the ease or difficult with which it can be machined. Machinability of a material may be assessed by tool life, metal removal rate, cutting forces and surface finish [1,2]. Titanium alloys are now being constituted in modern aerospace, marine, automotive, atomic power plant reactor, medical instruments and chemical industry due to their strength to weight ratio that can be maintained at elevated temperatures, excellent corrosion and fracture resistance and low modulus of elasticity [3-5]. However, machining of titanium and its alloys can be considered very difficult to cut materials due to its highly chemical reactivity and tendency to weld to the cutting tool, which resulted in edge chipping and rapid tool failure [4]. The advancement in the development of the cutting tools for the past few decades showed little improvement in the machinability of titanium alloys. Most of the cutting tools developed so far, including diamond ceramics and Cubic boron nitride, are highly reactive to titanium alloys, causing rapid wear especially at higher cutting speeds [6-8]. Titanium alloy, Ti-6Al-4V is known to be the workhorse for aerospace and non-aerospace applications. Previous study by [9] indicated that Ti-5Al-4V alloy possessed superior machinability in both drilling and turning tests when compared to Ti-6Al-4V.

Studies by [10-12]reported that, when machining titanium alloys, straight tungsten carbide (WC/Co) cutting tools have proved their excellence in almost all processes, except of the tool wear.

According to [4] cutting tool materials used for machining titanium alloys usually have short tool life and most react with the titanium work materials. This disadvantage is due to the generation of high temperatures closer to the cutting edge of the tool. Most of the problems associated with conventional machining of alloys have been dogged by high consumption of cutting tool materials due to excessive tool wear as a result of high-temperature generation at the cutting interfaces.

To over came this drawback, [4] suggested high-pressure jet-assisted cooling technology during machining of super alloys is the temperature reduction at the cutting interface due to improved access of coolant closer to the tool cutting edge. This proved significant improvement in tool life because of lower tool wear rates. The main tool failure criteria reported in the literature is rake and flank face wear which resulting from two wear mechanisms: dissolution-diffusion and attrition.

P. A. Dearnley *et al.* [11] found that ion implantation of chlorine and indium ions in tungsten carbide tools were successful in improving the life of the tools. The performance of mixed WC grades has been found to be poorer due to the high diffusion rates of TiC and TaC, leading to preferential attrition of these carbides from the tool [13].

Emmanuel [14] have investigated the high pressure coolant softens the machining surface by making the interference temperature low.

The aim of this present work is to investigate the influence of water soluble coolant in machining of titanium alloys (Ti-6Al-4V) with and without the application of coolant. The effect of cutting conditions on tool flank wear and surface roughness are also evaluated.

2. Experimental Procedure

Work piece samples of 40 mm diameter bar of 250 mm length were obtained from Kalpakam atomic power plant, Chennai. Work materials are being used in the power plant reactors. Turning tests were carried out on work material using water-soluble oil designated as Servo cut S lubricant oil. It forms a milky emulsion with water and contains rust inhibitor to impart anti-rust, anti-corrosion properties and a biocide to prevent bacterial growth in the emulsion. It has superior cooling and lubricating properties which impart excellent surface finish and minimizes tool wear. It is used in different concentration based on the type of operation. From open literature, to obtain best performance, stable emulsion oil should be added to 75% water. It meets BIS: 1115 1986 specification and is recommended for a variety of cutting operations on ferrous and non-ferrous metals.

All the tests were carried out on medium duty lathe with 2 kW spindle power. Cutting inserts used were DNMG 120408, TTI 15 ceramic insert with top rake angle of 0 degree. It has density of 4.2 g/cm^3, Vickers's hardness 2200, and toughness 4 $MN/m^{2/3}$ .It is suitable for high speed cutting due to superior hardness at high temperature, low chemical affinity to workpiece gives better finish and machining accuracy, high wear resistance and ensures longer tool life. Tool holder used was PCLNR 25 × 25 M 12 Chemical composition is shown in **Table 1**. Mechanical properties are shown in **Table 2**. **Figure 1** shows the experimental set up with work material Titanium Alloy (Ti-6Al-4V).

Experimental investigations were carried out on the work material with TTI 15 ceramic insert. **Table 3** summarizes the experimental conditions and **Table 4** shows the tool specifications of TTI 15 ceramic insert. Experiments were conducted and analyzed the machining parameters and graphs were drawn. Out of this, the best trial was investigated with the help of surface integrity.

The best trial was found as Cutting speed 135 m/min, feed 0.10 mm/rev and depth of cut 0.5 mm. using this parameter tool wear study was performed for time dura-

Table 1. Chemical composition of work material.

Alloy	Al	V	Fe	C	Ti
Ti-6Al-4V	6.40%	3.89%	0.16%	0.002%	Balance

Compositions are given in % by weight

Table 2. Mechanical Properties of Ti-6Al-4V.

Hardness (HRA)	70
Hardness Knoop	363
Tensile strength, Ultimate	950 MPa
Elongaion	14%
Poisson's ratio	0.342
Modulus of Elasticity	113 GPa
Density	4.43 g/cm^3

Figure 1. Experimental set up with Titanium alloy.

Table 3. Experimental conditions.

Cutting speed	45, 90 and 135 m/min
Feed	0.10, 0.20 and 0.32 mm/rev
Depth of Cut	0.5 and 0.75 mm
Cutting condition	With and without coolant
Machine	Medium duty lathe

Table 4. Tool specification of TTI 15 ceramic insert.

Composition	80% Al_2O_3 and 20% TiC
Grain Size	3.0 μm
Transverse Rupture Strength	551-786 MPa
Average density	3.90-3.99 g/cm^3
Youngs Modulus	641 GPa
Hardness	91-94 HRA
Coefficient of Thermal expansion	Good

tion of 45 minutes.

Average surface roughness (Ra) was measured with the help of surface roughness tester (Model : Mitutoyo - 301) with the cut-off 0.8 mm, Tool flank wear ($V_{Bmax} =$ 0.4 mm) as per ISO 3685:1993 was measured with the help of toolmakers microscope. Wear images were captured by Scanning Electron Microscope (SEM) and Optical microscope. **Table 5** shows the experimental readings

3. Results and Discussions

3.1. Surface Roughness

Figure 2(a) shows the influence of cutting speed on machined surface obtained while machining Ti-6Al-4V with coolant (wet machining) and **Figure 2(b)** shows the same graph without coolant (Dry machining). In general surface roughness of machined component decreased with the increase in cutting speed. It may be suggested that adherence of the work piece material to the tool at higher cutting speeds are less pronounced, perhaps due to the high temperature generated [5]

Surface roughness obtained was very good with coolant compared with dry machining. Similar trend was observed for other depth of cut also for dry and wet machining (**Figures 3(a,b)**). In both machining, surface roughness decreases as cutting speed increases. In dry machining, due to formation of Built-up-Edge (BUE), machined surface gets damaged by insert dragging over the surface of machined component with BUE [4,5]. In wet machining the coolant prevents the formation of BUE and also reduced the heat generated in the interface.

Figures 3(a,b) shows the influence of coolant on machining the workpiece with ceramic insert. From **Figure 3**, it is clearly understood that, surface roughness decrease with increasing the cutting speed. It is obvious that, ceramic insert has high hardness at high temperature. Analysis on surface roughness proved that an increase in feed rate produced a general trend towards higher surface

roughness [15] the machined surfaces consist of uniform feed marks in perpendicular to the tool feed direction. Surface damages are also observed after machining. Some feed marks also observed, this is attributed by plastic flow of material during the cutting process. Plastic flow of material on machined surfaces results in higher surface roughness values [16].

Generally surface finish of the machined component is good, with TTI 15 ceramic insert under wet machining. In 0.75mm depth of cut surface roughness is slightly higher at higher cutting speeds, because of influence of depth of cut on surface finish. In TTI 15 ceramic insert, surface roughness gradually decreases from maximum to minimum, because of formation of BUE, which affects the surface quality of the machined component at lower cutting speeds; at higher cutting speeds formation of BUE is not noticed. As a result good surface finish is obtained. It is suggested to machine the Titanium alloy under wet machining which soften the machined surfaces under coolant. It makes the interface to efficient cooling by coolant to get optimum life [14]

3.2. Effect of Feed Rate

Figures 4(a,b) show the effect of feed rate on average surface roughness on machining Ti-6Al-4V alloy with coolant for depth of cut 0.5 mm and 0.75 mm respectively. It is observed that, feed rate has more influence on surface finish. It increases from minimum to maximum. Good surface finish is obtained at higher cutting speed with low feed rate. At higher cutting speed tool wear is more. Confirming that cutting speed has the largest influence on tool wear/tool life, as commonly reported [7,15]

Table 5. Experimental data.

Expt.Nº	Cutting Speed (m/min)	Feed Rate (mm/rev)	Depth of Cut (mm)	Average Surface Roughness (μm) With Coolant*	Average Surface Roughness (μm) Without Coolant*
1	45	0.1	0.5	1.30	1.75
2	90	0.1	0.5	1.15	1.65
3	135	0.1	0.5	0.80	1.20
4	45	0.2	0.5	2.80	3.00
5	90	0.2	0.5	2.35	2.50
6	135	0.2	0.5	1.90	2.20
7	45	0.32	0.5	4.60	5.05
8	90	0.32	0.5	4.50	4.85
9	135	0.32	0.5	4.15	4.45
10	45	0.1	0.75	1.60	1.75
11	90	0.1	0.75	1.30	1.55
12	135	0.1	0.75	1.00	1.15
13	45	0.2	0.75	2.70	3.15
14	90	0.2	0.75	2.55	2.65
15	135	0.2	0.75	2.20	2.45
16	45	0.32	0.75	5.15	5.25
17	90	0.32	0.75	4.75	4.55
18	135	0.32	0.75	4.15	4.25

Average of three repetitions

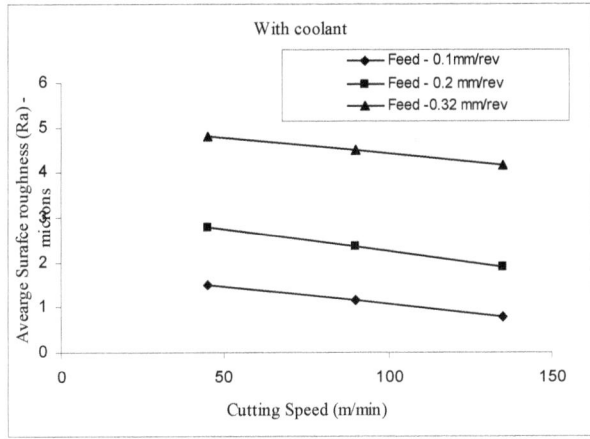

(a)

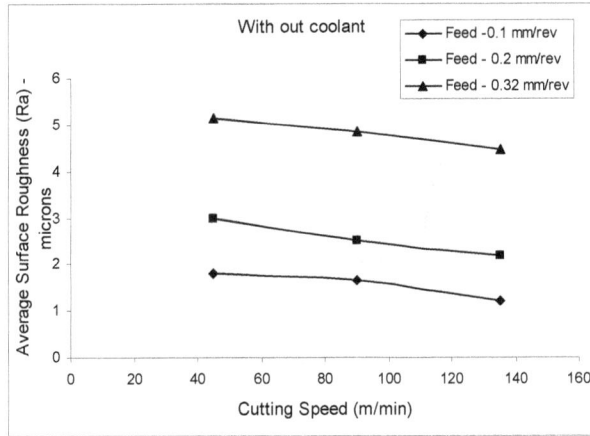

(b)

Figure 2. (a,b) Surface roughness (Ra) Vs Cutting Speed (depth of cut – 0.5 mm).

Similar trend exists for other machining condition (wet machining). Surface roughness values are lesser than that of dry machining under similar machining parameters. This is happened due to the presence of coolant in the interface which reduces the heat generation and avoids the adhesion nature of the work material.

3.3. Tool Wear

Main cause of the tool wear mechanism in machining Ti-6Al-4V was found to be diffusion, due to high temperature in the interface. Adhesive and attrition wear also take place due to micro hardening of the material [14]. From **Figure 5**, it was observed that tool flank portion in TTI 15 ceramic insert is subjected to uniform wear under wet machining. This could be reduced by high pressure jet assisted cooling in the tool interface in order to reduce the temperature [4]. The wear in TTI 15 insert attained only 0.1 mm after 45-minute duration.

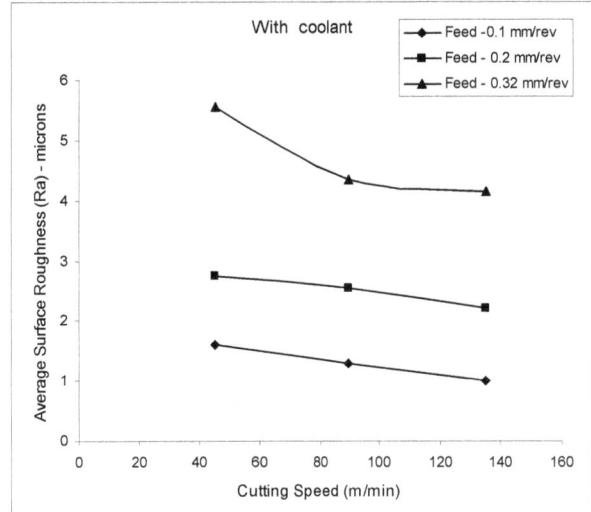

(a)

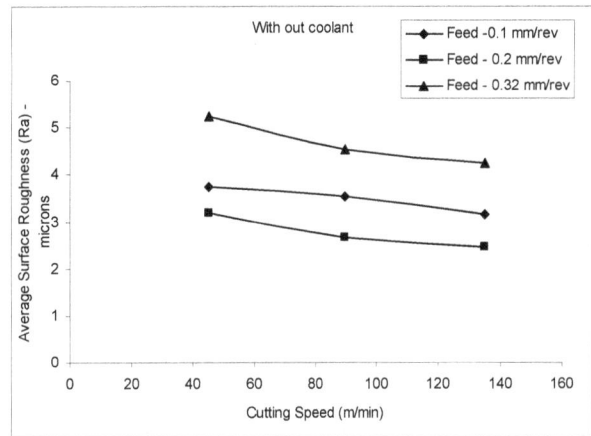

(b)

Figure 3. (a,b) Surface roughness Vs Cutting Speed (Depth of cut – 0.75 mm).

Tool wear is less at TTI 15 ceramic insert at the beginning. It was observed that uniform diffusion/adhesive wear is existing in the flank portion for the first 15 minutes duration, After that, it was noticed that, sudden increase in the wear, this is happened due to shock loading associated with the machine and micro chipping of the insert was found at the nose region. **Figure 6**, shows the rake and flank portion of TTI 15 ceramic insert. In **Figure 7**, built up edges are seen on the flank portion of the insert. It was also one of the reasons for tool wear [15].

When machining with TTI15 ceramic insert, there was as increase in the hardness of the surface of Ti-6Al-4V. This hardness may also believe to increase the adhesive wear [17].

When machining with dry condition, it was observed that tool diffusion/Adhesion wear is not uniform. There was a sudden increase in the wear, because of the heat

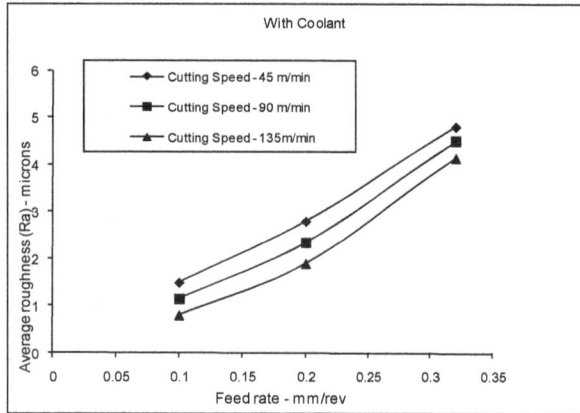

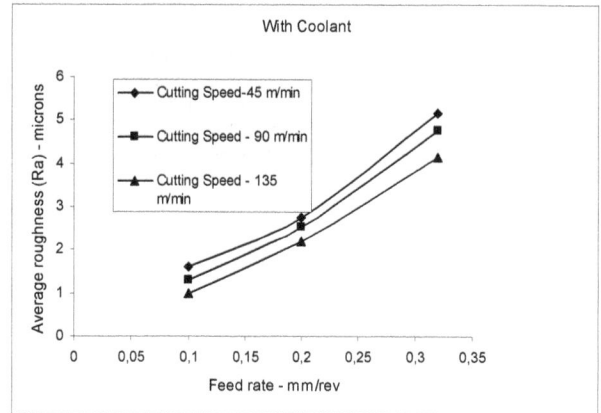

(a)   (b)

Figure 4. (a,b) Effect of feed rate on surface roughness.

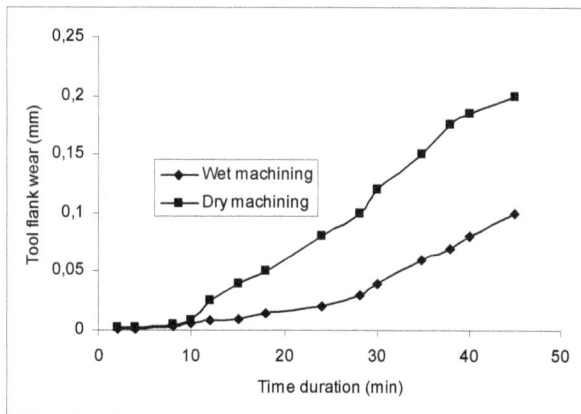

Figure 5. Tool wear versus machining time.

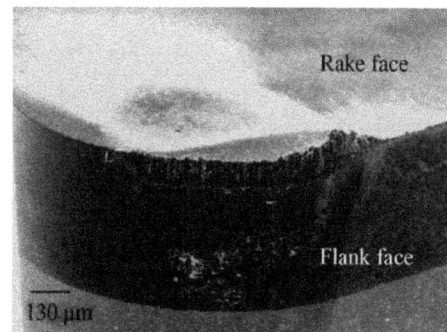

Figure 6. Tool wear with coolant.

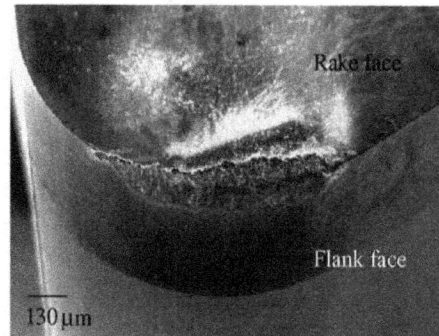

Figure 7. Tool wear without coolant.

generated in the interference zone was critical [10] The sharp cutting edge in the insert became blunt and BUE formation was more at the cutting edge because of the adhesion. At the end of 45 minute duration, tool wear on the insert under dry machining is approximately 100% more than that of the insert under wet machining.

4. Conclusions

The following conclusions are drawn, when machining Ti-6Al-4V with TTI 15 ceramic insert at various cutting speeds are as follows

1) Machining with coolant gives good surface finish compared with dry machining.

2) Tool wear is less for TTI 15 ceramic insert under the selected cutting conditions under wet machining.

3) Tool life is improved by 30% when machining with coolant.

4) Diffusion /adhesive wear is in the flank portion which was the dominant tool wear for TTI 15 ceramic insert.

5) Surface finish improves generally, with increase in cutting speed with and without coolant

6) Adhesive wear is observed at the flank portion of the insert on both dry and wet machining. Diffusion wear is more in dry machining

7) Attrition wear also presented in the insert.

REFERENCES

[1]   E. M. Trent, "Metal Cutting," Butterworth-Heinemann,

Oxford, 1991.

[2] E. M. Trent and P. K. Wright, "Metal Cutting," 3rd Edition, Butterworth Heinemann, Boston, 2000.

[3] W. Koning, "Applied Research on the Mach Inability of Titanium and Its Alloys," *Proceedings of the 47th Meeting of AGARD*, London, September 1978, pp 1-10.

[4] A. R. Machado and J. Wallbank, "Machining of Titanium and Its Alloys: A Review," *Proceedings' of the Institution of Mechanical Engineers Part B: Management and Engineering Manufacture*, Vol. 204, No. 11, 2005, pp. 53-60.

[5] E. O. Augur and Z. M. Wang, "Titanium Alloys and Their Machinability—A Review," Journal of Material Processing Technology, Vol. 68, No. 3, 1997, pp. 262-274.

[6] R. Komanduri and B. F. Von Turkovich, "New Observations on the Mechanisms of Chip Formation When Machining Titanium Alloys," *Wear*, Vol. 69, No. 2, 1981, pp. 179-188.

[7] P. D. Hartung and B. M. Kramer, "Tool Wear in Titanium Machining," *CIRP Annals—Manufacturing Technology*, Vol. 31, No. 1, 1982, pp. 75-80.

[8] N. Narutaki, A. Murakoshi, S. Motonishi and H. Takeyama, "Study on Machining Titanium Alloys," *CIRP Annals—Manufacturing Technology*, Vol. 32, No. 1, 1982, pp. 65-69.

[9] Y. Kosaler, J. C. Fanning and S. P. Fox, "Development of Low Cost High Strength Alpha/Beta Tiatanium Alloy with Superior Machinability," 10th *World Conference on Titanium*, Hamburg, July 2003, pp. 3028-3034.

[10] P. A. Dearnley and A. N. Grearson, "Evaluation of Principal Wear Mechanisms of Cemented Carbides and Ceramics Used for Machining Titanium Alloy IMI 318,"

Materials Science and Technology, Vol. 2, 1987, pp. 47-58.

[11] P. A. Dearnley, A. N. Grearson and J. Aucote, "Wear Mechanisms of Cemented Carbides and Ceramics Used for Machining Titanium Alloys," *High-Tech Ceramics*, Vol. 38, 1987, pp. 2699-2712.

[12] Z. M. Wang and E. O. Ezugwu, "Performance of PVD Coated Carbide Tool When Machining Ti-6Al-4V," *Tribology Transactions*, 1997, Vol. 40, pp. 81-86.

[13] X. Yang and C. R. Liu, "Machining Titanium and Its Alloys," *Machining Science and Technology*, Vol. 3, No. 1, 1999, pp. 107-139.

[14] E. O. Ezugwu, J. Bonneya, R. B. O. C da Silvab and O. Çakir, "Surface Integrity of Finished Turned Ti-6Al-4V Alloy with PCD Tools Using Conventional and High Pressure Coolant Supplies," *International Journal of Machine Tools & Manufacture*, Vol. 47, No. , 2007, pp. 884-891.

[15] J. F. Kahles, D. Eylon, F. H. Froes and M. Field, "Machining of Titanium Alloys," *Journal of Metals*, Vol. 37, No. 4, 1985, pp. 27-35.

[16] L. Zhou, J. Shimizu, A. Muroya and H. Eda, "Material Removal Mechanism beyond Plastic Wave Propagation Rate," *Precision Engineering*, Vol. 27, No. 2, 2003, pp. 109-116.

[17] J. I. Hughes, A. R. C. Sharman and K. Ridgway, "The Effect of Cutting Tool Material and Edge Geometry on Tool Life and Workpiece Surface Integrity," *Proceedings of the Institution of Mechanical Engineers, Part B: Journal of Engineering Manufacture*, Vol. 220, No. 2, 2006, pp. 93-107.

3

Finite Element Analysis of the Material's Area Affected during a Micro Thermal Analysis Applied to Homogeneous Materials

Yoann Joliff*, Lénaïk Belec, Jean-François Chailan

MAPIEM, EA 4323, Institut des Sciences de l'Ingénieur de Toulon et du Var, Cedex, France

ABSTRACT

Micro-thermal analysis (μ-TA), with a miniaturized thermo-resistive probe, allows topographic and thermal imaging of surfaces to be carried out and permits localized thermal analysis of materials. In order to estimate the effective volume of material thermally affected during this localized measurement, simulations, using finite element method were used. Several parameters and conditions were considered. So, thermal conductivity was found to be the driving physical parameter in thermal exchanges. Indeed, the evolution of the heat affected zone (HAZ) versus thermal conductivity can well be described by a linear interpolation. Therefore it is possible to estimate the HAZ before experimental measurements. This result is an important progress especially for accurate interphase characterization in heterogeneous materials.

Keywords: *Micro-Thermal Analysis, Localized Thermal Analysis, Heat Affected Zone, Thermal Conductivity, Finite Element Method*

1. Introduction

Micro thermal analysis is a characterization technique of thermal and mechanical behaviours of materials at submicroscopic scale [1]. This technique is therefore dedicated to numerous application domains dealing with low scale characterizations such as distinguishing different constituents of a heterogeneous material, identifying the polymorphism or thermal history of polymers, identifying a contaminant or surface film, characterizing interfaces and interphases [2]. Recent technological advances for this technique mainly concern the development of new thermal probes with tip edges that are reduced to the maximum in order to ensure localized measurements [3-5]. The present study aims at estimating the area thermally affected during a localized measurement with the thermal probe (Wollaston wire Pt/Rh). In fact, few works are concerned with this subject which is however crucial for the validity of results, particularly when polymer reinforced composite materials are considered. So, we were interested in developing numerical models that are able to quantify the heat affected zone (HAZ) and to identify parameters that influence its size during

localized measurements. The finite element method is used in order to quantify the volume of matter thermally stressed on virtual samples representative of real monophased materials. Therefore, it seems essential to identify physical phenomena (mainly thermal here) that occur during characterization.

The present work describes the technique and discusses the results obtained by this means. Thermal phenomena here above mentioned that have been widely studied in literature these last years will be presented. Then, results obtained from finite element method for the study of the HAZ will be detailed, analyzed and discussed.

2. Micro-Thermal Analysis Process

2.1. Description of the Micro-Thermal Analysis (μ-TA)

The μ-TA experimental device is a Scanning Force Microscope using a miniaturized thermo-resistive probe. The system is composed of four major elements: the thermal probe, a piezo-scanner to control its position in three directions, a thermal control unit and a system for

Finite Element Analysis of the Material's Area Affected during a Micro Thermal Analysis Applied to Homogeneous Materials

15

data acquisition. The scanner is linked to a feedback mechanism to maintain a constant force and the probe temperature is controlled by a wheatstone bridge so that an isoforce topographic image and an isothermal apparent thermal conductivity image can be simultaneously recorded. The thermoresistive probe derived from Dinwiddies *et al.*'s work [6] comprises a Wollaston lever arm at the end of which the Platinum thermoresistive element is fixed. The heating tip is a simple filament of Platinum/10% Rhodium curved in V-shaped (**Figure 1**) [7].

After positioning the probe on the sample's surface, a scan is performed at constant force and temperature. Mallarino *et al.* [8] show an example of apparent thermal conductivity image obtained after scanning a unidirectional glass fibre reinforced composite in transverse direction. Localized thermomechanical measurements intended to determine melting point and glass transition temperature in polymer or composite can then be performed at high temperature rate (10 - 20 K/s) on selected points of the scan. The high rate limits the HAZ dimensions for a given material diffusivity. The probe displacement is then followed versus temperature [8].

In this example, sensor position successively increases showing sample thermal expansion at glassy state, and decreases drastically around 100°C showing sample softening after glass transition temperature [8].

2.2. Heat Exchange between the Probe and the Sample

The quality of measurements is closely related to the nature of the contact zone between the platinum probe and the surface of the sample. It is therefore important to well-understand the thermal exchanges associated to this localized measurement in order to quantify heat transfers and consequently the HAZ. In a previous work, Shi *et al.*

conducted a detailed study of the thermal mechanisms that take place in the contact area between the thermocouple tip and a hot substrate [9]. Type and shape of the probe and sample, atmospheric environment and distance between the probe and the sample are the parameters to be considered. Existing literature brings back a lot of works [9-11] describing the presence, below 100°C, of a water meniscus which appears at the periphery of the probe tip and in the cavities created between probe and sample surfaces in the contact zone. In her PhD work [12], Gomes states that the heat exchange through the water meniscus is predominant. The intensity of this exchange is directly linked to the thermal conductivity of the sample.

Consequently, three ways can be considered for describing heat transfers (**Figure 2**)

o Transfer by conduction (probe/sample and probe/water/sample);

o Transfer by natural convection (air/sample);

Transfer by radiation (probe/sample).

The involvement of radiation in the heat balance is not significant compared to the natural convection with air [13]. A modelling of the thermal behaviour of the probe of the Scanning Thermal Microscope (SThM) has recently been proposed by Grossel *et al.* [14]. The thermal exchange along the probe was modelled as a convection gradient of which the heat transfer coefficient value h is maximal at the periphery of the probe/sample contact area (with $h = 1200$ W/m²/K at the end of the probe in contact with the sample and $h = 60$ W/m²/K at the other end [15]). A similar study has proposed h values lying between 950 and 1300 W/m²/K [16].

2.3. Contact Area between the Probe and the Sample

The HAZ will closely be related to the size of the contact area between the probe and the sample. Using analytical models, several authors have been able to quantify the dimensions of this contact area. So, Gomes *et al.* [16]

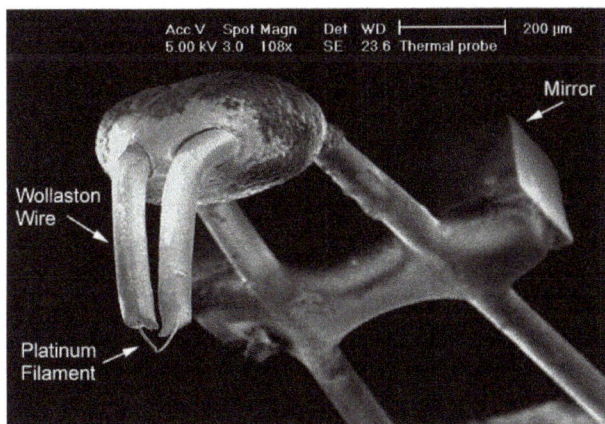

Figure 1. Micrograph of the thermal probe developed by Dinwiddie *et al.* [6,7]

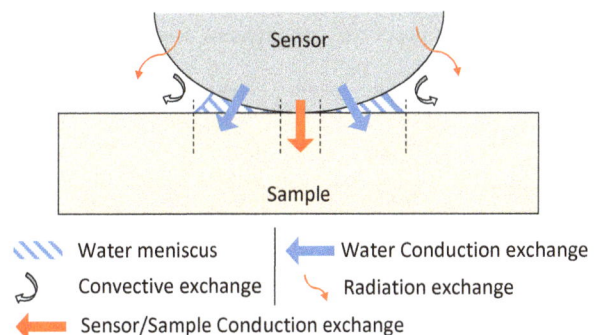

Figure 2. Different heat transfers mechanisms between the probe and the sample.

estimate a circular contact surface with a 80 nm radius and a concentric circular water meniscus surface with an outside radius of 306 nm. As for Lefèvre et al. [17-19], they have suggested contact radii values ranging from 200 nm to 1 μm depending on temperature and probe geometry.

3. Numerical Approach and Results

Numerical models developed here are based on a 500 nm radius and 1 mm height cylindrical sample (geometrical models have been described by an axisymmetric representation – 2D½). Although the approach presented here is dedicated to study polymer materials, metal and ceramic materials have been chosen for several reasons. First, below 30°C and 150°C, their thermal properties can be considered as independent of the temperature. Then, close to 110 - 130°C, mechanical properties of polymer materials decrease strongly and the thermal probe would push in the sample. So, the contact surface between the probe and the sample increases and must be considered in simulations. These phenomena are not considered here and will be taken into account in a future work where coupled thermomecanical simulations will be used.

Table 1 summarizes the type and properties of the materials that have been studied. Note that theses materials are supposed to be homogenous and isotropic.

The finite element models were realized with Abaqus code and the numerical models are meshed with 400 000 linear axisymmetric heat transfer quadrilateral elements. A gradient element size (small: 16 nm × 16 nm / large: 6 × 6 μm) was realized to have a fine discretization under the probe zone.

A transient heat transfer analysis is applied, and heat conduction is assumed to be governed by the Fourier law (Equation (1))

$$f = -k\frac{\partial \theta}{\partial x} \qquad (1)$$

with k: matrix of conductivity, $k = k(\theta)$,
f: heat flux.

Table 1. Physical properties of materials used for modelling [20].

	ρ (kg/m³)	λ (W/m/K)	Cp (J/kg/K)
Alumina (dense)	3 900	30	900
Asbestos (fibre)	1 400	0.168	1 045
Copper	8 910	390	385
Diamond	3 508	2 600	502
Iron	7 860	80	444
Gold	19 300	317	129
Zinc	7 150	116	380

x: position.

The thermal probe is replaced by a simplified thermal condition applied to the sample surface: the temperature increase of 15 °C/s (thermal loading of the probe) is supposed to be applied on a small surface of the sample equivalent to the probe contact area. The radius of this area is fixed at 80 nm according to the results of Gomes et al. [16] and the presence of the water meniscus is not taken into account in the first step. The initial temperature of the sample is fixed by a prescribed temperature equal to 30°C. The heat flux q on the top of sample surface due to convection is governed by boundary convection:

$$q = -h(\theta - \theta^0) \qquad (2)$$

with q: heat flux across the surface,
    h: reference film convection coefficient,
    θ: temperature at this point on the surface,
    θ^0: reference sink temperature value.

Radiation heat exchange is not taken into account (not significant compared with convection exchange). **Figure 3** summarized boundary conditions and thermal loading applied to the model.

The sample is considered as thermally affected if the temperature of the area concerned reaches values higher or equal to 50°C. The HAZ has been defined as the distance between the centre of the probe and the thermal isovalue of 50°C in the material. This value of 50°C was chosen as reference because polymer properties are not affected below glass transition temperature (generally $T_g > 60$°C).

3.1. Influence of the Materials Physical Properties on the Zize of the HAZ

Solving the heat transfer equations related to a thermal problem only requires three materials properties: density, thermal conductivity and specific heat. In this part of the study, the aim is to identify the properties that are really influent on the size of the HAZ.

Figure 4 shows the thermal gradient at the end of thermal loading in two samples, a heat insulation (asbestos) and a thermal conductor (copper). The quantity of material thermally impacted is more important in the radial direction than in the longitudinal one. For example, in the case of asbestos, a gap nearing 10% can be calculated between the radial and the longitudinal lengths whereas for copper, the difference is only about 3%. Considering this first result, the shape of the HAZ can therefore be described by a hemispherical shape for thermal conductor materials. While for thermal insulator materials, a hemi-elliptical shape is more appropriate. Later, in this paper, only the radial length on the top surface of the sample is studied.

Finite Element Analysis of the Material's Area Affected during a Micro Thermal Analysis Applied to Homogeneous Materials

17

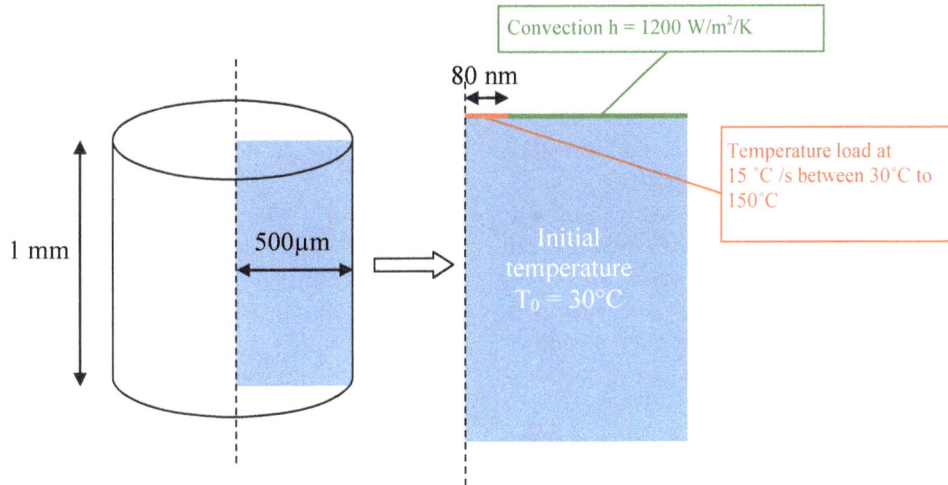

Figure 3. Boundary conditions and thermal load for numerical model.

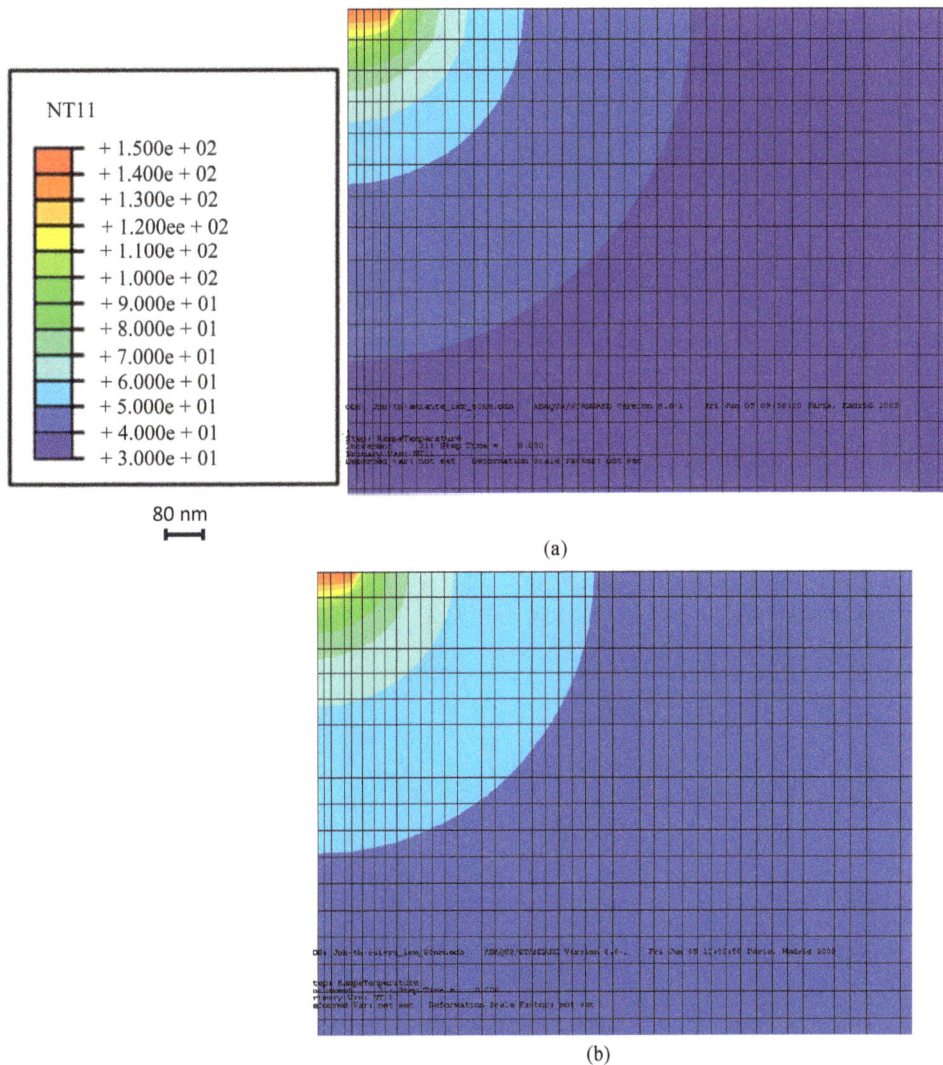

(a)

(b)

Figure 4. Thermal gradient evolution in the sample at the end of thermal loading, *i.e.* $T_{probe} = 150°C$ for asbestos (a) and copper (b) (zoom in the contact zone with the thermal probe).

Figure 5 presents, at the end of thermal loading (T_{probe} = 150°C), temperature variations along the horizontal upper surface (that is in contact with the probe) of the different model samples. Except for diamond, samples dimensions (1 mm × 1 mm) are enough to define a satisfying representative elementary volume (REV) for modelling μ-TA process. The HAZ is increases with the diffusivity (**Figure 6**). Moreover, a linear evolution is found between its size and the value of diffusivity.

The evolution of the HAZ versus thermal conductivity can also be well described by a linear interpolation as shown on **Figure 7**.

The other two parameters, specific heat and density, have been studied as a function of the HAZ (**Figure 8** and **Figure 9** respectively). It is difficult to use these results to quantify the HAZ since they could not be interpolated by a mathematical function.

3.2. Influence of the Contact Zone

Calculations were carried out by considering different sizes of contact zone for different materials with different

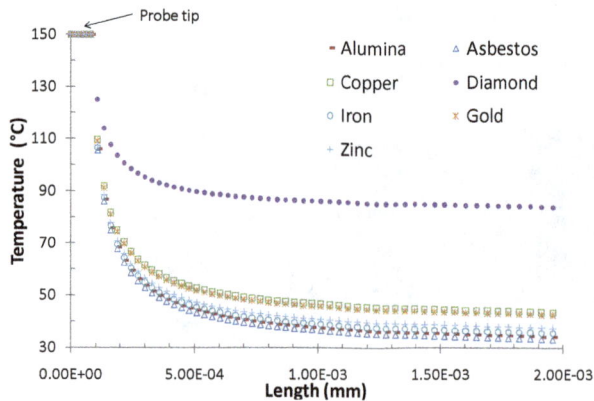

Figure 5. Temperature evolution along the horizontal surface of the sample at the end of thermal loading; *i.e.* T$_{probe}$ = 150°C (zoom in the contact zone with the thermal probe).

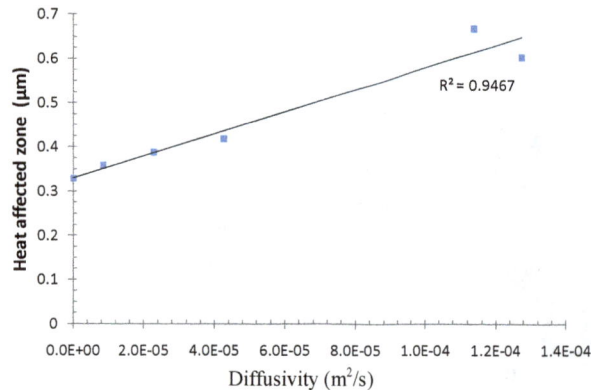

Figure 6. Evolution of heat affected zone versus diffusivity.

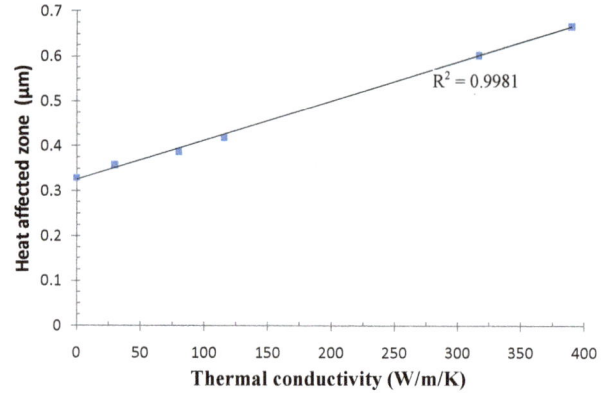

Figure 7. Evolution of heat affected zone versus thermal conductivity.

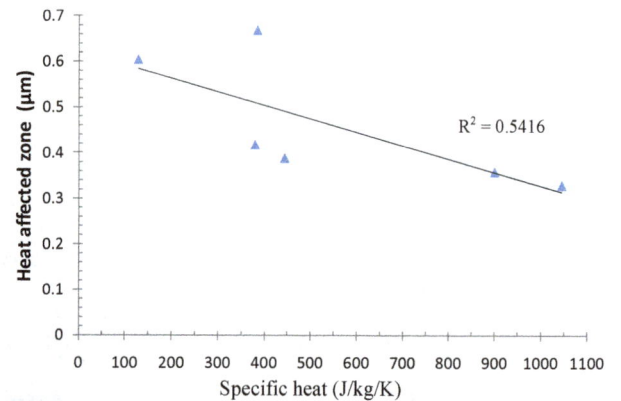

Figure 8. Evolution of heat affected zone versus specific heat.

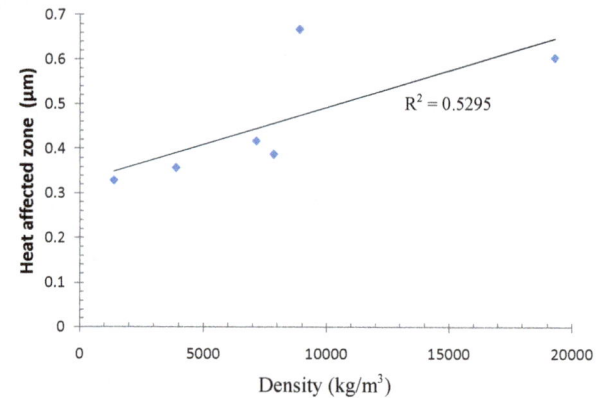

Figure 9. Evolution of heat affected zone versus density.

heat conductivities (copper, zinc, alumina and asbestos). **Figure 10** presents the numerical model used for these calculations.

Results are summarized on **Figure 11**. For thermal insulator materials (asbestos), the increase of the contact

Finite Element Analysis of the Material's Area Affected during a Micro Thermal Analysis Applied to Homogeneous
Materials

19

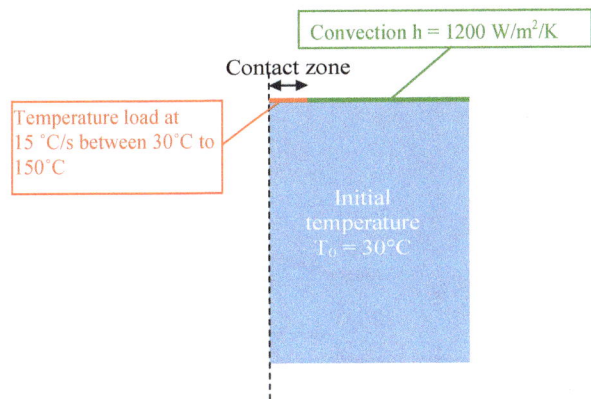

Figure 10. Numerical model used for study of the influence of the contact zone.

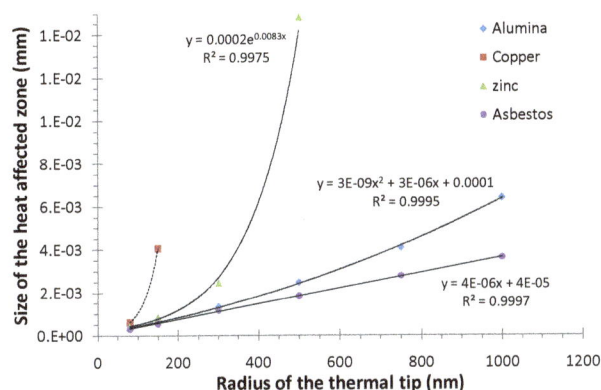

Figure 11. Evolution of heat affected zone versus contact radius between the probe and the sample.

radius leads to a linear increase of the size of the HAZ in the studied temperature range. For heat conductor materials (zinc), the change in HAZ size with temperature is exponential.

3.3. Influence of Water Meniscus

The shape of the probe influences the development of the water meniscus. Three angles are arbitrary fixed for the probe in the study: 15°, 45° and 85° (**Figure 12**). The length of the contact zone between the meniscus and the sample is constant and equal to 206 nm whatever the angle of the probe is [16]. The thermal condition used in previous simulation was applied on a new probe contact area (the sample surface and the water meniscus surface in contact with the thermal probe – **Figure 12**).

The contact zone between the sample and water meniscus is perfect and no dissipation is allowed. Using the same type of elements as previously, a new meshing was realized. A particular attention was paid to the meshing of the contact zone with thermal probe, and that of water meniscus. Their discretization has been made with small sized elements in order to optimize the results. Boundary

conditions and thermal loading applied to numerical models are summarized in **Figure 13**.

Figure 14 presents the results of the corresponding numerical calculations. From a certain value of thermal conductivity, the role of water meniscus becomes less important. Taking into account the water meniscus in calculations leads to an increase of the HAZ of 27% for alumina and only less than 5% for zinc. On the other hand, when the material has a low thermal conductivity (as for asbestos), the presence of water meniscus leads to an increase of the HAZ from 3 to 5 times depending on the angle of the tip.

4. Discussion

The finite element method enables to determine the influence of intrinsic or external parameters on the size of the zone really thermally affected during a micro thermal analysis (HAZ). The development of simple model representative of matrices with very different thermal properties is enough to define influent parameters such as thermal conductivity. From above results (**Figure 7**), it should be possible to estimate the size of the HAZ from thermal conductivity measurements after calibrating the µTA on several reference samples of well known thermal conductivities. Concerning specific heat which is difficult to quantify experimentally, no influence has been found on final result.

As expected, the initial contact area of the probe is a critical parameter on final HAZ size as observed on Figure 11. The interest of FEM analysis here is to quantify the influence of contact size on HAZ depending on thermal conductivity. For heat conducting materials, the exponential dependence between contact size and HAZ size requires the smallest possible contact areas. The technological development of micro thermal analysis then needs a miniaturization of this contact zone. As an example, Depasse et al. [21] refer to some 30 nm radius for this zone. This study however deals with an 80 nm radius according to Gomes et al.'s work [10,12,16] that are more successfully completed in regard to the analysis and the understanding of classical micro thermal analysis measurements with standard probes.

Since the probes are handmade, their shapes are heterogeneous, and the results obtained in displacement are therefore dispersed. Moreover, during experiments, these probes are often submitted to irreversible deformations that rapidly modify their initial shapes. The thermal tip usually tends to flatten, leading to the widening either of the angle of the thermal tip or of the angle between the probe and the sample. As mentioned above, this change has a little influence on heat transfer. However, it will be greatly influent during the development of water meniscus that drastically increases the size of the HAZ for heat

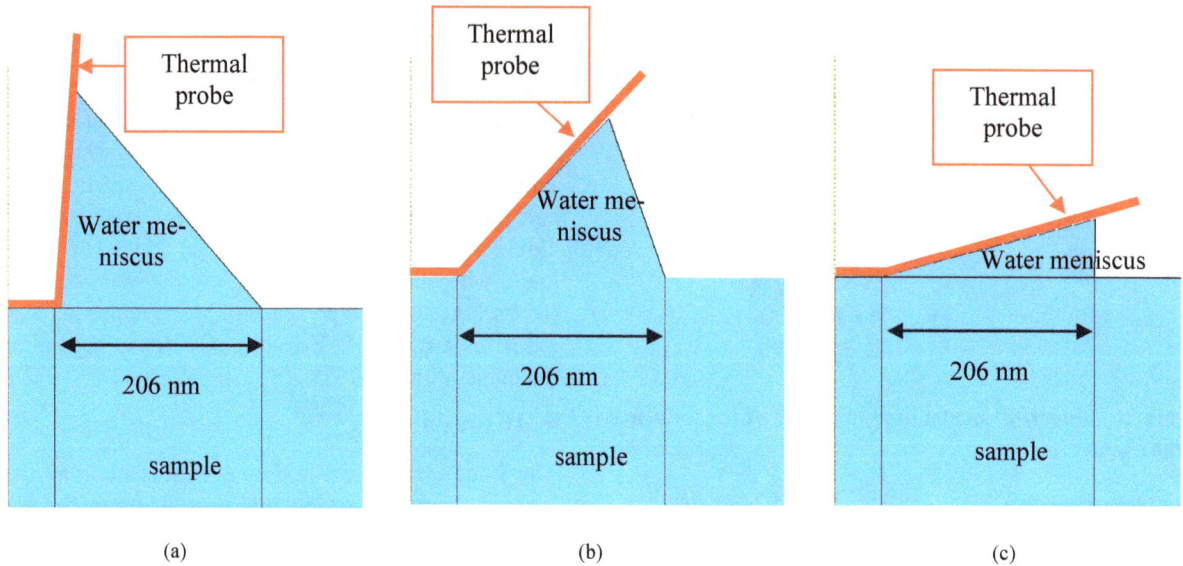

Figure 12. Probes and water meniscus profiles for the different numerical models: probe angle of 85° (a), 45° (b) and 15° (c).

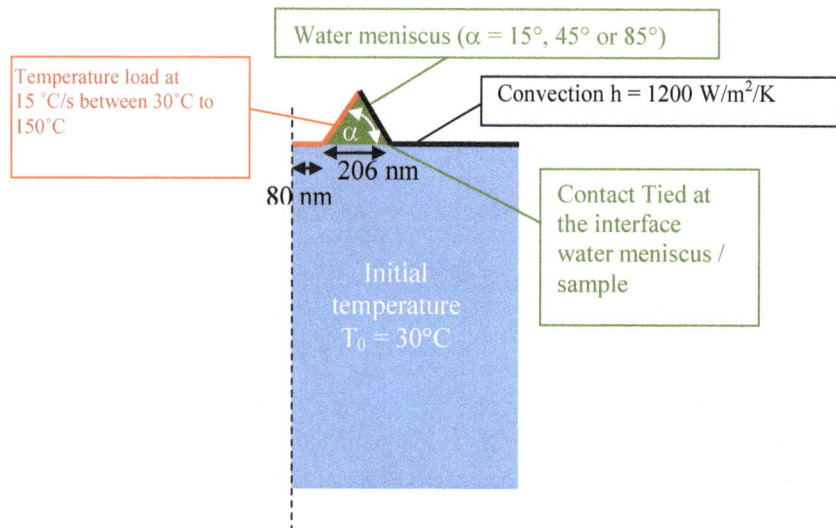

Figure 13. Numerical model used for study of the influence of the water meniscus.

insulator materials (such as asbestos). The thermal conductivity of water is in fact more important than that of the heat insulator material, what interferes with localized measurements. It is therefore fundamental to take this phenomenon into account for this materials family.

5. Conclusions

Making localized characterization is usually difficult to obtain. The physical parameters and the behaviours of materials are in fact modified compared to the macroscopic scale. So, it is an important challenge to be able to quantify at this small scale the thermal and mechanical stresses occurring on materials. Using numerical simulation and considering the heat insulator or conductor ma-

terials, this work led to the quantification of the heat affected zone during a localized micro thermal analysis. From the study of several parameters and conditions, the thermal conductivity has been found to be the driving physical parameter in thermal exchanges. It has therefore been possible to rapidly estimate the heat affected zone. Models showed that the water meniscus that appears at the interface between the probe and the sample has only little influence on the size of the heat affected zone for heat conductor materials. On the contrary, for heat insulator materials, this meniscus can lead to an increase of the size of the heat affected zone from 3 to 5 times. The contact angle has a limited influence on conductor materials and a higher influence on insulators. Finally, owing

Finite Element Analysis of the Material's Area Affected during a Micro Thermal Analysis Applied to Homogeneous
Materials

21

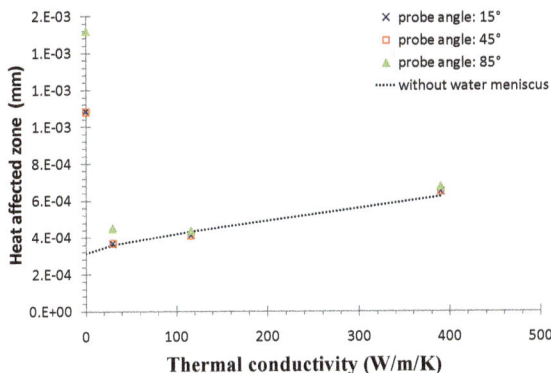

Figure 14. Evolution of heat affected zone for different shapes of probe and water meniscus.

to their varying shapes coming from diversified technological developments, the probes will probably modify the water meniscus and the size of the heat affected zone.

This work is just a first study on the topic of heat affected zone. In a near future, it will be necessary to complete it by defining new models that would describe multiphased materials and to quantify the corresponding heat affected zones surrounding different phases.

REFERENCES

[1] E. Gmelin, R. Fischer and R. Stitzinger, "Sub-Micrometer Thermal Physics—An overview on STHM Techniques," *Thermochimica Acta*, Vol. 310, No. 1-2, 1998, pp. 1-17.

[2] H. M. Pollock and A. Hammiche, "Micro-Thermal Analysis: Techniques and Applications," *Journal of Physics D: Applied Physics*, Vol. 34, 2001, pp. R23-R53.

[3] A. Altes, R. Tilgner and W. Walter, "Numerical Evaluation of Miniaturized Resistive Probe for Quantitative Thermal Near-Field Microscopy of Thermal Conductivity," *Microelectronics Reliability*, Vol. 46, No. 9-11, 2006, pp. 1525-1529.

[4] I. W. Rangelow, T. Gotszalk, N. Abedinov, P. Grabieca and K. Edingerb, "Thermal Nano-Probe," *Microelectronic Engineering*, Vol. 57-58, 2001, pp. 737-748.

[5] D.-W. Lee and I.-K. Oh, "Micro/Nano-Heater Integrated Cantilevers for Micro/Nano-Lithography Applications," *Microelectronic Engineering*, Vol. 84, No. 5-8, 2007, pp. 1041-1044.

[6] R. Dinwiddie, R. Pylkki and P. West, "Thermal Conductivity Contrast Imaging with a Scanning Thermal Microscope," *Thermal Conductivity*, Vol. 22, 1994, pp. 668-677.

[7] P. G. Royall, V. L. Kett and C. S. Andrews and D. Q. M. Craig, "Identification of Crystalline and Amorphous Regions in Low Molecular Weight Materials Using Microthermal Analysis," *The Journal of Physical Chemistry B*, Vol. 105, 2001, pp. 7021-7026.

[8] S. Mallarino, J. F. Chailan and J. L. Vernet, "Interphase Investigation in Glass Fibre Composites by Micro-Thermal Analysis," *Composites Part A: Applied Science and Manufacturing*, Vol. 36, No. 9, 2005, pp. 1300- 1306.

[9] L. Shi and A. Majumdar, "Thermal Transport Mechanisms at Nanoscale Point Contacts," *Journal of Heat Transfer*, Vol. 124, 2002, pp. 329-337.

[10] S. Gomés, N. Trannoy and P. Grossel, "D.C. Thermal Microscopy: Study of the Thermal Exchange between a Probe and a Sample," *Measurement Science and Technology*, Vol. 10, No. 9, 1999, pp. 805-811.

[11] S. Lefèvre, "Modélisation et Élaboration des Métrologies de Microscopie Thermique à Sonde Locale Résistive," Ph.D. Dissertation, Poitiers University, Poitiers, 2004.

[12] S. Gomes, "Contribution Théorique et Expérimentale à la Microscopie Thermique à Sonde Locale: Calibration d'Une Pointe Thermorésistive, Analyse des Divers Couplages Thermiques," Ph.D. Dissertation, Reims University, Reims, 1999.

[13] S. Lefèvre, S. Volz and P. O. Chapuis, "Nanoscale Heat Transfer at Contact between a Hot Tip and a Substrate," *International Journal of Heat and Mass Transfer*, Vol. 49, No. 1-2, 2006, pp. 251-258.

[14] P. Grossel, O. Raphaël, F. Depasse, T. Duvaut and N. Trannoy, "Multifrequential AC Modeling of the SThM Probe Behaviour," *International Journal of Thermal Sciences*, Vol. 46, No. 10, 2007, pp. 980-988.

[15] V. T. Morgan, "The Overall Convective Heat Transfer from Smooth Circular Cylinders," *Advances in Heat Transfer*, Vol. 11, 1975, pp. 199-264.

[16] S. Gomes, N. Trannoy, P. Grossel, F. Depasse, C. Bainier and D. Charraut, "D.C. Scanning Thermal Microscopy: Characterization and Interpretation of the Measurement," *International Journal of Thermal Sciences*, Vol. 40, 2001, pp. 949-958.

[17] S. Lefèvre, S. Volz, J.-B. Saulnier, C. Fuentes and N. Trannoy, "Thermal Conductivity Calibration for Hot Wire Based DC Scanning Thermal Microscope," *Review of Scientific Instruments*, Vol. 74, No. 4, 2003, pp. 2418-2423.

[18] S. Lefèvre, J.-B. Saulnier, C. Fuentes and S. Volz, "Probe Calibration of the Scanning Thermal Microscope in the AC Mode," Vol. 35, No. 3-6, 2004, pp. 283-288.

[19] S. Lefèvre and S. Volz, "3ω-Scanning Thermal Microscope," *Review of Scientific Instruments*, Vol. 76, No. 3, 2005, pp. 033701-033701-6.

[20] F. Cardarelli, "Materials Handbook—A Concise Desktop Reference," 2nd Edition, Springer-Verlag London, 2008.

[21] F. Depasse, P. Grossel and N. Trannoy, "Probe Temperature and Output Voltage Calculation for the SThM in A.C. Mode," *Superlattices and Microstructures*, Vol. 35, No. 3-6, 2004, pp. 269-282.

4

The Optical Parameters of $Zn_xCd_{(1-x)}Te$ Chalcogenide Thin Films

Umeshkumar P. Khairnar[1], Sulakshana S. Behere[2], Panjabrao H. Pawar[2]

[1]Department of Physics, S.S.V.P.S. ACS College, Shindkheda, Dhule, India; [2]Thin Film Laboratory, Department of Physics, Zulal Bhilajirao Patil College, Dhule, India.

ABSTRACT

A procedure to make optical quality thin films of $Zn_xCd_{(1-x)}$ Te by use of thermal evaporation of the ternary compound has been investigated. Structural and optical properties of $Zn_xCd_{(1-x)}$ Te solid solution with x = 0.1 to 0.5 were synthesized, from the resulting ZnTe and CdTe composition used in preparation of thin films. Structural investigation indicates they have polycrystalline structure. Composition was confirmed from EDAX while SEM picture shows homogeneity in films. Plots of $(\alpha h v)^2$ versus (hv) yield straight line indicating direct transition occurs with optical band gap energies in the range 1.7 - 2.3 eV. It is also found with increase Zn content the band gap of the films increases. Refractive indices and extinction coefficients have been evaluated in the spectral range (200 - 2500 nm).

Keywords: Thermal Evaporation, EDAX, XRD, Optical Band Gap

1. Introduction

Solid solution formation in semiconductors has been of interest for a number of years. An important question regarding ternary zinc-blende compound semiconductors is concerned with the structural and dynamic changes that can occur upon replacement of either cations or anions in the binary base material The II-VI compounds semiconductors and solid solutions based on them are promising source for various types of thin film devices such as thin film transistors [1], Solar cells [2] and photoconductors [3].

Thin films of $Zn_xCd_{(1-x)}Te$ were prepared by variety of techniques, such as, two source vacuum evaporation [6], molecular beam epitaxy [7], chemical vapour deposition and closed space vapour transport [8,9], physical vapour transport (PVT) [10], vertical Bridgman growth [11]. In the recent study $Zn_xCd_{(1-x)}Te$ thin films are deposited by thermal evaporation at substrate temperature (373 K) and the films are annealed and then characterized by energy dispersive X-ray analysis (EDAX) and scanning electron microscopy (SEM) technique for composition and surface morphology of the films. Optical properties of the films were studied by optical transmittance and reflectance measurement.

2. Experimental Details

For the preparation of ternary semiconductors, $Zn_xCd_{(1-x)}$ Te the constituent compounds ZnTe (Purity - 99.999%, Aldrich Co. Make, USA) and CdTe (Purity - 99.99+%, Aldrich Make, USA) have been taken in molecular stiochoimetry proportional weights and crushed and mixed homogenously. The different sets of samples of varying compositions (x = 0.1 to 0.5) were deposited via sublimation of the compound in vacuum higher than 10^{-5} mbar under controlled growth conditions of various compositions onto the amorphous precleaned glass substrates at the temperature of 373 K. The thicknesses of films were controlled by using quartz crystal thickness monitor model No.DTM-101 provided by Hind-High Vac. The deposition rate was maintained 10-15 Å/sec constant throughout sample preparations. The source to substrate distance was kept constant (15 cm) and substrate was kept at constant temperature (373 K). Deposited samples were kept under vacuum overnight. All the samples are deposited under the similar optimized condition. These samples were annealed at reduced pressure of 10^{-5} mbar for the duration of 3 hours at the temperature of 573 K and maintain carefully. These samples were then used for various characterizations. X-ray diffraction (XRD) studies were carried out using a Rigaku, Miniflex,

Japan X-ray diffractometer. The XRD patterns were recorded in the 2θ range of $20°$ - $80°$ glancing angle $30°$ using CuKα radiation ($\lambda = 1.5418$ Å). The morphology of the Zn$_x$Cd$_{(1-x)}$Te thin films was examined using Scanning Electron Microscope (SEM) (model 501, Philips, Holland with EDAX attachment) using acceleration voltage variable from 1.6 KV to 30 KV. For this purpose thin layer of gold (50Å) wa s deposited on the film using physical vapour deposition. UV-Vis spectra of the samples were recorded on a HITACHI-MODEL-330 UV-Vis spectrophotometer in the wavelength range 200 - 2500 nm.

3. Results and Discussion

All the Zn$_x$Cd$_{(1-x)}$Te films prepared by the above technique were polycrystalline of multi phase structure indicating preferential of the film crystallites corresponding to textured (100)H and (220)C growth [14].

From the **Figure 1** X-ray diffractograms of various compositions it is observed that for $x = 0.1$ and $x = 0.2$ there are only two prime peaks which corresponds to (100) plane of hexagonal CdTe and (220) plane of cubic ZnTe [14]. Diffraction analysis suggests that all the samples of various compositions are polycrystalline nature. However the charactertics peak of hexagonal CdTe and cubic ZnTe changes their angular position and relative intensities for different compositions suggesting multi phases and inhomogeneity in the growth of films. The samples of compositions ($x = 0.1, 0.2$ and 0.4) exhibit

predominant diffraction lines corresponding to (100) plane of CdTe (H) may be attributed to the characteristics growth with (100) reflecting plane as a preferred orientation. While the sample ($x = 0.3$ and 0.5) exhibits pre- dominant diffraction lines corresponding to (111) plane of ZnTe (C) is again attributed to the charactertics growth with the (111) reflecting plane as preferred ori- entation. The shifting of peak positions of these promi- nent diffraction lines suggests the formation of solid so- lution corresponding to Zn$_x$Cd$_{(1-x)}$Te material from the basic starting compounds CdTe and ZnTe.

From the scanning electron micrographs, it is found that the thicknesses of the films samples are too small to observed structure patterns. However films surfaces are very smooth. In all the compositions it has been observed that the reflectivity of the film gradually increases, which is quite natural that the reflectivity of the film is expected to increases with the increase of 'x' composition. The same observation as above shows the straitions, which are exactly parallel equidistance and extend from one end to another end. These straitions indicate the oscillatory growth [15], which indicate at the one end and terminate at the other end. It may be said that the oscillatory growth in the film which is manifested by straitions has been initiated by the presence of small composition of Zn ($x = 0.1$) in **Figure 2** and these oscillatory growth has been found to reduce successively with the composition $x = 0.3$ and $x = 0.4$. It is about reduce with composition of $x = 0.5$. The composition of starting basic ingredients and film composition comparison is presented in the **Table 1** and expressed in atomic percentages. The atomic percentage of basic ingredient seems to be in agreements with that obtain from EDAX analysis. From this table it is remarkable point to note that for composition ($x = 0.3$) the atomic percentage of basic ingredient taken is very close to atomic percentage obtain from EDAX spectra.

The reflectance and transmittance spectra of these sam-

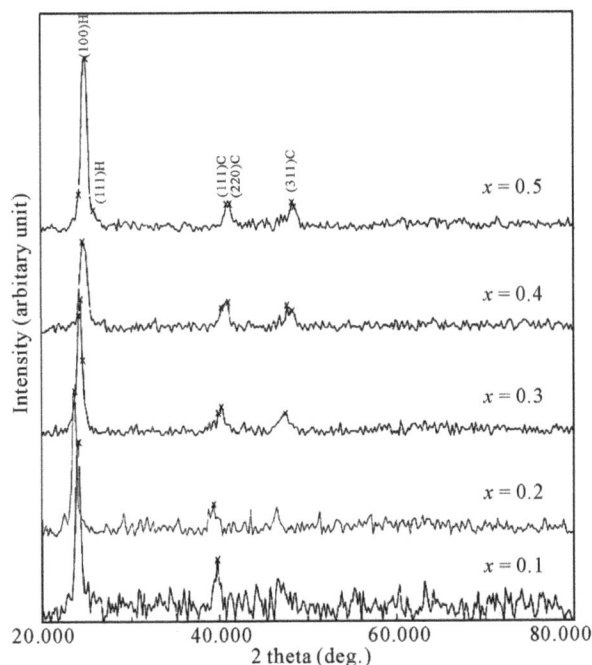

Figure 1. X-ray diffractograms of various Zn$_x$Cd$_{(1-x)}$Te structures.

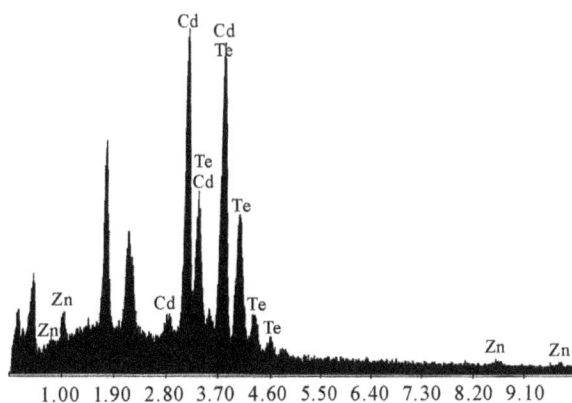

Figure 2. EDAX of ternary compound Zn$_x$Cd$_{(1-x)}$Te Thin Film. ($x = 0.1$).

Table 1. EDAX data for $Zn_xCd_{(1-x)}Te$ composite thin films.

	Basic Ingredient Taken			EDAX Composition		
'x'	At% Zn	At% Cd	At% Te	At% Zn	At% Cd	At% Te
0.1	5	45	50	7.07	40.52	52.41
0.2	10	40	50	13.42	37.78	48.81
0.3	15	35	50	15.59	34.46	49.96
0.4	20	30	50	28.20	25.90	45.90
0.5	25	25	50	31.49	20.95	47.56

ples were recorded using Hitachi Spectrophotometer model-330 in spectral region 200 - 2500 nm. Using these data, the absorption coefficient α has been calculated by applying the relation [12]. Absorption coefficients have been evaluated using percentage transmittance data as a function of wavelength presented in **Figure 3** for the samples of different compositions. The plot of $(\alpha h v)^2$ versus hv are plotted and shows clearly linear dependence for the value of $p = \frac{1}{2}$. This is attributed to an allowed and direct transition with direct band gap energies. The evaluated band gap energies are 1.7 eV, 2.05 eV, 2.2 eV, 2.3 eV and 1.5 eV for the samples of compositions x = 0.1 to 0.5 respectively clearly indicating dependence on compositions of films. Band gap energies are to be composition dependent. Band gap energy increases with the increasing 'x', this is as expected as band gap energy for ZnTe is 2.26 eV and band gap energy for CdTe is 1.5 eV [16].

CdZnTe thin films of thickness 450 - 1400 nm have been evaporated under vacuum onto unheated glass substrates, using a multilayer method [17]. He reported variation of optical band gap between 1.16 and 1.63 eV. Band-to-band transitions which give rise to the optical absorption in the visible region of the spectrum may be interpreted in terms of direct allowed transition with the band gap in the range of 2.05 - 1.92 eV. [18]. The band gap energy of the films measured by optical transmittance measurement is 1.523 eV [14]. Polycrystalline thin films of CdZnTe and CdMnTe have been grown by molecular beam epitaxy and metal-organic chemical vapor deposition [19]. He reported with band gaps of 1.65 - 1.75 eV for the top of a two-cell tandem design. P-i-n cells were fabricated and tested using Ni/p$^+$-ZnTe as a back contact to the ternary films. CdTe cells were also fabricated using both growth techniques.

Near normal incidence reflectance and transmittance data have been used to determine optical constant 'n' and 'k' [13]. The variation of refractive indices and extinction coefficients as a function of wavelength as represented in **Figure 4** for the samples of compositions

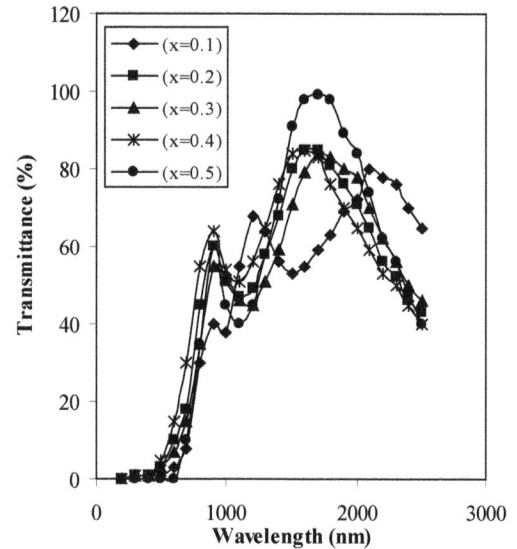

Figure 3. Spectral behaviour of transmittance with wavength.

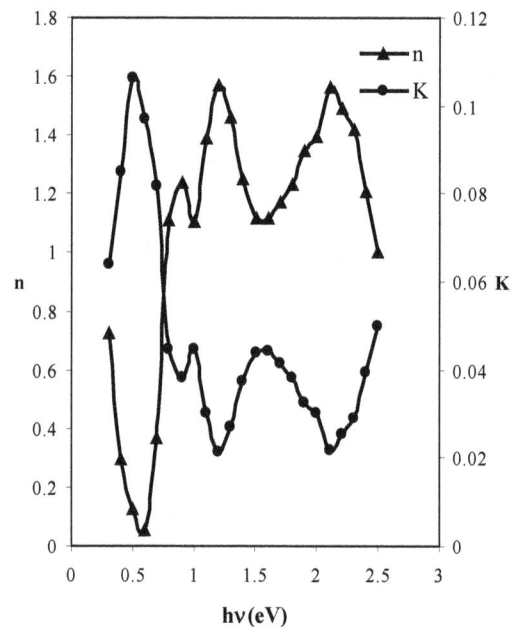

Figure 4. Variation of 'n' and 'k' with wavelength for $Zn_xCd_{(1-x)}Te$ ($x = 0.1$).

($x = 0.1$) It is found that variations in refractive indices and extinction coefficients are oscillatory in nature. Secondly variation in n and k seems to be complementary *i.e.* maxima of one and minima of the other at the same wavelength as estimated in **Table 2**.

Polycrystalline $Cd_{0.96}Zn_{0.04}Te$ thin films [20] were desited onto well-cleaned glass substrates kept at room temperature by vacuum evaporation. Optical properties of thin films were studied by optical transmittance meare-

Table 2. Well defined Maxima and Minima in variation of '*n*' and '*k*'.

Composition (x)	λ (μ)	Maxima		Minima	
		'*n*'	'*k*'	'*n*'	'*k*'
	0.6	-	0.10	0.055	-
	0.9	1.23	-	-	0.038
	1	-	0.044	1.10	-
0.1	1.2	1.57	-	-	0.021
	1.5	-	0.044	1.11	-
	2.1	1.56	-	-	0.021
	0.4	-	0.076	1.84	-
	0.9	1.96	-	-	0.019
0.2	1.2	-	0.035	1.92	-
	1.6	1.97	-	-	0.010
	0.4	-	0.065	1.86	-
	0.9	1.96	-	-	0.019
0.3	1.2	-	0.034	1.93	-
	1.7	1.97	-	-	0.010
	0.4	-	0.078	1.84	-
	0.9	1.96	-	-	0.017
0.4	1.1	-	0.031	1.93	-
	1.6	1.97	-	-	0.010
	0.9	1.95	-	-	0.021
0.5	1.1	-	0.046	1.90	-
	1.7	1.99	-	-	0.0007

ment and spectroscopic ellipsometry (SE). The spectra of various optical constants obtained from the SE $\varepsilon(E)$ data ($\varepsilon 1$, $\varepsilon 2$, R, n, k and α) revealed three distinct critical points (E1, E1+Δ1 and E2). The band gap energy of the films determined by transmittance measurement was 1.523 eV at room temperature. He reported refractive indices '*n*' varies between 2.4 to 2.6 and extinction coefficient '*k*' varies 0.5 to 1.

Optical properties of $Zn_xCd_{(1-x)}$Se films [21]. He reported variation in n and k in the wavelength range 600 - 1000 nm is very close matching with present work. They also reported the band gap energy increases with increasing Zn component in $Zn_xCd_{(1-x)}$Se films. At higher wavelengths the experimental results T and R satisfy the relationship T + R = 1. This indicates that neither absorp-

tion nor scattering of light occurs beyond the absorption edge. The appearance of maxima and minima results from interference effect and their number increases with increases film composition. These results are satisfactory and as theoretically expected.

4. Conclusions

- Homogeneous polycrystalline of multi phase structure of the thin films of $Zn_xCd_{(1-x)}$Te have been successfully deposited by thermal evaporation technique using basic ingredient ZnTe and CdTe elemental starting materials.
- EDAX composition seems to be closely matched with starting basic ingredients.
- The dependence of the optical parameters of the films on the light energy supports the direct character of the interband transition through an optical band gap in the range 1.7 - 2.3 eV.
- The variation in optical constants as a function of wavelength is oscillatory in nature having well defined maxima and minima, which depends on the composition of the thin films.

5. Acknowledgements

The authors are thankful to Prof. Dr. M. V. Patil, Principle, S. S. V. P. S. ACS College, Shindkheda. The authors are also grateful to Prof. Dr. P. P. Patil, Head, Department of Physics, North Maharashtra University, Jalgaon. One of the authors U. P. K. thanks to the University Grants Commission, New Delhi, for financial supports through the U.G.C. Minor Research Scheme No.F. 47 - 1275/09 (WRO) Pune.

REFERENCES

[1] T. H. Weng "Flash Evaporated Films of Indium-Doped CdS and CdS_xSe_{1-x}," *Journal of Electrochemical Society*, Vol. 117, No. 5, 1970, pp. 725-726.

[2] A. Rohatgi, S. A. Ringel, "Growth and Process Optimization of CdTe and CdZnTe Polycrystalline Films for High Efficiency Solar Cells," *Solar Cells*, Vol. 30, No. 1-4, 1991, pp. 109-122.

[3] S. I. Radautsan, O. V. Kulikova and O. G. Maksimova, "InSb-ZnxCd(1-x)Te Photosensitive Thin Film MIS – Structures," *Solar Energy Materials*, Vol. 20, No. 1-2, 1990, pp. 37-41.

[4] N. G. Patel, "Some Observations on the Switching and Memory Phenomenon in ZnTe-Si," *Journal of Materials Science*, Vol. 21, No. 6, 1986, pp. 2097.

[5] J. J. Kennedy, P. M. Amirtharaj and P. R. Boyd, "Growth and Characterization of $Cd_{1-x}Zn_xTe$ and $Hg_{1-y}Zn_yTe$,"

Journal of Crystal Growth, Vol. 86, No. 1-4, 1990, pp. 93-99.

[6] R. Weil, M. Joucla, J. L. Loison, M. Mazilu, D. Ohlmann, M. Robino and G. Schwalbach, "Preparation of Optical Quality ZnCdTe Thin Films by Vacuum Evaporation," *Applied Optics*, Vol. 37, No. 13, 1998, pp. 2681-2686.

[7] S. Mac kowski, G. Karczewski, F. Kyrychenko, T. Wojtowicz and J. Kossut, "Influence of MBE Growth Conditions on Optical Properties of CdTe/ZnTe Quantum Structures," *Thin Solid Films*, Vol. 367, No. 1-2, 2000, pp. 210-215.

[8] S. A. Ringel, R. Sudharsanan, A. Rohatgi, M. S. Owens and H. P. Gillis, "Influence of Thermal Annealing on the Structural and Optical Properties of Polycrystalline $Cd_{0.96}Zn_{0.04}Te$ Thin Films," *Journal of Optoelectronics and Advanced Materials*, Vol. 7, No. 3, 2005, pp. 1483-1491.

[9] A. Bansal and P. Rajaram, "Electrochemical Growth of CdZnTe Thin Films," *Materials Letters*, Vol. 59, No. 28, 2005, pp. 3666-3671.

[10] A. Mycielski, A. Szadkowski, E. usakowska, L. Kowalczyk, J. Domagaa, J. Bk-Misiuk and Z. Wilamowski, "Parameters of Substrates–Single Crystals of ZnTe and $Cd_{1-x}Zn_xTe$ ($x < 0.25$), Obtained by Physical Vapor Transport Technique (PVT)," *Journal of Crystal Growth*, Vol. 197, No. 5, 1999, pp. 423-426.

[11] V. M. Lakeenkov, V. B. Utimtsev, N. I. Shmatov and Y. F. Schelkin, "Numeric Simulation of Vertical Bridgman Growth of $Cd_{1-x}Zn_xTe$ melts," *Journal of Crystal Growth*, Vol. 197, No. 3, 1999, pp. 443-448.

[12] J. J. Pankove, "Optical Processes in Semiconductors," Prentice Hall, Englewood Cliffs, 1971.

[13] Goswami, "Thin Film Fundamentals," New Age International Publisher, New Delhi, 1996, p. 442.

[14] M. Sridharan, S. K. Narayandass, D. Mangalaraj and H. C.

Lee, "Optical and Opto-Electronic Properties of Polycrystalline $Cd_{0.96}Zn_{0.04}Te$ Thin Films," *Crystal Research Technology*, Vol. 38, No. 6, 2003, pp. 479-487.

[15] M. S. Joshi and A. S. Vagh, "Role of Spirals in the Growth of Prism Faces of Cultured Quartz," *Journal of Applied Physics*, Vol. 37, No. 1, 1966, pp.315-318.

[16] U. P. Khairnar, D. S. Bhavsar, R. U. Vaidya and G. P. Bhavsar, "Optical Properties of Thermally Evaporated Cadmium Telluride Thin Films," *Materials Chemistry and Physics*, Vol. 80, No. 2, 2003, pp. 421-427.

[17] G. G. Rusu, M. Rusu and M. Girtan, "Optical Characterization of Vacuum Evaporated CdZnTe Thin Films Deposited by a Multilayer Method," *Vacuum*, Vol. 81, No. 11-12, 2007, pp. 1476-1479.

[18] K. Prabakar, S. Venkatachalam, Y. L. Jeyachandran, S. K. Narayandass and D. Mangalaraj, "Optical Constants of Vacuum Evaporated $Cd_{0.2}Zn_{0.8}Te$ Thin Films," *Solar Energy Materials and Solar Cells*, Vol. 81, No. 1, 2004, pp. 1-12.

[19] A. Rohatgi, S. A. Ringel, R. Sudharsanan, P. V. Meyears, C. H. Liu and V. Ramanathan, "Investigation of Polycrystalline CdZnTe, CdMnTe, and CdTe Films for Photovoltaic Applications," *Solar Cells*, Vol. 27, No. 1-4, 1989, pp. 219-230.

[20] M. Sridharan, S. K. Narayandass, D. Mangalaraj, H. C. Lee, "Optical and Opto-Electronic Properties of Polycrystalline $Cd_{0.96}Zn_{0.04}Te$ Thin Films," *Crystal Research Technology*, Vol. 38, No. 6, 2003, pp. 479-487.

[21] P. Gupta, B. Maiti, A. B. Maity, S. Chaudhari, A. K. Pal, "Optical Properties of $Zn_xCd_{1-x}Se$ Films," *Thin Solid Films*, Vol. 260, No. 1, 1995, pp. 75-85.

5

Effect of Nitriding on Wear Behavior of Graphite Reinforced Aluminum Alloy Composites

Bhujang Mutt Girish, Bhujang Mutt Satish, Hanyalu Ramegowda Vitala

Research and Development Center, Department of Mechanical Engineering, East Point College of Engineering and Technology, Bangalore, Karnataka, India.

ABSTRACT

The paper evaluates the effect of nitriding on the wear behavior of graphite reinforced aluminum 6061 alloy composites. The composites were prepared using the liquid metallurgy technique. The content of graphite in the composites was varied from 3 to 7% (by weight) in steps of 2%. The nitriding process was carried out at 500°C for 24 hours. Three categories of specimens, namely, the nitrided composites, non-nitrided composites, as well as the alloy specimens were tested for their wear behavior. Pin-on disc equipment was used for wear testing. X-Ray Diffraction technique (XRD) was used to confirm the implantation of nitrogen in the composites. It was observed that the nitrided composites have better wear resistance than the non-nitrided composites.

Keywords: Composites, Nitriding, Surface Engineering, Wear

1. Introduction

Aluminum (Al) and its alloys posses numerous advantages such as low specific weight, high strength-to-weight ratio and low cost, and hence find wide applications in the automotive, aviation and space industries [1]. However, the main limitation of their wider application for commercial purpose is the lack of surface hardness, mechanical strength and wears resistance, together with low thermal and chemical stability. Therefore, a surface modification with the aim to improve the above properties is necessary for further industrial use. This can be achieved by formation of aluminum nitrite (AlN) on Al surface, obtained by nitrogen insertion into the surface of bulk Al as AlN exhibits excellent tribological properties combined with high thermal and chemical stability [2].

The use of aluminum for light weight machine parts and car components has recently increased, although non-metallic materials, such as resin, are also used for this purpose. Al alloys have advantages over non-metallic materials such as light weight, corrosion-resistance, and are workable and have good thermal conductivity. However, the hardness, wear and seize resistance of Al alloys is lower than that of the steel, and hence there is a limit on the application as sliding parts. Thus, research has been carried out on surface modification technology

to increase the applicability of Al alloys as sliding parts [3].

MMCs (Metal Matrix Composites) have received increasing attention in the recent decades as engineering materials. MMCs are primary candidate materials for industrial applications in the aerospace, automotive and power utility industries. However, their properties such as strength, toughness, and wear and corrosion properties depend to a great extent on several factors of which matrix properties are very important [4].

A more widespread use of light metal alloys in tribological applications like guide bars, bearing plates, seat supports or bushings demands powerful functional surface coatings to provide wear protection as well as compressive strength. The direct contact of uncoated light metal substrates with sliding or oscillatory counterparts results in severe wear, seizing and high frictional coefficient, even under lubricated conditions.

There are different surface treatment processes available commercially for aluminum alloys to increase the corrosion and wear resistance, such as nitriding, anodic oxidation, electrolysis, nickel plating etc. Nitriding is one surface modification technique that is widely used to increase the fatigue, mechanical strength and wear resistance of machine parts of carbon or alloys steels. It is a process of introducing nitrogen in to the metals so as to modify the surface properties, and increase the hardness, wear resistance and corrosion resistance [5].

The present paper aims to report the evaluation of wear behavior of nitrided aluminum MMCs reinforced with graphite particles and the same is compared with non- nitrided MMCs.

2. Experimental Method

2.1. Preparation of Composites

Aluminum 6061 alloy with the chemical composition given in **Table 1** was used as the matrix material. The optical emission spectrometer was used determine the chemical composition of the alloy. The reinforcement material used was graphite which is a solid lubricant with adequate resistance to wear. The composites were fabricated by liquid metallurgy technique, the details of which are available elsewhere [6]. Graphite particles (0%, 3%, 5% and 7% by weight) were heated separately to a nominal temperature to remove moisture content if any, and introduced into the vortex of the effectively degassed Al 6061 molten alloy. The molten alloy was stirred rigorously using a stirrer having a ceramic coated steel impeller.

2.2. Nitriding of Specimens

Nitriding is a promising method for surface treatment to improve hardness, corrosion, wear and fatigue resistance of materials. Nitriding process can improve the tribological and mechanical properties by enriching the near surface region with nitrogen. Nitrides produced by the combination of nitrogen with alloying elements posses a higher hardness as compared with the iron nitrides. It is due to this fact that the nitriding process is primarily used to increase the hardness and wear resistance of the surface of several metals including steel.

The specimens prepared for wear testing were washed in distilled water, and then with acetone and later subjected to the nitriding process. Nitriding was accomplished in a muffle furnace where ammonia gas was introduced into the air-tight furnace chamber heated to a temperature of 500°C. The specimens were exposed for a duration of 24 hours and the ammonia feed rate was 20CFH (Cubic feed/hr).

Ammonia gas decomposes giving the nitrogen in nascent form or monatomic nitrogen, which is the only form capable of entering the material and diffusing in it.

2.3. Wear Test Equipment

The wear tests were conducted in accordance with ASTM G99 standards using a pin-on-disc sliding wear testing machine, which is similar to the one, used by Poonawala *et al.*, [7]. EN24 steel disc of diameter 200 mm and chemical composition: C-0.45, Si-0.35, Mn-0.70, Cr-1.40, Mo-0.35, Ni-1.80, S-0.05, P-0.05 in weight %, was used as the counter face on which the test specimens slide. Hardness of the steel disc was HRc 57 achieved by oil quenching at 850°C and tempering at 550°C for 2 hours. Arrangements were made to hold a specimen and also for application of the load on the specimen. The test specimen was clamped in a vertical sample holder and held against the rotating steel disc. In the present investigation, loads of 20-160 N in steps of 20 N were used. The rotational speeds employed were 200, 250, and 300 rpm, which at an average distance of 80 mm from the center gave corresponding linear speeds of 1.25, 1.56, and 1.87 m/s respectively.

2.4. Testing of Specimens

The 'weight loss' method was adopted in the present study in which the pins of the material under investigation were 6 mm in diameter and 15 mm in length. A fresh disc was used each time and before each test, the disc was cleaned with acetone to remove any possible traces of grease and other surface contaminants. The specimens were cleaned with ethanol and weighed before and after the tests using a balance accurate to ± 0.001g. The wear results were computed from weight loss measurements. The duration of each test was exactly 60 minutes. The wear volume was calculated from the ratio of weight loss to density and wear rate was calculated using sliding distance and wear volume. Usage of such relations to calculate the wear parameters is common and has been used by Rohatgi *et al.*, [8]. The data for the wear tests was taken from the average of three measurements. The standard deviation was about 5%. The surfaces of the worn specimens were cleaned thoroughly to remove the loose wear debris and then observed using a scanning electron microscope (SEM). Along with the test specimens, even the surface of the counterpace steel disc was subjected to the SEM analysis in order to draw as much vital information as possible about the wear behavior of the composite specimens.

Table 1. Composition of Al-6061 alloy.

Elements	Cu	Mg	Si	Fe	Mn	Ni	Zn	Pb	Sn	Ti	Cr	Al
% by Wt	0.36	0.99	0.80	0.12	0.02	0.01	0.01	0.07	0.05	0.01	0.12	Bal

3. Results and Discussion

Wear of composite materials although appears to be simple, the actual process of material removal is quite complex. This can be attributed to the fact that a large number of factors influence wear such as metallurgical factors (weight percentage, chemical composition, and size of the reinforcement) or service (*i.e.*, speed, pressure) or other contacting factors (lubrication, corrosion, etc.) The factors that have been considered in the present study to explain the wear behavior of the composites are the applied load, the applied speed, and the weight percentage of the reinforcement.

Scanning electron microscopy were carried out on samples which were ground to 600 grit SiC and polished using alumina powders up to 1 μ, and etched using Keller's reagent. The microphotograph of the 7% graphite composite showing uniform distribution of the reinforcement is shown in **Figure 1**. The **Figure 2(a)** is the EDX of a nitrided Al6061 alloy specimen. The nitrogen peak is clearly visible indicating the presence of nitrogen. The **Figure 2(b)** shows the thickness of the coating that was obtained and it was found to be 1.10 mm.

3.1. Effect of Load and Speed on the Wear Rate

The specimens tested fall into three categories, namely, the unreinforced and non-nitrided specimen, the non-nitrided composite, and the nitrided composite. Three pin specimens were tested from each category at each specified load and speed.

The results were averaged to obtain the final wear rate, which are presented graphically in **Figures 3-5**. In the graphs shown in **Figure 3-5**, 'UN' means unnitrided specimens, while 'nit' means nitrided specimens. The wear rate of the unreinforced alloy is also plotted in order to enable comparison with the other category specimens. It is found from the graphs that the wear rate of the specimens belonging to all the three categories increased with the applied load. The wear rate of the non-nitrided composite as well as the nitrided composite specimen reduced with the increase in graphite content. It is clearly evident from the graphs that there exists a certain load, *i.e.*, a transition phenomena at which there is a sudden increase in the wear rate of the specimens belonging to all the three categories. However, the transition loads for the nitrided composites were much higher than that observed for the non-nitrided composites, and that of the non-nitrided composites was much higher than the unreinforced alloy and also the transition load increased with the increase in graphite particle content.

The **Figure 3** presents the behavior of the specimens tested at a speed of 1.25 m/s and load ranging from 20 to 200 N. It was observed that the unreinforced matrix alloy

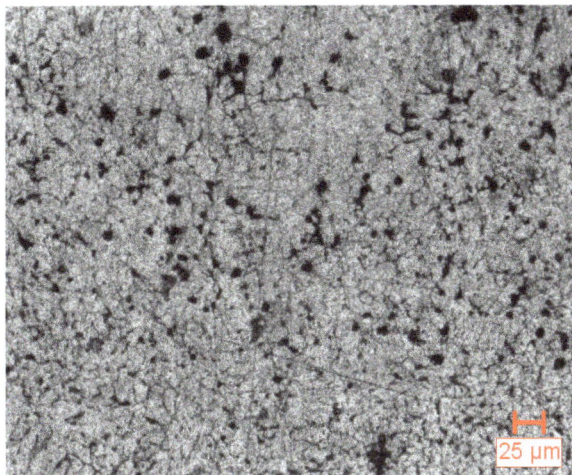

Figure 1. SEM showing uniform distribution of graphite particles in the alloy matrix.

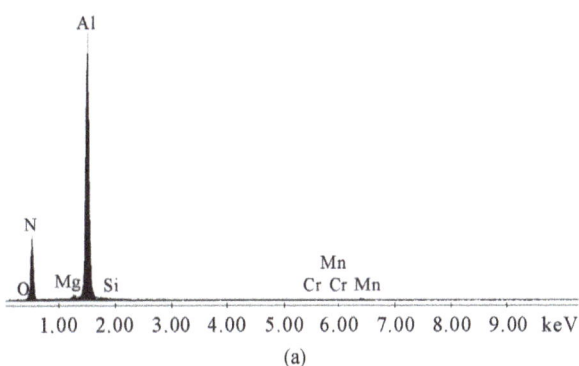

(a)

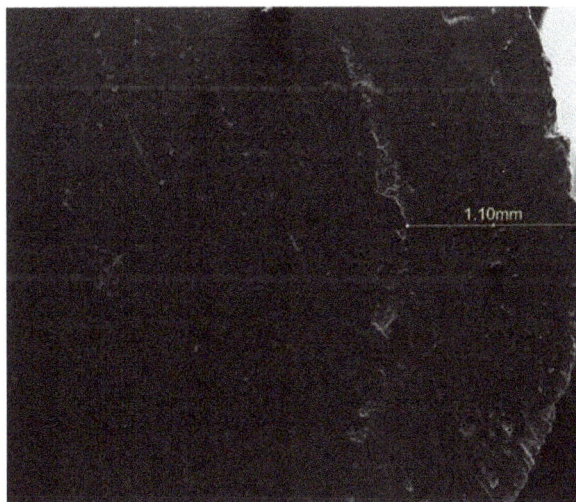

(b)

Figure 2. (a) EDX of a nitrided composite specimen; (b) SEM showing the thickness of the coating on the specimen.

showed a transition from mild to severe wear at a load of 60 N, while the 3 and 5% graphite reinforced composites showed a transition at 140 and 160 N respectively. The

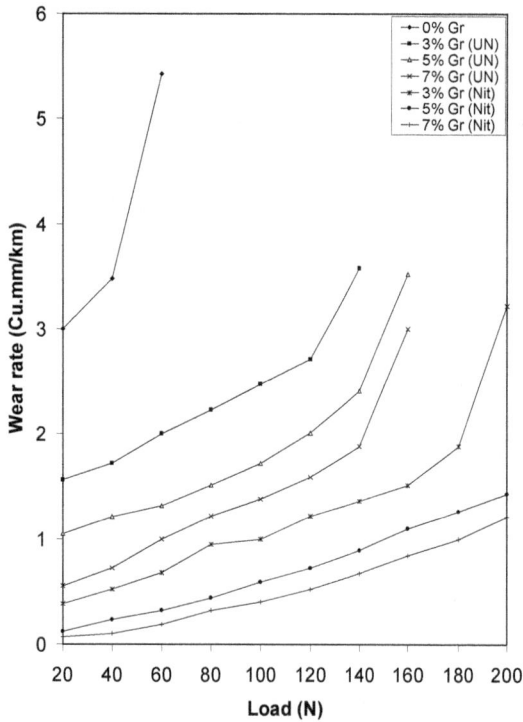

Figure 3. Graph of wear rate v/s load at a speed of 1.25 m/s.

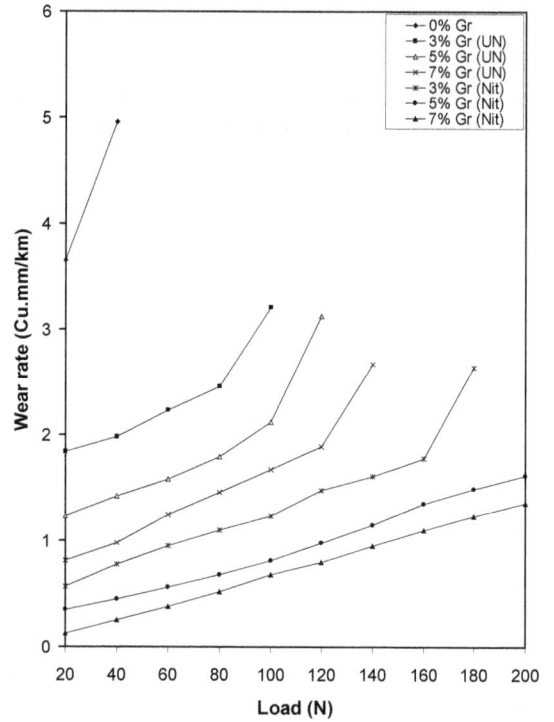

Figure 5. Graph of wear rate v/s load at a speed of 1.87 m/s.

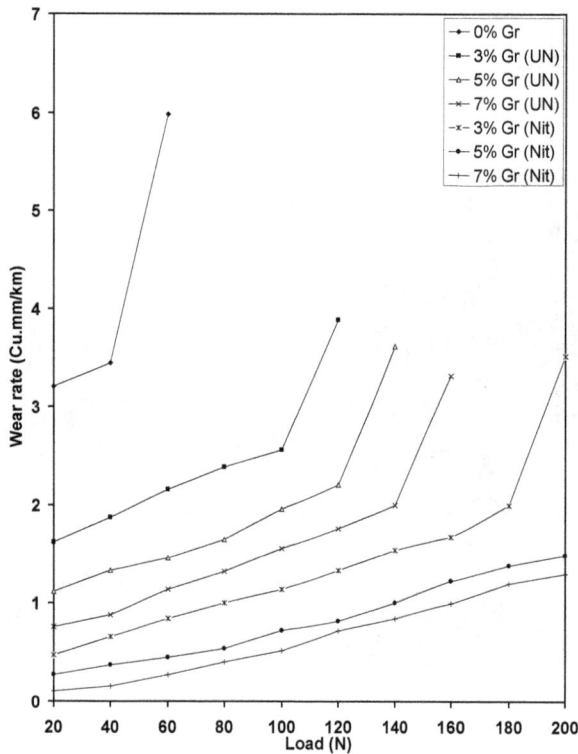

Figure 4. Graph of wear rate v/s load at speed of 1.56 m/s.

same observation was made at a load of 140 N in case of 7% graphite reinforced composites. This observation

which is evident from **Figure 3** indicates that the presence of graphite reinforcement delays the transition from mild to severe wear, and increases the transition load of the 7% reinforced composite by almost 2.5 times with respect to the unreinforced alloy. Interestingly, similar transition in the case of nitrided composites was observed at 200 N in those having a reinforcement content of 3% graphite, while the 5 and 7% reinforcement composites show no transition at all even at 200 N.

It follows from the results obtained that comparatively low wear rates exist at lower loads, thereby indicating the regime of mild wear. In this regime of mild wear, the composites demonstrate significant wear resistance than the alloy counterpart. At higher loads, the materials exhibit rapid increase in wear rate. At loads greater than the transition loads, severe wear occurs leading to seizure of the materials. The severe wear manifests itself by a rapid rate of material removal in the form of generation of coarse metallic debris, and also by massive surface deformation and material transfer to the counter face.

The composites behave very differently from the unreinforced alloy. The alloy shows a transition at 60 N when tested at 1.25 and 1.56 m/s, while the same is observed at 40 N in case of 1.87 m/s test. Similarly the composite with 3 % reinforcement shows a transition at 140 N when tested at 1.25 m/s, at 120 N when tested at speed of 1.56 m/s and 100 N at 1.86 m/s. Similar observations for 5 and 7% composites as well as the nitrided composites are clearly evident form the graphs shown in **Figures 4** and **5**.

The composite with 5% graphite shows a transition at 160 N when tested at 1.25 m/s, at 140 N at speed of 1.56 m/s, and the same is observed at 120 N in case of 1.87m/s test. The above observations clearly indicate that the sliding speeds employed have a significant effect on the wear rate transition on the materials. The transition in wear rate decreases with the increase in speed in all the materials. The results obtained are on par with the one obtained by Lee, *et al.*, [9] who have reported that the wear mechanisms are strongly dependent on the sliding speeds.

The mild wear of the alloy is oxidation dominated wear at low sliding speeds and loads. In the case of composites, due to the existence of graphite particles, the oxide film of the metal is not continuous and tenacious. It is removed by friction forces in the following sliding friction resulting in oxidation assisted mild abrasion wear. Hence it can be considered that the dominating wear mechanism is the removal and reproduction of the oxide film. This kind of wear is maintained until higher loads are employed, under which condition the wear mechanism transforms from mild to severe wear. The morphologies of wear surfaces of 7% graphite reinforced composites are shown in **Figure 6(a)-(c)**. **Figure 6(a)** shows the wear tracks for unreinforced alloy, while **Figure 6(b)** and **(c)** represent the wear tracks for composite and nitrided composite respectively. The wear tracks in **Figure 6(b)** and **(c)** show typical abrasive wear for the composites tested at low loads of 20 and 60 N respectively. Hence it can be concluded that the dominating wear mechanism is abrasive wear at low loads.

It was observed that at higher loads a transition occurs from mild to severe wear, and the wear rate quickly increases by tremendous rate. Due to the high loads employed, the friction and wear increased obviously. In this condition, the removal and formation of oxide films are faster than that of mild oxidation, thereby resulting in relatively higher wear rates.

3.2. Effect of Graphite Particles on Wear Rate

It follows from the observations that the graphite particles play a very strong role in enhancing the wear resistance of the composites. It was found that the transition load increases with the increase in graphite content and also the wear rate of the composites was lower than that of the base alloy without graphite. This is obviously due to the release of graphite particles by the composite specimens on to the mating surface during sliding which provides resistance to wear. The release of graphite onto the sliding interface causes formation of a thin film such that the relative movement of the mating surfaces pro-

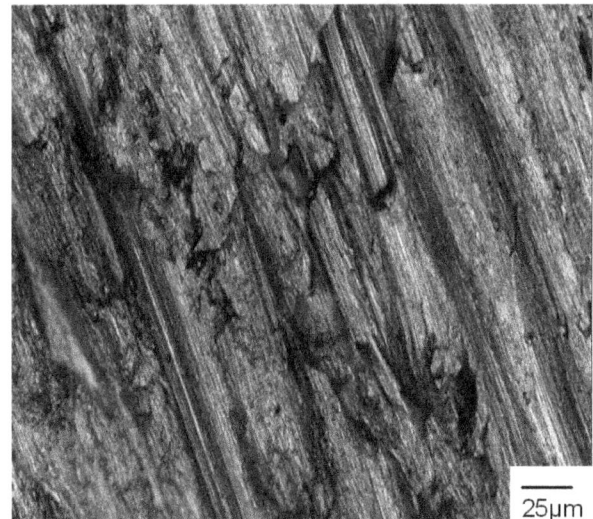

(a)

(b)

(c)

Figure 6. (a) SEM showing wear tracks in the unreinforced alloy; (b) SEM showing wear tracks in the 7% graphite reinforced non-nitrided composite; (c) SEM showing wear tracks in the 7% graphite reinforced and nitrided composite.

motes easy shear between the lamellar planes of graphite. The buildup of the film is a significant feature of graphite tribology. Graphite has a hexagonal layered structure and the bonds between the parallel layers are relatively weak (van der Walls type). The key to graphite's value as a self-lubricating solid lies in its layered lattice structure and its ability to form strong chemical bonds with gases such as water vapour. The adsorption of water vapour and other gases from the environment onto the crystalline edges weakens the interlayer bonding forces, resulting in easy shear and transfer of the crystalline platelets on to the mating surfaces. Graphite also performs well under boundary lubrication conditions because of its affinity fro hydrocarbon lubricants [10].

Hence it can be concluded that the ability of the sheared reinforcement layers to adhere to the sliding surface decides the effectiveness of the graphite particles in reducing the wear rate of the composite materials.

3.3. Effect of Nitriding on Wear Rate

Nitriding is a surface treatment process where nitrogen is supplied into the chamber of specialized equipment at relatively high temperature. A nitrided thin layer on the surface will be formed containing aluminum alloy based phases. The addition of graphite to the base alloy significantly improves the wear resistance. This is further improved by several folds in the case of nirtided composites. The hardness of the nitrided composites increases quite significantly as shown in the **Table 2**. The primary cause for the increase in hardness and hence wear resistance is the presence of AlN and nitrogen enriched aluminum at the surface. The presence of the phases is shown by the XRD analysis which is presented in **Figure 2(a)**.

3.4. SEM Analysis

In view of brevity and convenience, the SEM micrographs of only 7% composites at speed of 1.87 m/s have been presented. However, the explanation holds good even for the composites with 3 and 5% reinforcement as well. The SEM micrographs of a typical worn surface of the 7% graphite reinforced composites are presented in **Figure 6(a)**, **(b)** and **(c)** which shows the wear track morphology of the specimens tested at various loads.

It can be seen that a lot of parallel, continuous and

deeply ploughed grooves exist on the wear surface of the composites and there is an abrasion phenomenon observed at low loads. The parallel grooves suggest abrasive wear as characterized by the penetration of the graphite particles into a softer surface, which is an important contributor to the wear behavior of composites. The worn surfaces in some places reveal patches of material removed from the surface of the material during the course of wear and smeared on to the sliding surface.

4. Conclusions

The addition of graphite particles to the aluminum alloy improves the wear resistance of the composite in spite of the significant improvement to wear resistance. Nitriding which is a surface treatment process further improves the wear resistance by several folds. The primary cause for the increase in hardness and hence wear resistance is the presence of AlN and nitrogen enriched aluminum at the surface.

Table 2. Vickers micro hardness number for nitrided and non-nitrided specimens.

Specimen Type	VHN
Non-Nitrided Specimens	68
Nitrided Specimens	82

REFERENCES

[1] P. H. Chong, H. C. Man and T. M. Yue, "Microstructure and Wear Properties of Laser Surface Claded Mo-WC MMC on AA6061 Aluminium Alloy," *Surface and Coatings Technology*, Vol. 145, No. 1-3, 2001 pp. 51-59.

[2] L. Liu, W. W. Li, Y. P. Tang, B. Shen and W. B. Hu, "Friction and Wear Properties of Short Carbon Fiber Reinforced Aluminium Matrix Composites," *Wear*, Vol. 266, No. 7-8, 2009, pp.733-738.

[3] Y. Sahin and K. Ozdin, "A Model for Abrasive Wear Behaviour of Aluminium Based Composites," *Materials and Design*, Vol. 29, No. 3, 2008, pp. 728-733.

[4] A. T. Alpas and J. Zhang, "Effect of Microstructure and Counterface Materials on the Sliding Wear Resistance of Particulate Reinforced Aluminium Matrix Composites," *Metallurgical Transactions*, Vol. 25, 1994, pp. 969-983.

[5] D. D. Nolan, S. W. Huang, V. Leskovsek and S. Braun, "Sliding Wear of Titanium Nitride Thin Film Deposited on Ti-6Al-4V alloy by PVD and Plasma Nitriding Processes," *Surface and Coatings Technology*, Vol. 200, No. 20-21, 2006, pp. 5698-5705.

[6] S. C. Sharma, B. M. Girish, D. R. Somashekar, B. M. Satish and R. Kamath, "Sliding Wear Behavior of Zircon Particles Reinforced ZA-27 Alloy Composite Materials," *Wear*, Vol. 224, No. 1, 1999, pp. 89-94.

[7] N. S. Poonawala, A. K. Chakrabarti and A. B. Chattopadhyay, "Wear characteristics of Nitrogenated Chromium Cast Irons," *Wear*, Vol. 162, 1993, pp. 580-588.

[8]   P. K. Rohatgi, S. Ray and Y. Lin, "Tribological Proper-
      ties of Metal Matrix Graphite Particle Composites," *In-
      ternational Materials Review*, Vol. 37, 1992, pp. 129-138.

[9]   P. P. Lee, T. Savaskan and E. Laufer, "Wear Resistance
      And Microstructure of Zn-Al-Si and Zn-Al-Cu alloys,"

Wear, Vol. 117, No. 1, 1987 pp. 79-84.

[10] Y. Sahin, "Tribological Behaviour of Metal Matrix and
      Its Composites," Materials and Design, Vol. 28, No. 4,
      2007, pp. 1348-1352.

6

Analysis of Mobile Phone Reliability Based on Active Disassembly Using Smart Materials[*]

Zhifeng Liu, Liuxian Zhao, Jun Zhong, Xinyu Li, Huanbo Cheng

School of Mechanical and Auto Engineering, Hefei University of Technology, Hefei, China.

ABSTRACT

When using shape memory materials into active disassembly of actual electronic products, because the elastic modulus of shape memory materials is affected by the temperature is relatively large, therefore, the main difference of environmental reliability between active disassembly products and common products is the impact of collision and vibration under different temperature. Establishing three-dimensional analysis model, comparing the impact of collision and vibration of mobile phone shells which are made up of PVC materials after casting & radiation and PC/ABS materials under different temperature. Analyzing the reliability of mobile phone under different temperature and optimizing its structure according to data of testing.

Keywords: *Active Disassembly, Mobile Phone, Environmental Reliability, Analysis of Collision, Analysis of Vibration*

1. Introduction

With the rapid development of electronic technology, it is growing demand for electronic products and the pace of elimination of these products is accelerating which result in the serious pollution of environment. Therefore, Design for Disassembly (DfD) of products is getting more and more attention in order to meet the requirements of environment protection [1,2]. However, the current disassembly process of products is only one to one; that is to say, worker can only disassemble one product at one time. It is very difficult to achieve the automation and assembly line of disassembly process of products. What is more, the disassembly process can pose threat to human health. So, it is a growing concern of people about how efficiently to recycle and reuse these obsolete electronic products [3-5].

In recent years, with the deep research and extensive application of smart materials, the method of Active Disassembly using Smart Materials (ADSM) is getting more and more concerns [6-10]. The technology of ADSM was put forward by Dr. Chiodo in Brunel University at 1997. Active disassembly using smart material (ADSM) is a method using shape memory material to replace fastener, when it is heated to the stimulated temperature, the product can be disassembled actively. The

R&D center of Nokia, Helsinki University of Technology and Helsinki University of Art and Design started to research the active disassembly of mobile phone in cooperation at 2000-2001, which aimed to develop a thermal stimulated mechanical structure for simplifying the recycling process of mobile phone. By testing, the average disassembly time of mobile phone using ADSM is 2s, which is only 1/50 of the average disassembly time using artificial. Therefore, the method of ADSM can improve the efficiency of disassembly and achieve the economic and efficient recycling of mobile phone. The active disassembly tests of mobile phone and radio are shown in **Figure 1** [11].

Any brand of mobile phone needs to go through rigorous testing to verify whether the mobile phone can meet the required reliability norms before sales in the market. Environment is one of the most important factors which affect the reliability of mobile. According to some materials, it is about 52% accidents of electronic products are caused by environment. Including 40% are caused by temperature, 27% are caused by vibration, 19% are caused by humidity and the remaining accidents are caused by dust, smoke etc [12].

The main difference between active disassembly mobile phone and common mobile phone is that active disassembly mobile phone is made up of smart drives or active disassembly fasteners using shape memory alloy

---

*National Natural Science Foundation of China (50775064).

Figure 1. The active disassembly tests of mobile phone and radio [11].

(SMA) or shape memory polymer (SMP). And put the drives and fasteners into the product when design and assembly. Because properties of SMA and SMP are easy to be affected by temperature, so, the main difference of reliability between active disassembly mobile phone and common mobile phone is caused by collision and vibration under different temperature.

2. Constructing the CAD Model of Active Disassembly Mobile Phone

Constructing active disassembly mobile phone model used three-dimensional modeling software according to the structure of bar phone. As shown in **Figure 2**.

In order to be simple and can explain the problem, this phone model consists five parts. Number the parts according to the order from the outside to the inside, 1 - front cover, 2 - battery compartment cover, 3 - LCD screen and the circuit board, 4 - battery, 5 - back cover. The active disassembly fasteners of front cover and back cover are made up of PVC material which receives ir-radiation after casting, the dose of radiation is 4kGy (SMP for short). There are two L shaped slot in the area of back cover in order to place Ni-Ti memory alloy films which act as drive of disassembling the battery, the drive can be changed by changing the size of memory alloy films, shown in **Figure 3**.

Here select the stimulation temperature of Ni-Ti memory alloy films as 75°C. The size of phone model is 105 mm × 45 mm × 8 mm, the screen is 2 inch on standard, and the thickness of shell is 1 mm. The shell of common mobile phone is made up of PC/ABS. The properties of the shell of active disassembly mobile phone and common phone are shown in **Table 1**.

Table 1. The properties of SMP and PC/ABS.

Name	Density kg/m³	Elastic modulus GPa	Tensile strength MPa	Poisson's ratio
SMP	1000	3.7	70	0.4
PC/ABS	1140	2.5	80	0.3

3. Analysis of Collision of Active Disassembly Mobile Phone

Simulating the free falling body of mobile phone using computer simulation in order to observe the response of mobile during the process of free falling body.

3.1. Establishing the Equation of Collision

Establishing the equation of collision before analyzing the collision of mobile phone using Abaqus. The equation of collision can be described as below under the overall coordinate system.

$$Ma + Cv + Kd = F^{ex},   (1)$$

in which M means the mass matrix of the structure, C means the damping matrix of the structure, K means the stiffness matrix of the structure, a means the vector of the acceleration, v means the vector of the velocity, d means the vector of the displacement, F^{ex} means he vector of the external forces including the force of collision.

If order $F^{in} = Cv + Kd$ and suppose $F^{re} = F^{ex} - F^{in}$, then equation of collision can be described as below.

$$Ma = F^{re},   (2)$$

If use concentrated mass, the mass matrix becomes the diagonal matrix, and then each equation of degree of

freedom is independent. That is,

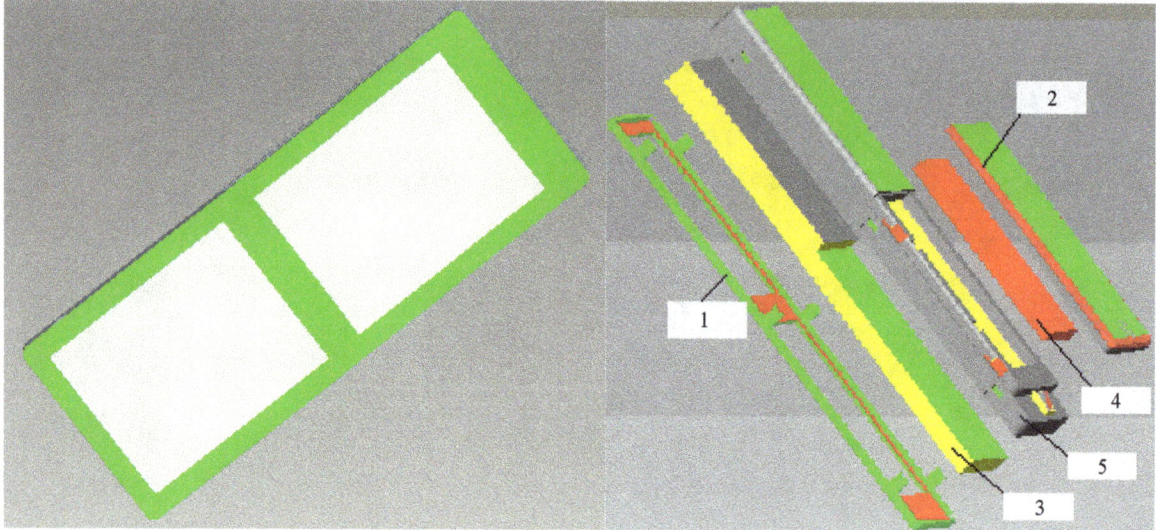

Figure 2. The structure of mobile phone.

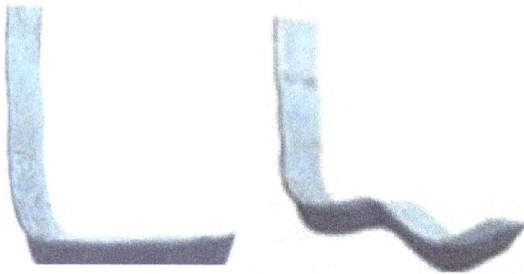

Figure 3. The Ni-Ti memory alloy films before and after deformation.

$$M_i a_i = F_i^{re}, \quad (i = 1, 2, \cdots), \qquad (3)$$

Solving the equation of collision using display method, firstly obtain the following equation by Equation (3).

$$a_i = F_i^{re} / M_i , \qquad (4)$$

Then obtaining the velocity (v_i) by integration of time, obtaining the displacement (d_i) by integration of time once again. Here, the integration of time using display format of central difference. The display format of central difference is described as below.

$$\begin{cases} v_{n+1/2} = v_{n-1/2} + a_n \left(\Delta t_{n+1/2} + \Delta t_{n-1/2} \right)/2 \\ d_{n+1} = d_n + v_{n+1/2} \cdot \Delta t_{n+1/2} \\ \Delta t_{n+1/2} = \left(\Delta t_n + \Delta t_{n+1/2} \right)/2 \end{cases} \qquad (5)$$

Therefore, obtaining the displacement, velocity and acceleration of each discrete point of time in the overall domain of the time used the above recurrence formula.

The display format of integration does not require solution and reverse solution of matrix, it is not necessary to solve the simultaneous equations and there is no problem of convergence. Therefore, the speed of computation is much fast; the criteria of stability can automatically control the step of time and ensure the speed of integration [13].

3.2. The Model of Finite Element Analysis

The experimental condition of the example is shown in **Figure 3**. The shell of mobile phone collides with the rigid plane from a height of 2 mm by free falling body. The element type of mobile phone shell is C3D8R, eight-node, reduced integration, linear, solid element, and the rigid plane uses discrete rigid body for simulation. Selecting the type of parts as discrete rigid body when constructing the model, therefore, it is not necessary to define the properties of materials. Selecting Explicit solver in Abaqus [14]. The properties of materials of mobile phone shell are shown in **Table 1**.

3.3. Analysis of the Result of Collision Simulation

Simulating the mobile phone collides with the rigid plane by Abaqus finite element method. As shown in **Figure 4**.

It can been seen from **Figure 4** that when the temperature is 20°C - 60°C, the maximum stress of mobile phone shell made up of SMP is 67.1 MP when it falls from a height of 2 m , and it does not exceed the tensile strength of SMP which is 70 MP. While the maximum stress of mobile phone shell made up of PC/ABS is 119 MP under the same condition, which exceeds the tensile strength of PC/ABS. Therefore, the mobile phone made up of PC/ABS can be destroyed falling on the rigid plane from a height of 2 m at room temperature. Thus, the re-

liability of mobile phone made up of SMP is higher than

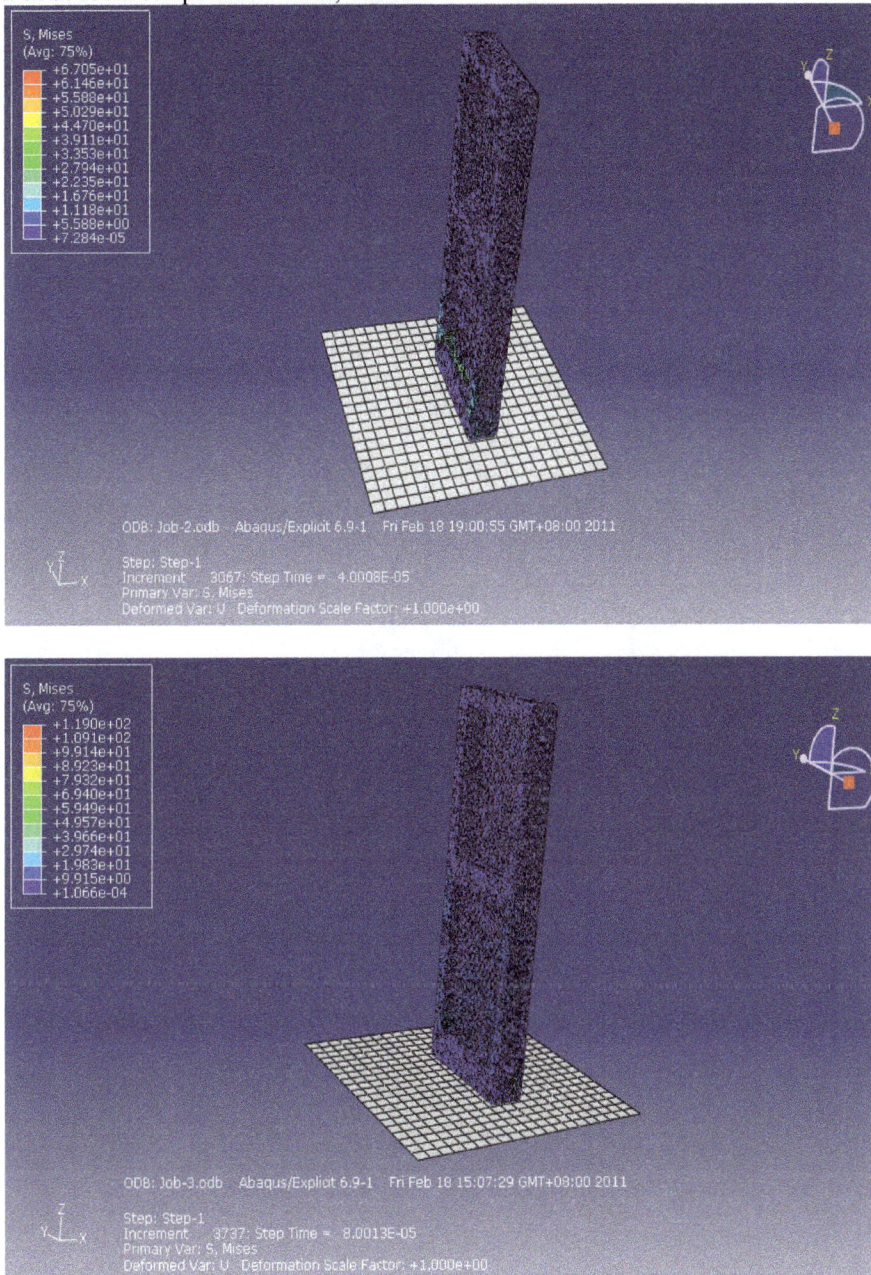

Figure 4. The collision experiment of mobile phone using SMP and PC/ABS.

the reliability of mobile phone made up of PC/ABS.

The maximum stress of collision of active disassembly mobile phone and the common mobile phone under different temperature are shown in **Table 2** and **Table 3**.

4. Analysis of Vibration of Active Disassembly Mobile Phone

4.1. The Analysis Model of Mobile Phone

The analysis of vibration of mobile phone mainly analyzes whether the vibration can cause the fasteners between front cover and back cover to detach. Therefore, just analyzing the displacement of fasteners between front cover and back cover is enough. The properties of materials of mobile phone shell are shown in **Table 1**. The parameters of mobile phone motor are shown in **Table 4**. The vibration time of motor is 5 s.

4.2. Analysis of the Result of Vibration Simulation

The mobile phone can only move alternatively along the
Table 2. The maximum stress of collision of active disassembly mobile phone under different temperature.

Temperature ˚C	The maximum temperature MPa	Tensile strength MPa
20 - 60	67.1	70
70	36.5	12.9

Table 3. The maximum stress of collision of common mobile phone under different temperature.

Temperature ˚C	The maximum temperature MPa	Tensile strength MPa
20 - 40	119	80

Table 4. The parameters of motor.

Rated voltage V (DC)	Rated Current (mA)	Rated Speed R/M	Hammer length mm	Mass g
3.0	80	20000	5	2

Z direction using finite element method of Abaqus. Therefore, it is necessary to define the constraints of X, Y directions and three rotation directions. The mobile phone suffers a sinusoidal alternating load, the amplitude is 2 mm, the result is shown in **Figure 5**.

It can be seen from **Figure 5** that the maximum stress of front cover made up of SMP is 3.96 MP under the temperature of 20˚C - 60˚C, and the maximum displace-

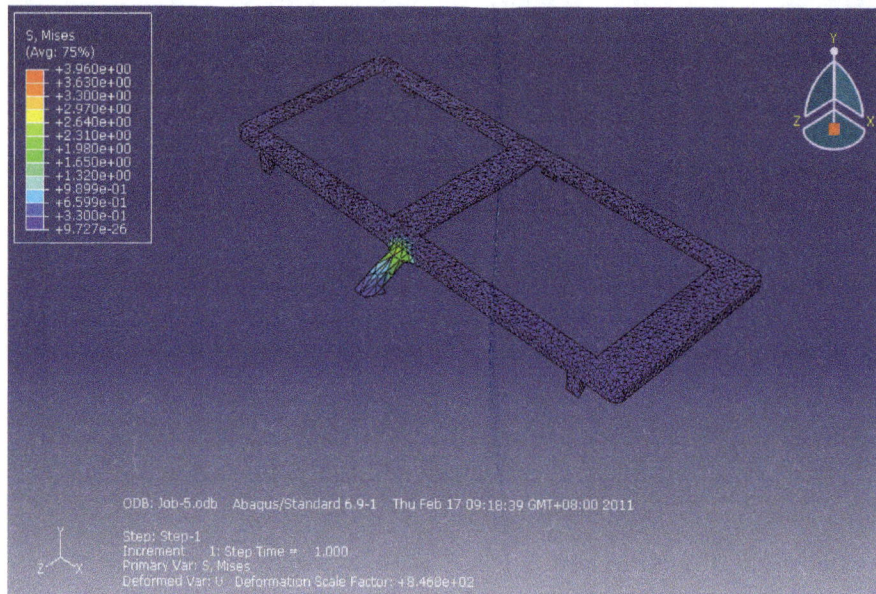

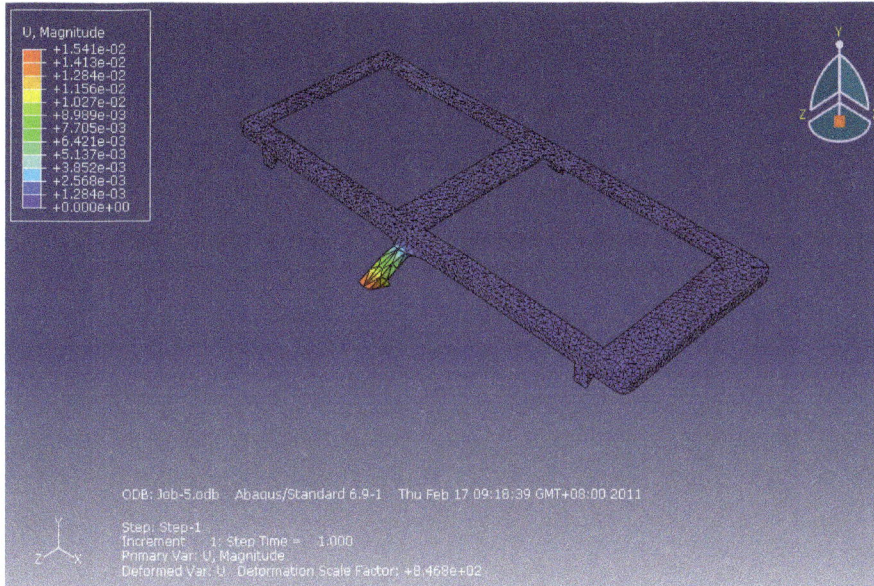

Figure 5. The collision experiment of mobile phone using SMP.

ment is 1.54×10^{-2} mm, while the length of connection between front cover and back cover is 0.8 mm, therefore, the fasteners between front cover and back cover do not detach and it ensures the reliability of vibration of mobile phone. The maximum stress of vibration of active disassembly mobile phone under different temperature is shown in **Table 5**.

For easy comparison, conducting the same analysis of mobile phone made up of PC/ABS, the result is shown in **Figure 6**.

It can be seen from **Figure 6** that the maximum stress of front cover made up of PC/ABS is 0.12 MP, and the maximum displacement is 2.74×10^{-2} mm, while the length of connection between front cover and back cover is 0.8 mm, therefore, the fasteners between front cover and back cover do not detach and it ensures the reliability of vibration of mobile phone. The maximum stress of vibration of mobile phone made up of PC/ABS under different temperature is shown in **Table 6**.

The vibration reliability of mobile phone made up of

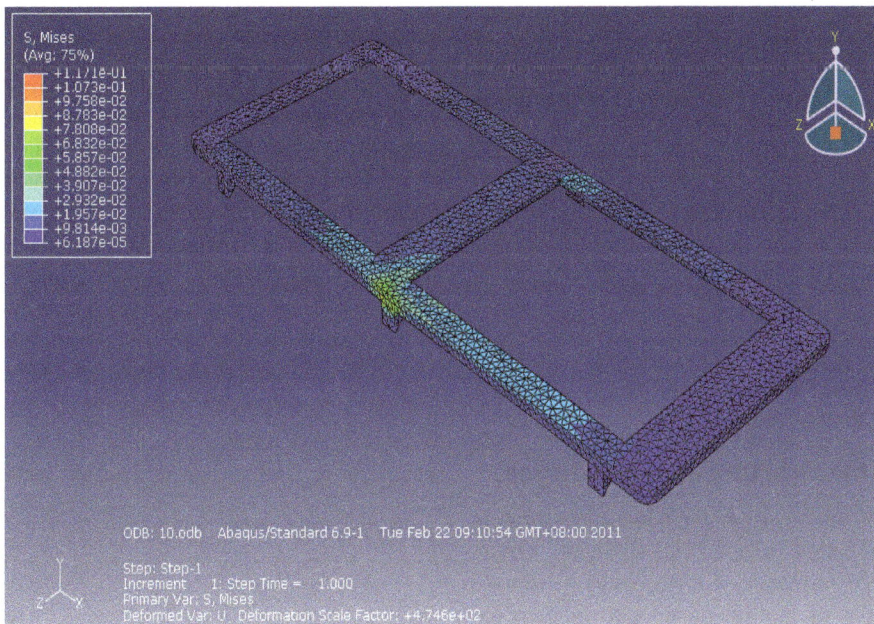

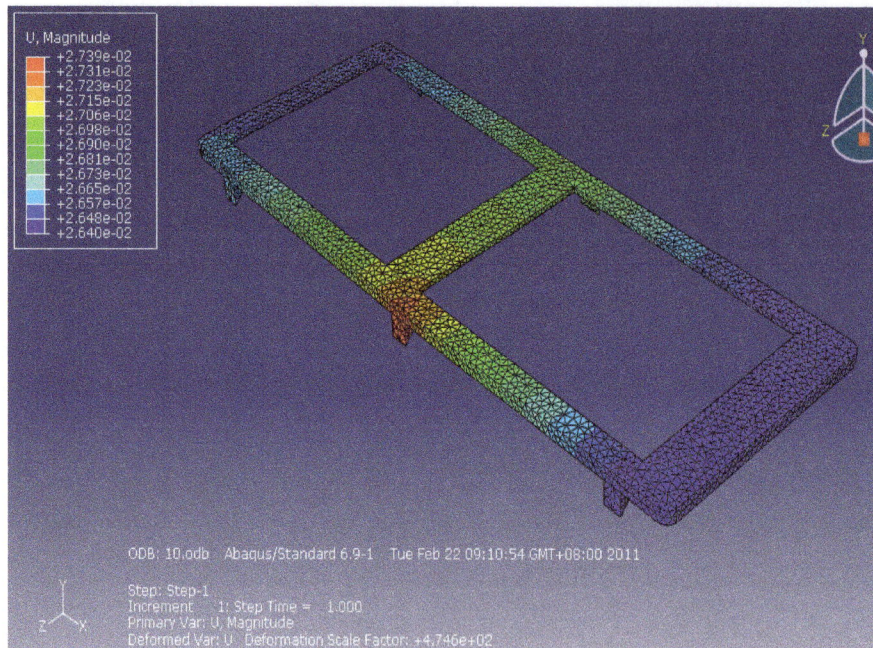

Figure 6. The collision experiment of mobile phone using PC/ABS.

Table 5. The maximum stress of vibration of active disassembly mobile phone under different temperature.

Temperature °C	Maximum displacement mm	The length of connection of fasteners mm
20 - 60	1.54×10^{-2}	0.8
70	7.15×10^{-2}	0.8
80	3.32×10^{-2}	0.8
90	1.54	0.8

Table 6. The maximum stress of vibration of mobile phone made up of PC/ABS under different temperature.

Temperature °C	Maximum displacement mm	The length of connection of fasteners mm
20-40	2.74×10^{-2}	0.8
90	1.18×10^{-2}	0.8

PC/ABS is higher than the vibration reliability of mobile phone made up of SMP by comparison.

4.3. Optimizing the Structure of Mobile Phone

In order to expand the stress of collision and vibration quickly to other locations around the shell and obtain better mechanical performance, it is necessary to replace the material of fillet of mobile phone into magnesium alloy material, as shown in **Figure 7** [15]. Currently, the price of magnesium alloy shell is about the double of plastic shell. However, the mass of magnesium alloy is light, the density is low, the resistance to stress is relatively strong, the hardness is several times more than traditional plastic, but its weight is only 1/3 of ordinary plastic. Therefore, it can improve the reliability of mobile phone using magnesium alloy shell. What is more, it can make the product more attractive by treating the surface of magnesium alloy. Therefore, magnesium alloy shell can be used in high-end mobile phones [15].

5. Conclusions

1) The mobile phone made up of SMP is not destroyed falling from a height of 2 m under the room temperature, while the mobile phone made up of PC/ABS is destroyed under the same conditions. Therefore, the collision reliability of mobile phone shell made up of PVC material which receives irradiation after casting is higher than the collision reliability of mobile phone shell made up of PC/ABS. And it can improve the collision reliability using magnesium alloy shell of mobile phone.

2) The mobile phone made up of SMP and the mobile phone made up of PC/ABS both can ensure the reliability

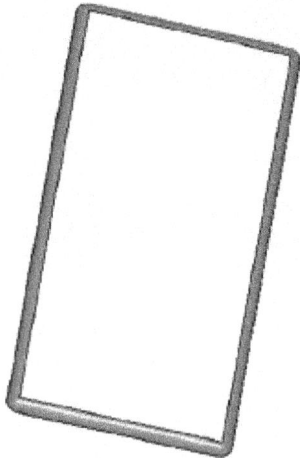

Figure 7. Magnesium alloy shell of mobile phone.

of vibration under room temperature. Because the property of SMP is easily affected by the temperature, the elastic modulus of SMP declines sharply under high-temperature condition which results in the reliability of mobile phone made up of SMP is relatively low. While the elastic modulus of PC/ABS affected by the temperature is not so large, therefore, the mobile phone made up of PC/ABS can ensure the reliability of vibration under the temperature of 90°C.

3) Mobile phone is used under the room temperature because the electronic components of mobile phone are easily destroyed under high temperature. The mobile phone made up of SMP can ensure the reliability of collision and vibration under room temperature, therefore, the mobile phone made up of SMP meets the environmental reliability

6. Outlook

It is necessary to analyze the reliability of multi-step active disassembly products, such as the impact of collision and vibration on multi-step active disassembly mobile phone. What is more, it is necessary to research the impact of active disassembly on the economy and environment of products.

7. Acknowledgements

Thanks to the support of NSFC item: 50775064.

REFERENCES

[1] W. Xu and J. H. Tao, "Research of Recycling Oriented Disassembly Technology," *Machine Tool & Hydraulics*, Vol. 37, No. 2, 2009, pp. 24-28.

[2] K. Lee and R Gadh, "Computer Aided Design for Disassembly: *A Destructive Approach*," *Proceedings of the 1996 IEEE International Symposium on Electronics & the Environment*, New Jersey, 1999, pp. 173-178.

[3] R. C. Li, J. J. Tang, X. Hong and K. G. Wu, "Harmful Influence to Environment and Recycling of Discarded Household Appliance," *Environmental Pollution and Control*, Vol. 25, No. 4, 2003, pp. 109-110.

[4] F. H. Lian, Q. P. Su, M. H. Wang, Y. Zhang and G. H. Tang, "Recycling and Disposal of the Old and Waste Household Appliances," *Environmental Pollution and Control*, Vol. 26, No. 1, 2004, pp. 67-69.

[5] J. Z. Li, P. Shrivastava and Z. Gao, "Printed Circuit Board Recycling: A State-of-the-Art Survey," *Proceeding of the 2002 IEEE International Symposium on Electronics and the Environment*, Vol. 27, 2004, pp. 234-241.

[6] K. Lee and R. Gadh, "Computer Aided Design for Disassembly: A Destructive Approach" *Proceedings of the 1996 IEEE International Symposium on Electronics & the Environment*, New Jersey, 1999, pp. 173-178.

[7] J. D. Chiodo, E. H. Billett and D. J. Harrison, "Preliminary Investigations of Active Disassembly Using Shape Memory Polymers," *Proceedings First International Symposium on Environmentally Conscious Design and Inverse Manufacturing*, Tokyo, 1999, pp. 590-595.

[8] S. Seelecke, "Shape Memory Alloy Actuators in Smart Structures: Modeling and simulation," *American Society of Mechanical Engineers*, Vol. 57, No. 1, 2004, pp. 23-46.

[9] Z. F. Liu, X. Y. Li and H. C. Zhang, "Research on Design Methods of Products Based on ADSM," *Chinese Journal of Mechanical Engineering*, 2009, Vol. 45, No. 10, pp. 192-197.

[10] Z. F. Liu, L. X. Zhao, X. Y. Li, H. C. Zhang and H. B. Cheng, "Research on Multi-Step Active Disassembly Method of products Based on ADSM," *Conference of Manufacturing Engineering and Automation,* Vol. 139-141, Guangzhou, 2010, pp. 1428-1432.

[11] J. D. Chiodo and C. Boks, "Assessment of End-of-Life Strategies with Active Disassembly Using Smart Materials," *The Journal of Sustainable Product Design*, Vol. 2, No. 1-2, 2002, pp. 69-82.

[12] J. Peng, "Humid and Hot Environment and Electronic Product Reliability," *Electronic Product Reliability and Environmental Testing*, Vol. 5, 2003, pp. 57-60.

[13] J. P. Zu and C. Q. Xu, "The Simulation and Investigation for Free - Drop of Mobile Phone," *Manufacture Information Engineering of China*, Vol. 35, No. 11, 2006, pp. 68-70.

[14] J. Ding, G. Li, C. Q. Xue and L. Yu, "A Study on Anti-impact Property of the Bottom Shape of Plastic Mobile Phone Case," *Electro-Mechanical Engineering*, Vol. 25, No. 1, 2009, pp. 42-48.

[15] B. Y. Peng, J. B. Xie, Q. H. Feng and G. F. Yin, "Digitalized Analysis of Open-faced Mobile Phone Damage Caused by Falling," *Journal of Xihua University*, Vol. 28, No. 1, 2009, pp. 74-77.

7

Modeling of Adsorption of Bi(III) from Nitrate Medium by Impregnated Resin D2EHPA/XAD-1180

Nasr-Eddine Belkhouche[*], Nacera Benyahia

Laboratory of Separation and Purification Technologies, Department of Chemistry-Faculty of Sciences, Tlemcen University, Algeria.

ABSTRACT

Di(2-ethylhexyl)phosphoric acid (D2EHPA) in acetone was supported on the Amberlite XAD-1180 polystyrene divinyl-benzene copolymer resin. The use of XAD-1180 impregnated with D2EHPA for the extraction of bismuth(III) from nitrate medium was carried out using batch technique. Various parameters affecting the uptake of this metal ion were described in the previous paper Reference [1] and the capacity of the impregnated resin for bismuth(III) was found to be 490.7 mg/g of resin. Effect of temperature on the values of distribution equilibrium was studied to evaluate the changes in standard thermodynamic quantities. A comparison of Langmuir forms I, II and Freundlich sorption isotherms was realized and the kinetic models applied to the adsorption rate data were evaluated for Lagergren first order, the pseudo second order and Morris–Weber models. From the results, the adsorption of Bi(III) onto D2EHPA/XAD-1180 resin shown the exothermic character and followed the Langmuir form II isotherm. Thus, the capacity of monolayer adsorption of Bi(III) was equal to 769.23 mg/g of resin. Both the Lagergren pseudo first order and film-diffusion models were found to best describe the experimental rate data.

***Keywords**: Bismuth, XAD-1180 Resin, D2EHPA, Sorption Isotherms, Kinetic Models*

1. Introduction

The previous paper [1] was devoted to study the kinetics of bismuth(III) extraction from nitrate medium by solvent impregnated resin technique (SIR) using the Amberlite XAD-1180 resin as support for di(2-ethylhexyl) phosphoric acid (D2EHPA) as organophosphorus extractant, in order to know the best operating conditions of later selective extraction from other metals such as lead, copper and tin. The bismuth(III) was fixed at 490.7 mg/g of XAD-1180 resin, at 295 K. The extractant impregnated resin (EIR) of bismuth(III) was studied in function of the experimental parameters such as: Amberlite XAD-1180 impregnation, D2EHPA/XAD-1180 ratio, Contact time and stirring speed, pH of Bi(III) solution, Concentration of Bi(III), Aqueous phase volume, NaCl electrolyte and elution of Bi(III) from loaded EIR. The results were used to determine the constants of polynomial model which described the experimental data of bismuth (III) extraction process.

In this present paper we are interested to study the thermodynamic parameters of the distribution equilibrium of Bi(III) sorption process for evaluated the changes in standard thermodynamic quantities. The sorption isotherms such as: Langmuir forms I, II and Freundlich were tested for experimental data of Bi(III) sorption onto D2EHPA/XAD-1180 resin. Also, the kinetics models as the Lagergren first order, the pseudo second order and Morris–Weber were applied for modeling the adsorption rate data.

2. Results and Discussion

2.1. Effect of the Temperature

The study of the temperature effect on the bismuth(III) sorption from nitrate medium onto 15 mmol of D2HPA/g of XAD-1180 resin was carried out by using 250 ppm of the concentration of metal ion at pH 3.6 with v/m ratio equal to 50 ml/g.

The distribution coefficient (K_d) of metal ion between the aqueous bulk phase and the resin phase was calculated from the Equation (1):

$$K_d = \frac{C_0 - C_e}{C_e} x \frac{V}{M} \qquad (1)$$

Figure 1 shows the variation of the distribution coefficient (K_d) of bismuth(III) sorption in function of different temperatures. From where, an increasing of the temperature from 22 to 60°C decreased the adsorption of the bismuth(III). The van't Hoff relation [2] given by Equation (2) can be used to calculate the enthalpy changes associated with the adsorption process of the bismuth(III).

$$\log K_d = -\frac{\Delta H^0}{2.303R} \times \frac{1}{T} + C \qquad (2)$$

From the plots of K_d vs. $1/T$ (**Figure 1**), a straight line was observed, from which ΔH^0 (the enthalpy variation) can be deduced according to the Equation (3):

$$\Delta H^0 = -2,303R \times Slope \qquad (3)$$

The free energy variation ΔG^0 was also calculated based on the logarithmic value of the distribution ratio $Log K_d$ at 22°C according to the Equation (4):

$$\Delta G^0 = -2.303RT Log K_d \qquad (4)$$

Also, the entropy variation ΔS^0 was obtained from ΔG^0 and ΔH^0 with the Equation (5):

$$\Delta S° = \frac{\Delta H° - \Delta G°}{T} \qquad (5)$$

The thermodynamic parameters of the sorption of bismuth (III) were given in **Table 1**. The negative sign of the enthalpy variation value showed the exothermic character of the liquid-solid extraction and sorption process. This result is similar to the previous paper [3]. While the negative sign of the free energy variation value

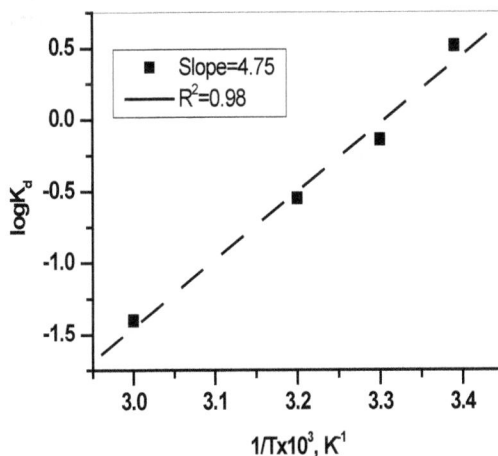

Figure 1. Variation of log K_d with $1/T$ for the sorption of Bi (III) ion from nitrate medium by D2EHPA/XAD-1180 resin.

Table 1. Thermodynamic parameters for the adsorption of bismuth (III) from nitrate medium by D2EHPA/XAD-1180 resin.

Metal ion	ΔH^0 (kJ/mol)	ΔG^0 (kJ/mol)	ΔS^0 (J/mol·K)
Values (KJ/mol)	−91.01	−2.91	−298.64

indicated the spontaneous phenomenon of bismuth(III) sorption and the value sign of the entropy variation suggested that the system exhibit a disorder.

As reported in literature [4], the process of solvent impregnated resin (SIR) can be evaluated as film-diffusion controlled when $E_a < 16.7$ kJ/mol, particle diffusion controlled when $E_a > 42$ kJ/mol and reaction controlled when $E_a = 50.2$ kJ/mol. The activation energy (E_a) of sorption reaction of bismuth(III) by the XAD-1180 resin impregnated with D2EHPA was calculated by applying the Arrhenius relation where E_a was be found to 1.14 kJ/mol which confirmed that the sorption was governed by the film-diffusion.

2.2. Sorption Isotherm

The experimental results obtained for the adsorption of bismuth(III) by D2EHPA impregnated onto XAD-4 resin at temperature equal to 295 K under the optimum conditions [1] were tested for Langmuir form I, II and Freundlich adsorption isotherms. The Langmuir isotherm can be written under the Equation (6) and Equation (7), form I and II respectively as:

$$\frac{C_e}{q_e} = \frac{1}{bQ_0} + \frac{C_e}{Q_0} \quad (Form \ I) \qquad (6)$$

$$\frac{1}{q_e} = \frac{1}{bQ_0 C_e} + \frac{1}{Q_0} \quad (Form \ II) \qquad (7)$$

where

$$q_e = (C_0 - C_e) x \frac{V}{M} \qquad (8)$$

The Langmuir isotherms Form I and II for sorption of bismuth (III) ions on the impregnated resin were presented in **Figure 2** and **Figure 3** respectively.

The representation of Langmuir isotherm form II for experimental data of bismuth(III) sorption by D2EHPA impregnated in XAD-1180 resin (**Figure 3**) showed a good linear fitting ($R^2 = 0.99$) compared with that given in **Figure 2**. From where the fit of experimental data using Langmuir isotherm form I was equal to 0.92. Thus, the sorption of Bi(III) onto D2EHPA impregnated in XAD-1180 was expected as a monolayer adsorption and that all active sites are similar and have the same energy [5]. Parameters of Langmuir model form II are given in **Table 2**.

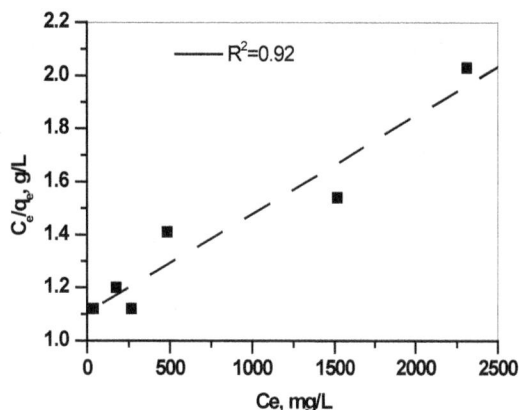

Figure 2. Langmuir isotherm (Form I) for sorption of Bi(III) onto impregnated D2EHPA/XAD-1180 resin.

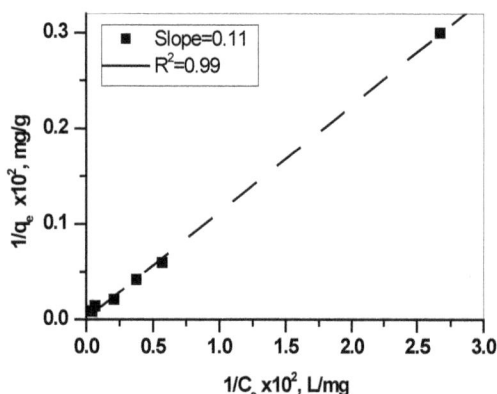

Figure 3. Langmuir isotherm (Form II) for sorption of Bi(III) onto impregnated D2EHPA/XAD-1180 resin.

Table 2. Parameters of Langmuir isotherm for sorption of Bismuth(III) by D2EHPA/XAD-1180 resin.

Metal ion	Parameters of Langmuir model form II	
	Q_0 (mg/g)	b(L/g)
Bi(III)	769.23	0.011

In fact, the fit of data using the equation of Freundlich isotherm was carried out. From the fitting factor ($R^2 = 0.97$) which was lower than that the Langmuir isotherm, we suggests that, the sorption process was restricted to one specific class of sites and assumes surface homogeneity.

2.3. Kinetic Modeling

Several kinetic models were tested to select the model that describes our experimental data for established to the appropriate mechanism of sorption of Bi^{3+} by D2EHPA impregnated onto XAD-1180 resin. The batch sorption process of bismuth(III) was analyzed using Lagergren

first order and the pseudo second order kinetics model [6,7]. The equation of Lagergren was widely used in liquid-solid extraction for sorption of solute from aqueous or organic solution [3]. The Legergren first order model was given by the Equation (9):

$$\log\left(q_e - q_t\right) = \log q_e - \frac{k_1}{2.303}t \qquad (9)$$

From the results shown in **Figure 4**, linear fit was observed for the metal ion, during the first 30 min of shaking time, the first order rate constant (k_1) was found to be approximately 0.10 min^{-1}. In fact, the experimental data were analyzed using the Legergren pseudo second order model but the data fitting was not better than the Legergren pseudo first order which confirm that it's appropriate to use this last model to predict the sorption kinetics of bismuth(III) ions onto D2EHPA impregnated in XAD-1180 resin.

The diffusion of the particles from the bulk solution into the sorbent pores can constitute a limiting step in the process of bismuth(III) sorption by D2EHPA impregnated in XAD-1180 resin. For this, the experimental data were used to study intraparticle diffusion. The equation of Morris–Weber model is given as Equation (10):

$$q_t = K_{ad}\sqrt{t} \qquad (10)$$

As shown in **Figure 5**, multi-linearity correlation of experimental data was obtained by plotting a graph of q_t vs. $t^{0.5}$. From theory [8,9], the preliminary conclusions indicated that the intraparticule diffusion cannot be involved in the sorption process (linear plot, $R^2 = 0.95$) and was not the rate controlling step because the fit line no pass through the origin.

In the way to verify the conclusions brought on the rate controlling step of bismuth(III) sorption, the graphs: $[-Ln(1 - F)] = kt$ and $[3-3(1 - F)^{2/3} - 2F] = kt$ plotted in

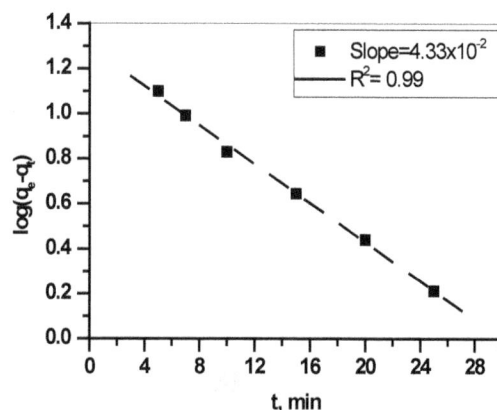

Figure 4. Legergren pseudo first order plot for the removal of Bi(III) ion from nitrate solution by D2EHPA/XAD-1180.

the cases of film-diffusion controlled and chemical reaction controlled respectively [10,11].

From the results shown in **Figure 6** and **Figure 7**, the

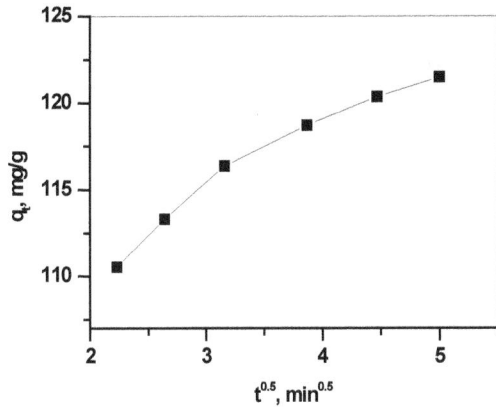

Figure 5. Morris–Weber plot for the adsorption of Bi(III) ion from nitrate solution by D2EHPA/XAD-1180 resin.

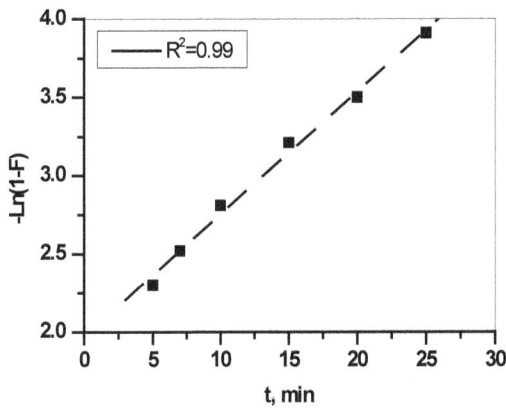

Figure 6. Plot [–Ln(1 – F)] vs. time for the adsorption of Bi(III) ion from nitrate solution by D2EHPA/XAD-1180 resin.

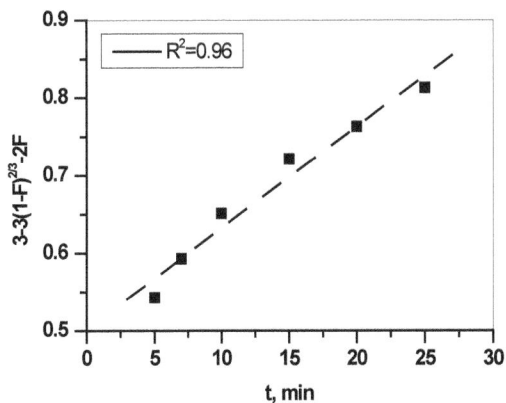

Figure 7. Plot [3-3(1 – F)²/³ – 2F] vs. time for the adsorption of Bi(III) ion from nitrate solution by D2EHPA/XAD-1180 resin.

linear correlation of experimental data was better ($R^2 = 0.99$) when plotting [–Ln(1 – F)] vs. time where the correlation factor was equal to 0.95 in the case of the plotting [3 – 3(1 – F)$^{2/3}$ – 2F] vs. time. Thus, the film-diffusion was the rate controlling step of bismuth(III) sorption by impregnated resin D2EHPA/XAD-1180. This result was similar to that found by the activation energy explanation.

The calculation of the diffusion coefficient (D_r) was given by the Equation (11) [10]:

$$D_r = k \frac{r_0^2}{\pi^2} \qquad (11)$$

where, the diffusion coefficient was equal to $5.84 \cdot 10^{-6}$ $cm^2 \cdot min^{-1}$.

3. Conclusions

The physical impregnation of D2EHPA in Amberlite XAD-1180 resin for the Bi(III) sorption from aqueous nitrate medium was carried in batch system. The thermodynamics values of extraction reaction of Bi(III) shown the exothermic process where the activation energy was be found to 1.14 kJ/mol. The equilibrium isotherms for sorption of the investigated metal ion were modeled successfully using the Langmuir isotherm (Form II) where the sorption capacity of monolayer was found equal to 769.23 mg/g of impregnated resin.

Also, the experimental data were tested for different kinetic model expressions and the data were successfully modeled using the Legergren pseudo first order where the first order rate constant was found to be approximately 0.1 min^{-1}. Besides, the pushed studies on the mechanism of the Bi(III) sorption showed that the rate controlling step was the fim-diffusion and the coefficient of diffusion was $5.84 \cdot 10^{-6}$ $cm^2 \cdot min^{-1}$.

REFERENCES

[1] N. Belkhouche and M. A. Didi, "Extraction of Bi(III) from Nitrate Medium by D2EHPA Impregnated onto Amberlite XAD-1180," *Hydrometallurgy*, Vol. 103, No. 1-4, 2010, pp. 60-67.

[2] C. H. Weng and C. P. Huang, "Adsorption Characteristics of Zn(II) for Dilute Aqueous Solution by Fly Ash," *Colloids and Surface. A: Physicochemical Engineering Aspects*, Vol. 247, No. 1-3, 2004, pp. 137-143.

[3] E. A. El-Sofany, "Removal of Lanthanum and Gadolinium from Nitrate Medium Using Aliquat-336 Impregnated onto Amberlite XAD-4," *Journal of Hazardous Materials*, Vol. 153, No. 3, 2008, pp. 948-954.

[4] K. J. Laidler, "Chemical Kinetics," Mc-Graw Hill, London, 1975, p. 11.

[5] A. M. El-Kamash, N. S. Awad and A. A. El-Sayed, "Sorption of Uranium and Thorium Ions from Nitric Acid Solution Using HDEHP-Impregnated Activated Carbon," *Arab Journal of Nuclear Sciences and Applications*, Vol. 38, No. 1, 2005, pp. 44-49

[6] Y. S. Ho and G. McKay, "Pseudo-Second Order Model for Sorption Processes," *Process Biochemical*, Vol. 34, No. 5, 1999, pp. 451-465.

[7] W. J. Weber and J. M. Morris, "Kinetics of Adsorption of Carbon from Solutions," *Journal of Sanitary Engineering Division American Society Engineers*, Vol. 89, 1963, pp. 31-60

[8] N. K. Lazaridis, T. D. Karapantsios and D. Georgantas, "Kinetic Analysis for the Removal of a Reactive Dye from Aqueous Solution onto Hydrotalcite by Adsorption,"

Water Research, Vol. 37, No. 12, 2003, pp. 3023-3033.

[9] M. Alkan, O. Demirbas, S. Alikcapa and M. Dogan, "Sorption of Red 57 from Aqueous Solution onto Sepiolite," *Journal Hazardous Materials*, Vol. 116, No. 1-2, 2004, pp. 135-145.

[10] F. J. Alguacil, "A Kinetic Study of Cadmium(II) Adsorption on Lewatit TP260 Resin," *Journal of Chemical Research*, Vol. 2003, No. 3, 2003, 144-146.

[11] R. Chiarizia, E. P. Horwitz and S. D. Alexandratos, "Uptake of Metal Ions by a New Chelating Ion Exchange Resin. Part 4: Kinetics," *Solvent Extraction and Ion Exchange*, Vol. 12, No. 1, 1994, pp. 211-237.

Symbols

b: Constant related to the free energy of adsorption, b ∞ exp($-\Delta G/RT$)

C: Constant

C_0: Initial concentration of metal ion in solution, mg/L

C_e: Equilibrium concentration of metal ion in solution, mg/L

$F = q_t/q_e$

k: Rate constant, min^{-1}

k_1: Pseudo first order rate constant, min^{-1}

K_{ad}: Rate constant of intraparticle transport, mg/g min$^{0.5}$

M: Weight of the adsorbent, g

Q_0: Monolayer adsorption capacity, mg/g

q_e: Amount of solute sorbed per unit weight of adsorbent at equilibrium, mg/g

q_t: Concentration of ion in the adsorbent at time t, mg/g

R: Universal gas constant (8.314 J·mol^{-1}·K^{-1})

r_0: radius of particles of resin (0.024 cm)

T: Absolute temperature, Kelvin

V: Volume of bulk solution, L

8

Effect of Self-Assembled Monolayers on the Performance of Organic Photovoltaic Cells

Hanène Bedis

[1]UMAO, Faculté des Sciences de Tunis, Campus Universitaire, Tunis, Tunisia; [2]ITODYS, 15 Rue Jean-Antoine de Baïf, Paris, France.

ABSTRACT

The improvement of the performance of organic photovoltaic cells (OPVCs) and the photogeneration process in these devices may occur via multiple mechanisms depending on their structure and/or architecture. For this purpose we investigate how self-assembled monolayers of thiol molecules ($C_{12}H_{25}SH$ and $3T(CH_2)_6SH$) and benzoic acid molecules (ABA and NBA) affect the efficiency and the photogeneration of free carriers in a sexithiophene based photovoltaic cells. Firstly, we provide the results of absorption spectra for samples with SAM of thiol that show there effect on orientation of 6T molecules on these structures and the organization degree of the thiol molecules on ITO substrate. Afterward, we describe from current vs. applied voltage after illumination, the enhancement of the performance of these cells. In the second, we study the effect of SAM of benzoic acids molecules on the photovoltaic behavior. A theoretical model is used for quantitative description of the open circuit voltage as a function of carrier's generation rates at the electrodes. The results of I-V characterization under illumination show that open circuit voltage as well as short circuit current is dramatically affected by the dipolar layer. The orientation and the magnitude of dipole moment of benzoic acid molecules are the crucial factors that affect the organic photovoltaic parameters.

Keywords: *SAMs, Oligothiophene, Photovoltaïc Cell, Photogeneration, Efficiency, Performance*

1. Introduction

Organic semiconductors materials are potential candidates in electronics industry, more diversified and less expensive. Indeed, the fact that is a synthetic materials contrary to the mineral semiconductors, allows to modeling the constitutive molecules to adjust a very precise property. It is by proceeding that we were able to conceive a multitude of electronic structures such as organic light-emitting diodes (OLEDs) which cover the totality of the visible spectral field and organic photovoltaic cells (OPVCs).

Organic photovoltaic cells are gaining considerable interest motivated by a steady enhance of their conversion efficiency during the last decade [1-3]. Nevertheless, their operating mechanisms are still subject of controversial theories. In fact, the observed photocurrent in OPVCs is an experimental evidence of free photocarrier production. Nonetheless, the origin of photogenerated carriers that are responsible of the observed photocurrent may be attributed to mainly three mechanisms. The first

is the ionization of tight bound excitons or the Frenkel-type excitons in the bulk which is described by Onsager theory [4,5]. In this mechanism Frenkel excitons lead to the formation of charge transfer excitons resulting in geminate pairs of photogenerated carriers lying on adjacent molecules. The geminate pairs escape the strong columbic attraction under the effect of built-in electric field E at a given temperature T with the probability P(E,T). The second mechanism occurs at the interface with the electrodes. In this mechanism one of the geminate pair carriers passes through the electrode and the other is left "free" to contribute to photocurrent. The last mechanism is attributed to detrapping of trapped carriers. In this mechanism photogenerated tight bound excitons transfer enough energy to trapped charges by colliding them. Actually, all of the three described mechanisms may occur simultaneously but in some cases one among them may be preponderant on the two others as for example the case of materials with low mobility, in which bulk photogeneration may be ignored because their short lifetime and photogenerated carriers in the bulk that are far from the electrodes disappear rapidly and minimizing the photocurrent.

The development of organic photovoltaic cells (OPVCs) is still a matter of research despite their low efficiency relatively to mineral ones which is precisely a crucial factor for their commercialization. This is due to mainly two reasons. The first one may be qualified as historical and is related to the youthfulness of organic optoelectronics by comparison to minerals. The second one is physical and is due to mainly two reasons:

- The first one is the low dielectric constant of organic semiconductors (~3) relatively to that of inorganic ones (~10). This property makes photo-excited electron-hole geminate pairs much more bound, due to columbic interaction, in organic materials indeed excitons in organic semiconductors are Frenkel-type with a binding energy in the order of some tenths of eV [5].

- The second one is the low mobility of carriers in organic semiconductors which inhibits short lifetime carriers from attaining elec trodes [6].

The relatively high binding energy prevents these photogenerated species from dissociating which reveals low free carrier production efficiency. The most probable sketch to account for photocurrent observed experimentally in organic solar cells is that free carriers are produced from dissociation of those tightly bound geminate pairs by interaction with the metal/Semiconductor interface [5].

Self Assembled Monolayers (SAMs) which were widely used in OLEDs are supposed to induce modification of metallic electrode work function leading to significant improve of carrier injection into the organic semiconductor. Another advantage in using SAM layers at electrode interfaces is to improve the adhesion of the organic film onto the metal or the oxide electrode [7]. Then, the quality of the interface reflected by the degree of the structural order, has been improved substantially by using self-assembled monolayer (SAM) at the interface either in the organic transistor [8], the organic diode [9] or the organic light emitting diode [10].

We are interested in this work on photovoltaic cells based of the α-sexithiophene (6T) molecules (**Figure 1(a)**). The characteristics of these molecules apart their stability and facility of synthesis, they are essentially motivated by their structural order, the degree of organization and the big purity which offers to oligomeres contrary to polymers. These molecules present by their simplicity a better understanding of the optoelectronics phenomena and allow to modeling the behavior of polythiophenes.

We present in this article the methods for improving performance of sexithiophene based photovoltaic cell

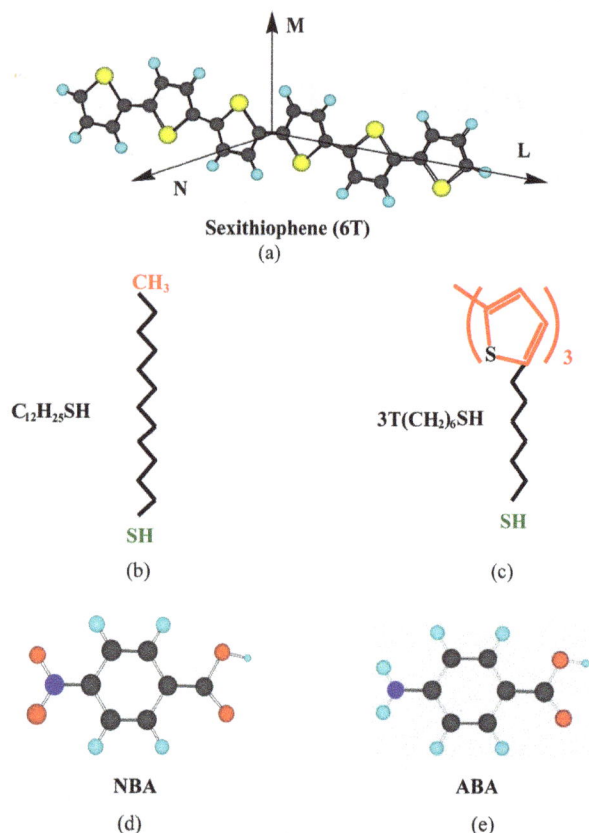

Figure 1. Chemical structure of: (a) sexithiophene, (b) Dodecanthiol ($C_{12}H_{25}SH$), (c) Thiol with head group the terthiophene ($3T(CH_2)_6SH/6T$), (d) p-Nitro-benzoic acid NBA and (e) p-Amino-benzoic acid ABA.

through introduce a self-assembled monolayer of thiols molecules differed with functional groups ($C_{12}H_{25}SH$ and $3T(CH_2)_6SH$) (**Figure 1(b)** and **Figure 1(c)**) and benzoic acid molecules (Amino-Benzoic Acid and Nitro-Benzoic Acid) (**Figure 1(d)** and **Figure 1(e)**) on ITO in order to control photocarrier generation at the interface ITO/organic and to provide an increase in device efficiency. In this study, the evaporate 6T both on SAMs devices were characterized electrically after illumination. We describe from current vs. applied voltage, the enhancement of devices efficiencies and we can estimate their photovoltaic parameters. A theoretical model is used for quantitative description of the open circuit voltage as a function of carrier's generation at the electrodes, and can explain the effect of orientation and the magnitude of dipole moment of SAM of benzoic acid on the photogeneration rate of free carriers and their effect on the organic photovoltaic parameters.

2. Experimental

2.1. Preparation of ITO

The purpose of the preparation of the ITO is to eliminate

the impurities and to increase its surface reactivity and the interactions with the grafted molecules. For the whole of our experimental work, the glassmaking and the whole set of our manipulating tools are carefully cleaned in order to avoid to the maximum any contamination which may trouble grafting of SAM.

Prior to SAMs grafting ITO substrates, provided by SOLEMS, were carefully cleaned according to the following protocol: first the samples are immerged in an ultrasound bath of ultrapure water for 30 mn, second they are rinsed in pure NaOH (30%) during 15 mn and third, they are immerged in pure sulphuric acid (98%) during 1mn. After that the samples are rinsed again in an ultrasound bath of ultra pure water during 1mn. This chemical treatment is used in order to render ITO surface more hydrophilic and reactive with respect to grafted SAM.

2.2. Preparation of SAM and Devices Structure

SAMs are obtained by solution dipping of cleaned ITO substrates according to the following procedure: we prepared two solutions with the molecule shown in **Figure 1**.

To deposit SAMs on the cleaned ITO substrates, we used in the first for thiol molecules a pure solution of 2ml of Dodecanethiol: $C_{12}H_{25}SH$ (**Figure 1(a)**), and a prepared solution of $3T(CH_2)_6SH$ (**Figure 1(b)**) in benzene with a concentration of 10^{-3} mol/l. The samples of cleaned ITO substrates were left for one week in these solutions in order to maximize the SAM graftin g [11]. Afterward the samples were rinsed in pure ethanol (C_2H_6O (96%)) in an ultrasonic bath, then dried with argon. All the experimental steps were carried out at room temperature. For acid molecules, we used pure solution of 2 ml of 4-Aminobenzoic acid (ABA), 4-Nitro benzoic acid (NBA) **Figure 1(d)** and **Figure 1(e)**. All the experimental steps were performed at room temperature.

In order to obtain comparable results either between SAM coated ITO and bare ITO we performed both sexi-

thiophene (6T) and aluminum (Al) coatings in the same conditions. We deposited 80 nm of 6T on five substrates: four of them were coated with SAMs (thiols and benzoic acids) the last one is just for bare cleaned ITO. A semi-transparent 20 nm thick aluminum cathode was finally deposited through a shadow mask under 10^{-6} torr leading to final structures ITO/SAM/6T/Al (**Figure 2**) and ITO/6T/Al. Each sample consisted of two pixels each of which had a rectangular shape of 2 mm/4 mm.

2.3. Electrical and Optical Characterization

Electrical measurements were performed with a Keithley 4200-SCS Semiconductor Characterization System. For I-V measurement under illumination, the first irradiating light source was a 150 W Tungsten lamp for the first series of samples with SAM of thiols and the second irradiating light source was a 150 W Xenon arc lamp (model 150 W/1 XBO, Osram).

Absorption spectra have been recorded with a Carry 500 UV-VIS-NIR spectrophotometer that directly gives the variation of optical density D.O or absorbance vs. wave length of thin film of 6T. All measurements are taken in the ambient air and at room temperature on the optical seat by illumination of the cell through the semi-transparent Aluminium electrode.

3. Theory

In this section we calculate the open voltage for the sample geometry described above. We assume that for applied voltages $V = V_{oc}$ the charge density inside the sample is low enough that it does not cause any band bending, it remains localized at the contact, the photocurrent is equal to zero and the electric field remains uniform and equal to $(V_{bi} - V)/d$, where d is the thickness of the poly- mer layer.

The carrier conduction of in this case is limited of hole and current densities J_h of hole respectively is:

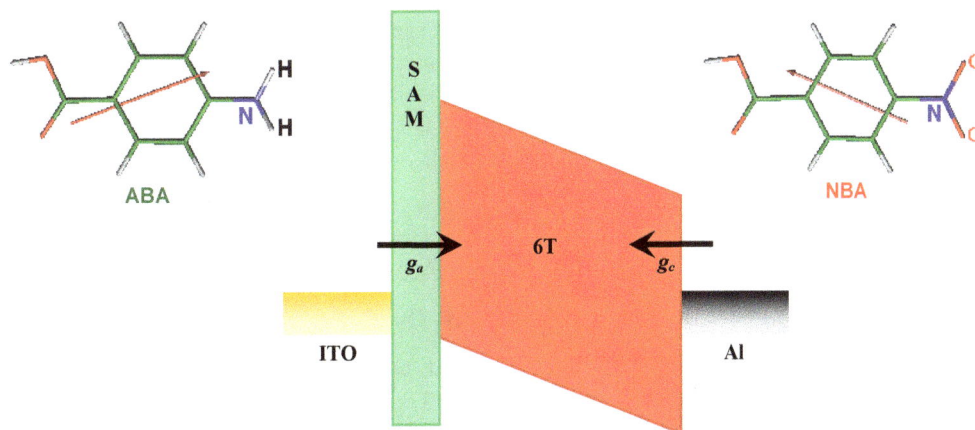

Figure 2. Energy diagram of ITO/SAM/6T/Al structure.

$$J = -e\mu_h p(x)\frac{V_{bi}-V}{d} + eD_h\frac{\partial p(x)}{\partial x} \quad (1)$$

where e is the elementary charge, μ_h is the hole mobility, D is the diffusion coefficient and p is the hole density. Solving the Equation (1) we get the hole density [12]:

$$p(x) = \frac{p_c - p_a}{e^{-qd}-1}e^{-qx} + \frac{p_a e^{-qd} - p_c}{e^{-qd}-1} \quad (2)$$

where the subscripts "a" and "c" denote charge concentrations at the anode $(x=0)$ and the cathode $(x=d)$, respectively, and $q = \frac{1}{d}\frac{e(V_{bi}-V)}{kT}$.

Using the Einstein relation between the mobility and the diffusion coefficient one get the dark current:

$$J_D = e\left(\frac{V_{bi}-V}{d}\right)\frac{(\mu p_a - \mu p_c e^{qd})}{e^{qd}-1} \quad (3)$$

The total current density under illumination is in the following form:

$$J_t = J_D + J_L \quad (4)$$

where the steady state current density in the dark is:

$$J_D = e\left(\frac{V_{bi}-V}{d}\right)\frac{(\mu p_a - \mu p_c e^{qd})}{e^{qd}-1} \quad (5)$$

Since exctions in organic materials are tightly bound, the basic charge generation mechanism for photoconductivity is believed to involve dissociation of the excited state via transfer of charge to the metal electrode, leaving the other charge free inside the organic layer [12]. Thus, in order to calculate the current density under illumination one can neglect bulk photogeneration and assume that only photogenerated carriers at the electrodes contribute to photocurrent. Therefore, J_L is obtained by substituting respectively p_a and p_c in J_D by $g_a I$ and $g_c I$, where I stand for the illumination intensity and g_a and g_c are the density of photogenerated holes respectively at the anode and the cathode. Thus J_L is written as:

$$J_L = e\left(\frac{V_{bi}-V}{d}\right)\frac{(\mu g_a I - \mu g_c I e^{qd})}{e^{qd}-1} \quad (6)$$

Then, the total current density is written in the form:

$$J_t = e\mu_h\left(\frac{V_{bi}-V}{d}\right)\left[\frac{p_a + g_a I - (p_c + g_c I)e^{qd}}{e^{qd}-1}\right] \quad (7)$$

Or at $V = V_{oc}$, the total current density equals zero then we get the relation between the built-in potential and the open circuit voltage as follows:

$$V_{oc} = V_{bi} - \frac{kT}{e}\ln\left(\frac{p_a + g_a I}{p_c + g_c I}\right) \quad (8)$$

Since generally organic semiconductors are known to be excellent photoconductors that is photocurrent is orders of magnitude greater that dark current, then one may neglect p_a and p_c with respect to $g_a I$ and $g_c I$ respectively leading to the expression:

$$V_{oc} = V_{bi} - \frac{kT}{e}\ln\left(\frac{g_a}{g_c}\right) \quad (9)$$

On the other hand, the built-in potential V_{bi} in conjugated non-doped thin film devices is proportional to the workfunction difference between the cathode and the anode.

$$V_{bi} = \frac{1}{e}(\phi_{ITO} - \phi_{Al}) \quad (10)$$

where ϕ_{ITO} and ϕ_{Al} are the workfunctions of the anode (ITO) and of the cathode (Al). In the case of SAM coated electrodes one should take into account the workfunction shift due to dipole layer introduced by dipolar character of SAM molecules. Then the Potential energy shift owing to a dipolar SAM layer at the ITO surface is given by [13]:

$$\Delta\phi = \frac{e\Gamma\mu_D}{\varepsilon_0\varepsilon} \quad (11)$$

where Γ is the number of molecules by unit surface (2×10^{18} m^{-2} for ABA molecule and 1.3×10^{18} m^{-2} for NBA molecule [14,15], μ_D is the dipole moment of the individual molecule, ε_0 is the vacuum permittivity and ε is the dielectric constant of SAM molecules, ($\varepsilon = 5.3$) [14]. Dipole moments of both molecules are calculated using Gaussian program and they are also verified in the literature [16]. Thus, the built-in potential of SAM based devices as /SAM/6T/Al may be written as follows:

$$V_{bi} = \frac{1}{e}((\phi_{ITO} - \phi_{Al}) + \Delta\phi) \quad (12)$$

Note that the built in potential V_{bi} can be reduced or enhanced depending on the sign of μ_D i.e. on the orientation of the dipole moment.

Finally one can write the expression of the open circuit voltage as:

$$V_{oc} = \frac{1}{e}((\phi_{ITO} - \phi_{Al}) + \Delta\phi) - \frac{kT}{e}\ln\left(\frac{g_a}{g_c}\right) \quad (13)$$

4. Results and Discussion

4.1. Devices with SAM of Thiol

4.1.1. Optical Characterization

The UV-Visible absorption spectra have been record to a Carry 500 scan/UV-VIS-NIR spectrophotometer. **Figure 3(a),** shows the optical absorbance at normal incidence of

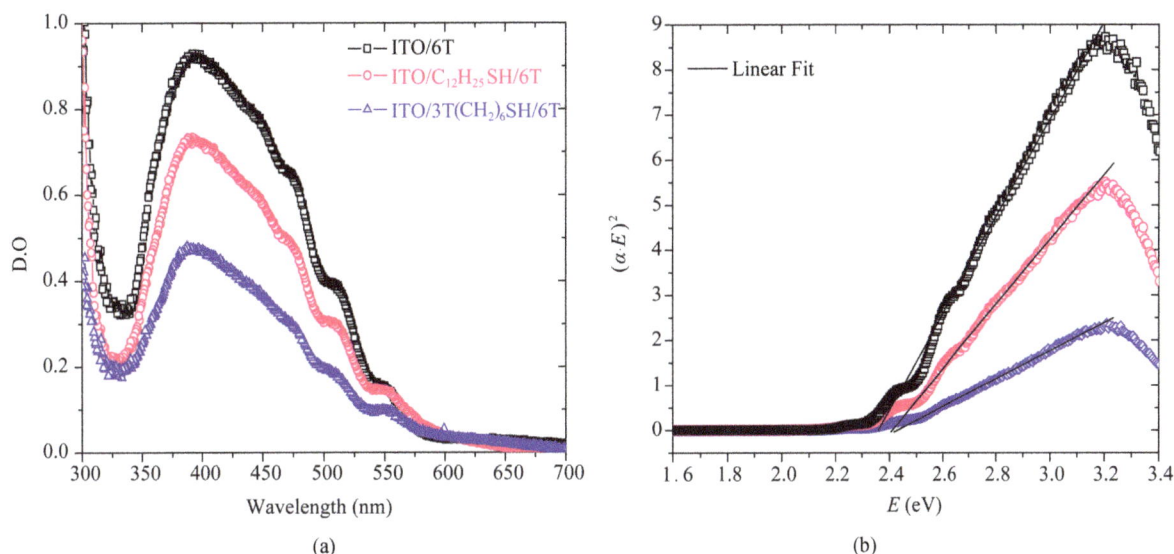

Figure 3. UV-visible absorption spectra of 6T thin film deposited on SAM coated ITO and bare ITO substrates.

both samples: 6T on bare ITO and SAM coated ITO . Although the 6T thin film layer is deposited simultaneously on both substrates, we notice that for the sample with SAM coated ITO the optical density (O.D.) is reduced by a factor 2. This difference is not related to the film thickness bus may be accounted of the position of α-6T molecules. In fact, electronic spectra of oligothiophene thin films are dramatically affected by the orientation of crystallites in the layer. Indeed, since the molecules of oligothiophene are lengthened and since the $\pi\pi^*$ transition dipole moment μ of these molecule is quasi-parallel to their long axis L, then light absorption which is proportional to scalar product of the incident light electric field E by the transition dipole moment μ is maximum when E and μ are parallel and null when E and μ are perpendicular [16]. Therefore the difference between the tree spectra may be accounted for by a better orientation of 6T molecules in the case of $C_{12}H_{25}SH$ and $3T(CH_2)_6SH$ than in the case of bare ITO . Indeed, 6T thin film seems to be much more organized with the SAM of $3T(CH_2)_6SH$ then with $C_{12}H_{25}SH$ and bare ITO , which can induce a quasi-epitaxial growth on terthiophene groups at the surface.

From the absorption spectra we can thus deduce the values of the band gap energy (E_g) for sample with and without SAM of thiol by using the Tauc model [17]. This model supposes a parabolic variation of edge of the absorption band with energy of the photons. The absorption coefficient α is then associated to the energy of the photons E and to the gap (E_g) by the following relation which allows to extrapolate the bandwidth.

$$\alpha E = A\left(E - E_g\right)^n \quad (14)$$

where $E = h\nu$, effective energy of photon, A is constant

and n is an index connect to the nature of electronic transition. We note according to **Figure 3(b)** that electronic transitions which appear in the absorption spectra are controlled by the vibrationnels levels from the molecule 6T. In fact, the permit transition of first excited state, correspond to the transition $^1a_g \rightarrow ^1a_u$ and contained in the plan (**LM**), allows us to deduce gap energy and the other transitions correspond of transitions towards the vibrationnels levels of the molecule 6T. This enabled us to adjust our spectrum on the first absorption band. We notice from adjustments results that the values of gap for n = 1/2 are in the same order of magnitude and are in agreement with that are deferred in the literature [16]. Moreover the best linear curve in the area of edge of absorption band corresponds to an average value of gap of the order 2.2 eV, and all the found values show that the gap of sexithiophene layer is independent of the functional surface.

Consequently, grafting of aliphatic self assemblies monolayer of thiols or with thiophene head group on ITO, emerge on the level of the orientation and the organization of the layers. This results are confirmed with the results of measurement of contact angle and the surface energy [11] that prove the better orientation of $C_{12}H_{25}SH$ molecules towards the surface of ITO and the layer is relatively dense, but they have tendency to disorganize with the time of adsorption. In addition $3T(CH_2)_6SH$ molecules are better organized on the surface and form a dipolar layer with the interface which would involve an electrical improvement of contact ITO.

4.1.2. Current-Voltage Characteristics

Figure 4 shows the tendency observed of ITO/6T/Al (REF) and devices with self assembled monolayer of

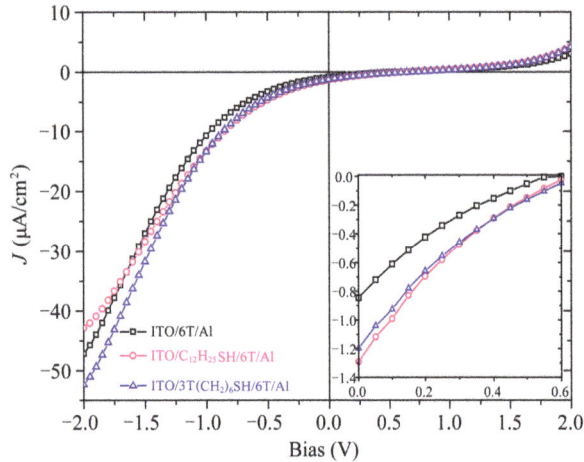

Figure 4. Current density characteristics at applied voltage illuminated with tungsten lamp for devices with SAMs of C$_{12}$H$_{25}$SH and (3T(CH$_2$)$_6$SH/6T), linear plot.

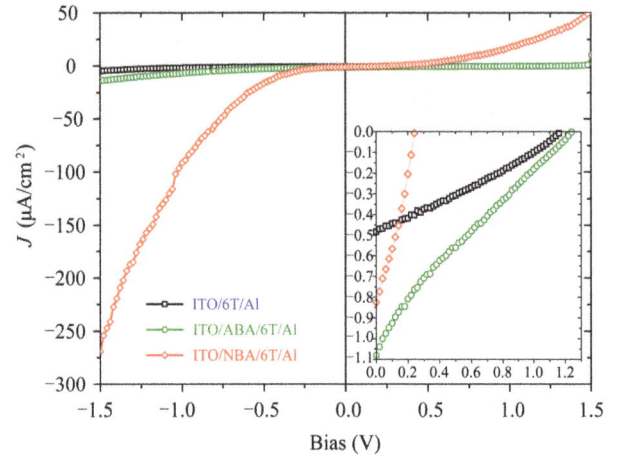

Figure 5. Current density characteristics at applied voltage illuminated with Xenon lamp for devices with SAMs of ABA and NBA, linear plot.

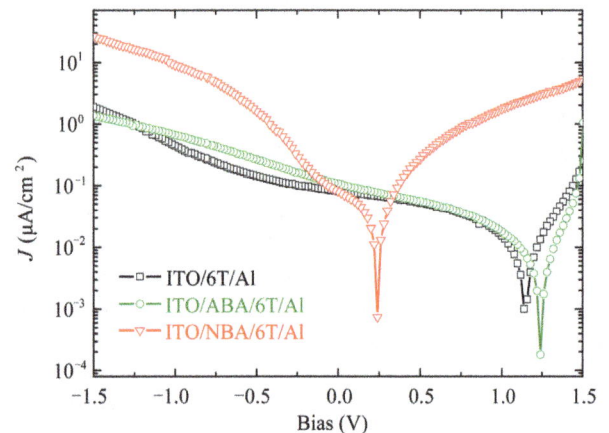

Figure 6. Current density characteristics at applied voltage illuminated with Xenon lamp for devices with SAMs of ABA and NBA, logarithmic plot.

thiols under illumination by a tungsten lamp inside the Al. We clearly observe on the **Figure 3** an effect of the light on I-V characteristics. It is a photocurrent that is strongly depends of the applied voltage. By observing photovoltaic parameters in **Table 1**, we remark clearly that the potential of open circuit and efficiency increased considerably for (ITO /SAM/6T/Al) compared to (ITO /6T/Al). Moreover, these results explain that SAM of thiol where the head groups are oligothiophenes present the best performances of the realized devices in spite of the fact that it absorbs a little luminous power that the two other devices. This would be show a better organization of the layer [11,18,19] and induced a stronger mobility of the charges, affect the device efficiency. We could thus assure that the effect of the SAM containing thiols is limited to the photogeneration in volume and that really there is no direct interface effect on the dissociation of excitons. Thus, for devices with SAM of thiols the improvement of efficiency is probably imputable at the photogenerate free carrier of factors which are much less significant with tungsten lamp. In the continuation we will study from J-V characteristic under Xenon illumination, the effect of dipolar SAM molecules of benzoic acid in photovoltaic conversion.

4.2. Devices with SAM of Benzoic Acid

4.2.1. Current-Voltage Characteristics
The current-voltage characterization under illumination has shown different characteristics depending on the SAM nature (**Figure 5, Figure 6**). We notice that reverse biased photocurrent in NBA coated ITO samples is by far greater than the ABA coated ITO and the bare ITO samples, whereas its open circuit voltage is clearly lower than both of the other samples. On the other hand ITO/

ABA/6T/Al cell shows the largest short–circuit current (J_{sc}) and also the largest open circuit voltage (V_{oc}) (**Table 2**) summarizes the values of photocurrent densities (J_{ph}), open circuit voltages (V_{oc}) and the efficiencies with bare ITO and SAM-coated ITO samples.

It turns out that the conversion efficiency is dramatically affected by the grafting of the acid molecules on ITO. Indeed the efficiency of ITO/NBA/6T/Al is reduced nearly to the half of that of bare ITO sample, whereas in ITO/NBA/6T/Al sample the efficiency in almost twice that of bare ITO sample. The improvement of efficiency in the case of ABA can be accounted for by an enhancement of interfacial charge carriers photogénération due to the interaction of excitons with the dipole layer altogether with a suitable orientation of the dipole moment, that is the electric field lying in the dipolar SAM layer has the same orientation than that of the intrinsic electric

Table 1. Open circuit voltage, current density and efficiency of conversion results of devices with SAM of thiols.

	V_{oc} (V)	J_{cc} ($\mu A/cm^2$)	FF(%)	$\eta \times 10^{-2}$ (%)
ITO/6T/Al	0.55	0.85	18.5	2.4
ITO/C$_{12}$H$_{25}$SH/6T/Al	0.60	1.29	18.7	4.0
ITO/3T(CH$_2$)$_6$SH/6T/Al	0,65	1.20	17.8	3.9

Table 2. Open circuit voltage, current density and efficiency of conversion results of devices with SAM of benzoic acids.

	V_{oc} (V)	J_{cc} ($\mu A/cm^2$)	FF(%)	$\eta \times 10^{-2}$ %
ITO/6T/Al	1.18	0.48	28.75	2.7
ITO/ABA/6T/Al	1.24	1.08	21.73	4.9
ITO/NBA/6T/Al	0.24	0.83	31.97	1.1

field. Also, we can express this result by orientation of dipole moment of SAM in the direction favor of increase of interfacial energy between the ITO and the SAM layer. After collection of carriers at the interfaces we will have a significant generation of photocurrent.

4.2.2. Orientation of the Dipole Moment on Photogeneration Carriers

The change of ITO work function is related to the molecules adsorbed on the surface that are supporting a dipole moment owing to the presence of partial charges on the functional groups. These dipoles are aligned on the surface and form a dipolar layer which can be seen as an effective layer lying in a planar capacitor formed by charges of opposite sign at the edges of SAM layer and in which lays an intrinsic electric field (**Figure 7**). The orientation of the dipole either facilitates or inhibits the extraction of electrons or holes by the surface which entrains a shift of the photogenerated carriers resulting in a shift of potential surface [20-23].

We note that SAMs of benzoic acids increase or decrease the work function of the anode with the additional potential barrier with the interfaces depending on the orientation of the dipole moment of grafted acid molecules [24,25].

Therefore, at a first insight one can say that grafting SAMs with permanent dipolar moment molecules enhances the creation of carriers at the interface which may improve the performances of organic photovoltaic cells. But, the orientation of dipole layer will increase the built-in potential if the dipole layer field is in the same direction of the built-in electric field and vice-versa.

The experimental parameters according to Equation (11), (12) and (13) for bare ITO and SAM-coated ITO devices are summarized in (**Table 3**). We notice that ITO/NBA/6T/Al cells exhibit the largest photocurrent but the lowest open circuit voltage, whereas for ITO/

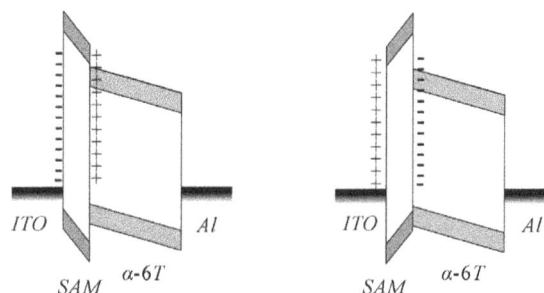

Figure 7. Photogeneration at interfaces.

ABA/6T/Al the J_{ph} and V_{oc} are slightly enhanced relatively to ITO/6T/Al. The NBA sample which has higher dipole moment than ABA is more efficient in creating photocarriers, but on the other hand it has a quite low open circuit voltage. The first feature can be interpreted by the strength of the dipolar moment which induces a quite strong field nearby the dipolar surface (**Table 3**); hence the dissociation rate of interacting excitons is considerably enhanced. Whereas the second feature is mainly due to the orientation of the dipolar moment, which results in an effective field oriented in the opposite direction of the built-in potential of the bulk sexithiophene thin film (**Figure 7(a)**). Thus the orientation of the dipolar moment of the SAM is a crucial factor in determining the open circuit voltage of a thin film photovoltaic cell. In fact, for ABA based samples in which the dipolar moment of the molecules is oriented such as the built-in potential is enhanced (**Figure 7(b)**), the open-circuit voltage as well as photocurrent are slightly enhanced compared to ITO/6T/Al samples.

In the NBA sample which has higher dipole moment than ABA is more efficient in creating photogenerated carriers, but on the other hand it has a quite low open circuit voltage. For NBA based samples, a remarkable decrease of built in potential that follow the magnitude

Table 3. Results of photocurrent density, Open circuit voltage, build in potential and free-standing dipole moment of SAM of the devices.

	Dipole moment (D) [13]	$\Delta\phi$ (eV)	J_{Ph} (μA/cm^2) (@−1V)	V_{OC} (V)	V_{bi} (eV)	g_a/g_c
Bare ITO	-	-	0.81	1.18	0.60	8.40·10^{-11}
ABA	+2.64	+0.38	3.31	1.24	0.98	2.55·10^{-5}
NBA	−5.94	−0.55	45.44	0.24	0.05	5.17·10^{-4}

and the orientation of dipole moment. We can assume that strength of the dipolar moment induces a quite strong field nearby the dipolar surface.

5. Conclusions

The grafting of self assembled monolayers with thiols molecules and dipolar molecules of benzoic acid on ITO may be a fashionable way to improve photovoltaic performance of organic cells. Analyze of UV-Visible absorption spectra shows an effect of thiols SAM on the orientation of the sexithiophene molecule on the substrate (gap 2.2 eV), what can be related to the degree of organization of the thin layer that is better with the molecules 3T(CH$_2$)$_6$SH. The current vs. applied voltage characterisation show an enhancement of device efficiency that confirm the effect of thiols molecules on the photogeneration of free carriers in the bulk i.e. far from electrodes and contribute to photocurrent, due to the conjunction between the low mobility of free carriers in organic materials and their short lifetime. Moreover, it may be worth remembering that dipolar benzoic acids derivative (ABA and NBA) increase the efficiency of photovoltaic cells. This increase is significant especially for oriented dipole molecule of ABA at ITO/6T interface (0.05%) reported with NBA. This improvement is affected by interaction of tightly bound Frenkel excitons with a surface dipole that may lead to efficient dissociation of geminate electron-hole pairs that has a significant effect on the photocurrent rate. However the contribution of photogenerated carriers to photocurrent is strongly dependent on dipole orientation. In fact the NBA compound has a large dipole moment but is oriented in the sight of reduction in photocurrent contrary to ABA. Then grafting strong dipole moment molecules oriented towards the cathode would provide a significant improvement of organic photovoltaic cells.

6. Acknowledgements

The author will like to thanks and express her gratitude to Dr. Fayçal Kouki and Prof. Habib Bouchriha, directors of research in UMAO (University El-Manar, Tunis) for their helpful and critical discussions to accomplish the study. I express also my thanks to Mr. Gill Horowitz and Mr. Philippe Lang, directors of research in ITODYS (University Paris7) for their assistance and support in experimental studies.

REFERENCES

[1]   G. Yu and A. J. Heeger, "Polymer Photovoltaic Cells: Enhanced Efficiencies via a Network of Internal Donor-Acceptor Heterojunctions," *Science,* Vol. 270, No. 5243, 1995, pp. 1789-1791.

[2]   P. Peumans and S. R. Forrest, "Very-High-Efficiency Double-Hetero-Structure Copper Pthalocyanine/C$_{60}$ Photovoltaic Cells," *Applied Physics Letters*, Vol. 79, No. 1, 2001, pp.126-128.

[3]   G. Li and Y. Yang, "High Efficiency Solution Processable Polymer Photovoltaic Cells by Self-Organization of Polymer Blends," *Nature Materials*, Vol. 4, No. 11, 2005, pp. 864-868.

[4]   C. L. Braun, "Electric Field Assisted Dissociation of Charge Transfer States as a Mechanism of Photocarrier Production," *Journal of Chemical Physics*, Vol. 80, No. 9, 1984, pp. 4157-4161.

[5]   F. Kouki and H. Bouchriha, "Photogeneration Process in Pristine Sexithiophene Based Photovoltaic Cells," *Organic Electronics*, Vol. 11, No. 8, 2010, pp.1439-1444.

[6]   S. F. Alvaradoand and D. D. C. Bradley, "Direct Determination of the Exciton Binding Energy of Conjugated Polymers Using a Scanning Tunneling Microscope," *Physical Review Letters*, Vol. 81, 1998, pp. 1082-1085.

[7]   X. Crispin, "Interface Dipole at Organic/Metal Interfaces and Organic Solar Cells," *Solar Energy Mater and Solar Cell*, Vol. 83, No. 2-3, 2004, pp. 147-168.

[8]   D.H. Kim and S. Young Oh, "Effects of ITO Surface Modification Using Self-Assembly Molecules on the Characteristics of OLEDs," *Ultramicroscopy*, Vol. 108, No. 10, 2008, pp. 1233-1236.

[9]   H. Sirringhaus and R. Friend, "Integrated Optoelectronic Devices Based on Conjugated Polymers," *Science*, Vol. 280, No. 5370, 1998, pp.1741-1744.

[10] S. Goncalves-Conto and L. Zuppiroli, "Interface Morphology in Organic Light-Emitting Diodes," *Advanced Materials*, Vol. 11, No. 2, 1999, pp.112-115.

[11] H. Bedis Ouerghemmi and F. Kouki, "Self-Assembled Monolayer Effect on the Characteristics of Organic Diodes," *Synthetic Metals*, Vol. 159, No. 7-8, 2009, pp. 551-555.

[12] G. G. Malliaras and J. C. Scott, "Photovoltaic Measurement of The Built-in Potential in Organic Light Emitting Diodes and Photodiodes," *Journal of Applied Physics*, Vol. 84, No. 3, 1998, pp. 1583-1587.

[13] S. G. Ray and H. Waldeck, "Organization-Induced Charge Redistribution in Self-Assembled Organic Monolayers on Gold," *Journal of Physical Chemistry B*, Vol. 109, No. 29, July 2005, pp. 14064-14073.

[14] H. Bässler and G. Vaubel, "Exciton-Induced Photocurrents in Molecular Crystals," *Discussions of the Faraday Society*, Vol. 51, 1971, pp. 48-53.

[15] L. Zuppiroli1 and M. Grätzel, "Self-assembled Monolayers as Interfaces for Organic Opto-Electronic Devices," *European Physical Journal B*, Vol. 11, No. 3, 1999, pp. 505-512.

[16] F. Kouki and F. Garnier, "Experimental Determination of Excitonic Levels in Alpha-Oligothiophenes," *Journal of Chemical Physics*, Vol. 113, No. 1, 2000, pp. 385-391.

[17] J. Tauc and A. Vancu, "Optical Properties and Electronic Structure of Amorphous Germanium," *Physica Status Solidi*, Vol.15, No. 2, 1966, pp. 627-637.

[18] N. Camaioni and Giovanna Barbarella, "Branched Thio-Phene-Based Oligomers as Electron Acceptors for Organic Photovoltaics," *Journal of Materials Chemistry*, Vol. 15, No. 22, 2005, pp. 2220-2225.

[19] H. Hoppe and N. Serdar Sariciftci, "Polymer Solar Cells," *Advances in Polymer Science*, Vol. 214, 2007, pp.1-86.

[20] J. W. King and S. P. Molnar, "Molecular Structural Index Control in Property-Directed Clustering and Correlation," *International Journal of Quantum Chemistry*, Vol. 80, No. 6, 2000, pp. 1164-1171.

[21] I. H. Campbell and D. L. Smith, "Electrical Impedance Measurements of Polymer Light Emitting Diodes", *Applied Physics Letters*, Vol. 66, No. 22, 1995, p. 3030.

[22] P. S. Davids and D. L. Smith, "Nondegenerate Continuum Model for Polymer Light – Emitting Diodes," *Journal of Applied Physics*, Vol. 78, No. 6, 1995, pp. 4244-4252.

[23] A. J. Cambell and D. G. Lidzey, "Space-Charge Limited Conduction with Traps in Poly (Phenylene Vinylene) Light Emitting Diodes," *Journal of Applied Physics*, Vol. 82, No. 12, 1997, pp. 6326-6342.

[24] M. Carrara and L. Zuppiroli, "Carboxylic Acid Anchoring Groups for the Construction of Self-Assembled Monolayers on Organic Device Electrodes," *Synthetic Metals*, Vol. 121, 2001, pp. 1633-1634.

[25] H. Bedis, F. Kouki and H. Bouchriha, "Effect of Self Assembled Monolayers on Carrier Photogeneration in Organic Photovoltaic Cells," *International conference ELECMOL'08 Grenoble-HMMNT-Minatec*, 2008.

9

Effect of Annealing on Structural, Morphological, Electrical and Optical Studies of Nickel Oxide Thin Films

Vikas Patil[1*], Shailesh Pawar[1], Manik Chougule[1], Prasad Godse[1], Ratnakar Sakhare[1], Shashwati Sen[2], Pradeep Joshi[1]

[1]Materials Research Laboratory, School of Physical Sciences, Solapur University, Solapur, India; [2]Crystal Technology Section, Bhabha atomic Research Centre, Mumbai, India.

ABSTRACT

Sol gel spin coating method has been successfully employed for the deposition of nanocrystalline nickel oxide (NiO) thin films. The films were annealed at 400°C - 700°C for 1 h in an air and changes in the structural, morphological, electrical and optical properties were studied. The structural properties of nickel oxide films were studied by means of X-ray diffraction (XRD) and scanning electron microscopy (SEM). XRD analysis shows that all the films are crystallized in the cubic phase and present a random orientation. Surface morphology of the nickel oxide film consists of nanocrystalline grains with uniform coverage of the substrate surface with randomly oriented morphology. The electrical conductivity showed the semiconducting nature with room temperature electrical conductivity increased from 10^{-4} to 10^{-2} $(\Omega cm)^{-1}$ after annealing. The decrease in the band gap energy from 3.86 to 3.47 eV was observed after annealing NiO films from 400°C - 700°C. These mean that the optical quality of NiO films is improved by annealing.

Keywords: *Nickel Oxide, Sol Gel Method, Structural Properties, Optoelectronic Properties*

1. Introduction

Metal oxides can adopt a large variety of structural geometries with an electronic structure that may exhibit metallic, semiconductor, or insulator characteristics, endowing them with diverse chemical and physical properties. Therefore, metal oxides are the most important functional materials used for chemical and biological sensing and transduction. Moreover, their unique and tunable physical properties have made themselves excellent candidates for electronic and optoelectronic applications. Nanostructured metal oxides have been actively studied due to both scientific interests and potential applications [1,2].

NiO is an important antifromagnetic p-type semiconductor with excellent properties such as gas-sensing, catalytic and electrochemical properties, and has been studied widely for applications in solid-state sensors, electrochromic devices and heterogeneous catalysts as well as lithium batteries. The nickel oxide thin films have been prepared using various techniques including thermal evaporation [3], spray pyrolsis [4], chemical vapor deposition [5], electrochemical deposition [6], sol-gel [7,8], sputtering [9-11], chemical solution deposition [12-16], *etc.* Among these, chemical solution deposition, also called as a chemical bath deposition, is an advantageous technique due to its low cost, low-temperature operating condition and freedom to deposit materials on a variety of substances. Verkey and Fort [14] deposited nickel oxide thin films using nickel sulfate and ammonia solution over the temperature range 330 - 350 K. Pramanik and Bhattacharya [12] prepared nickel oxide thin films from an aqueous solution composed of nickel sulfate, potassium persulfate, and ammonia atroom temperature. Han *et al.* [15] studied growth mechanism of nickel oxide thin films following Pramanik's chemistry. Banerjee *et al.* [16] obtained hexagonal mesoporous nickel oxide using dodecyl sulfate as a surfactant and urea as a hydrolyzing agent.

To the best of our knowledge, few works are available in the literature on the sol-gel synthesis and characterization of NiO-based nanosystems [7,8].

In the present study, we report new method of synthesis and characterization of nanocrystalline NiO thin films by simple and inexpensive sol-gel spin coating technique and effect of annealing on their structure, morphology, electric transport and optical properties.

2. Experimental Details

Nanocrystalline NiO thin films have been synthesized by a sol-gel method using Nickel acetate $Ni(CH_3COO)_2 \cdot 4H_2O$ as a source of Ni. In a typical experiment; 3.322 gm of nickel acetate was added to 40 ml of methanol and stirred vigorously at 60°C for 1 hr, leading to the formation of light green colored powder. The as prepared powder was sintered at various temperatures ranging from 400 - 700°C with a fixed annealing time of 1hr in an ambient air to obtain NiO films with different crystallite sizes. The nanocrystalline NiO powder was further dissolved in m-cresol and solution was continuously stirred for 11 h at room temperature and filtered. The filtered solution was deposited on to a glass substrate by a single wafer spin processor (APEX Instruments, Kolkata, Model SCU 2007). After setting the substrate on the substrate holder of the spin coater, the coating solution (approximately 0.2 ml) was dropped and spin-casted at 3000 RPM for 40 s in an air and dried on a hot plate at 100°C for 10 min. **Figure 1** shows the flow chart for the sol-gel synthesis of the NiO films prepared by using the spin-coating technique. The structural properties of the films were investigated by means of X-ray diffraction (XRD) (Philips PW-3710, Holland) using Cu Kα radiation (λ = 1.5406 Å). The surface morphology of the films was examined by scanning electron microscopy (SEM) (Model Japan), operated at 20 kV. The room temperature dc electrical conductivity measurements were performed using four probe techniques. The optical absorption spectra of the NiO thin films were measured using a double-beam spectrophotometer Shimadzu UV-140 over 200 - 1000 nm wavelength range. The thickness of the film was measured by using weight difference method and Dektak profilometer.

3. Results and Discussion

3.1. NiO Film Formation Mechanism and Thickness Measurement

The mechanism of NiO film formation by the sol gel spin coating method can be enlightened as follows:

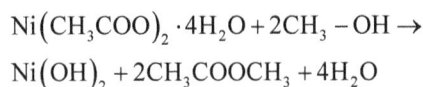

$$Ni(CH_3COO)_2 \cdot 4H_2O + 2CH_3 - OH \rightarrow$$
$$Ni(OH)_2 + 2CH_3COOCH_3 + 4H_2O$$

Since to improve crystallinity and remove hydroxide phase, films were annealed for 1 h pure NiO film is formed after air annealing by following mechanism:

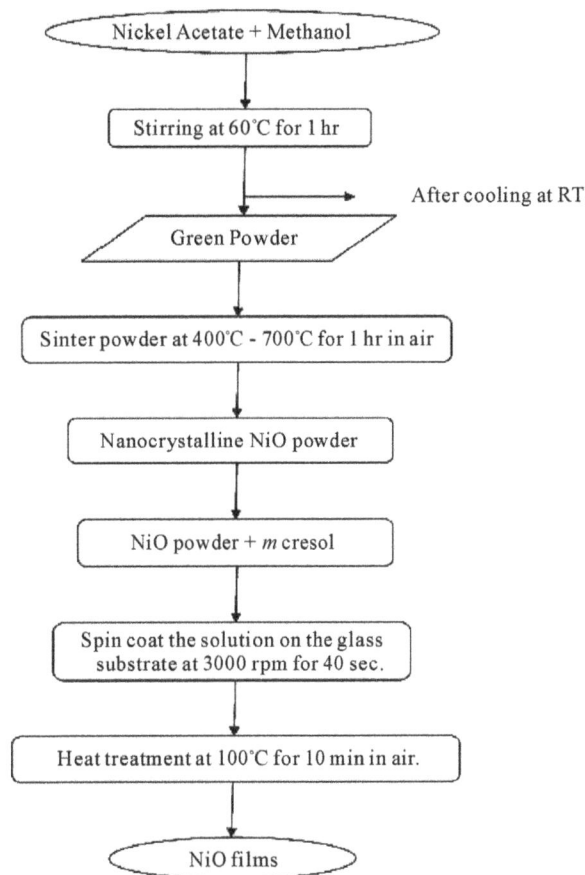

Figure 1. Flow diagram for NiO films prepared from the sol-gel process using the spin-coating method.

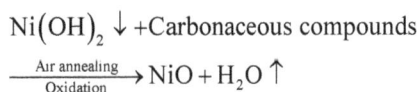

$$Ni(OH)_2 \downarrow + \text{Carbonaceous compounds}$$
$$\xrightarrow[\text{Oxidation}]{\text{Air annealing}} NiO + H_2O \uparrow$$

Thickness was calculated by weight difference method using formula:

$$t = m/A\rho \tag{1}$$

where t is film thickness of the film; m is actual mass deposited onto substrate; A is area of the film and is the density of nickel oxide (6.67 g/cc^2).

It was observed that increasing the annealing temperature resulted in a decrease in film thickness from 0.9061 μm (400°C annealing) to 0.4997 μm (700°C annealing). The NiO thin film thickness is also confirmed by using Dektak profilometer and is presented in **Table 1**.

3.2. Structural Analysis

Structural analysis of the NiO films annealed at 400°C - 700°C was carried out by using CuK$_\alpha$ radiation source of wavelength (λ = 1.54056 A°) and the diffraction patterns of films were recorded by varying diffraction angle (2θ) in the range 20° - 80°. **Figure 2** shows XRD pattern for

Table 1. Effect of annealing on NiO thin film properties.

Sr. No.	Annealing temperature °C	Crystallite size nm	Thickness μm	Energy gap E_g, eV	Activation energy, E_a, eV	
					HT	**LT**
1	400	41.55	0.9061	3.86	0. 110	0.082
2	500	43.20	0.7414	3.69	0. 236	0.086
3	600	46.80	0.6425	3.60	0.344	0.096
4	700	50.67	0.4997	3.47	0.481	0.143

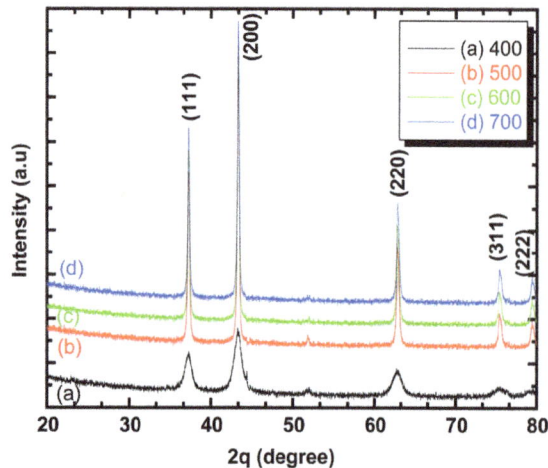

Figure 2. X-ray diffraction patterns of NiO film at different annealing temperatures.

the NiO films annealed at 400 - 700°C .The observed 'd' values are in good agreement with standard 'd' values and the diffraction peaks are indexed to the cubic phase of NiO with a = b = c = 4.1678 A° [Joint Committee on Powder Diffraction Standards (JCPDS) No. 73-1519]. It shows well-defined peaks having orientations in the (1 1 1), (2 0 0), (2 2 0), (3 1 1) and (222) planes. The absence of impurity peaks suggests the high purity of the nickel oxide. Compared with those of the bulk counterpart, the peaks are relatively broadened, which further indicates that the deposited material has a very small crystallite size [17]. The crystallite size (D) is calculated using equation as follows [18]:

$$D = 0.9\lambda/\beta\cos\theta \qquad (2)$$

where, β is the half width of diffraction peak measured in radians. The calculation of crystallite size from XRD is a quantitative approach which is widely accepted and used in scientific community [19-22]. The average crystallite size is increased with increasing annealing temperature revealing a fine nanocrystalline grain structure (**Table 1**).

3.3. Surface Morphological Studies

The two-dimensional high magnification surface mor-

phologies of NiO thin films annealed at 400°C -700°C were carried out using SEM images are shown in **Figure 3(a-d)**. From the micrographs, it is seen that the film consists of nanocrystalline grains with uniform coverage of the substrate surface with randomly oriented morphology and the crystallite size is increased from 40 - 52 nm as annealing temperature increases from 400°C - 700°C. The crystallite size calculated from SEM analysis is quite in good agreement with that of crystallite size calculated from XRD analysis. Similar results are also observed by Patil et al. [22] for sol gel derived TiO_2 films.

3.4. Electrical Transport Properties

3.4.1. Electrical Conductivity Measurement

The four-probe technique of dark electrical conductivity measurement was used to study the variation of electrical conductivity of film with annealing temperature. The variation of log σ with reciprocal temperature (1000/T) is depicted in **Figure 4**. After annealing, room temperature electrical conductivity was increased from 10^{-4} to 10^{-2} $(\Omega\cdot cm)^{-1}$, due to increase in the crystallite size and reduced scattering at the grain boundary. Similar type of increase in electrical conductivity has been observed by Patil et al. [22]. From **Figure 4** it is observed that the conductivity of film is increases with increase in annealing temperature, further it is observed that conductivity obeys Arrhenius behavior indicating a semiconducting transport behavior. The activation energies were calculated using the relation:

$$\sigma = \sigma_o \exp\left(-E_a/kT\right) \qquad (3)$$

where, σ is the conductivity at temperature T, σ_o is a constant, k is the Boltzmann constant, T is the absolute temperature and E_a is the activation energy. The activation energy represents the location of trap levels below the conduction band. From **Figure 4**, activation energy (HT) was increases from 0.110 eV, to 0.481 eV, when film annealed from 400°C - 700°C indicating no significant change.

3.4.2. Thermo-emf Measurement

The dependence of thermo-emf on temperature is de-

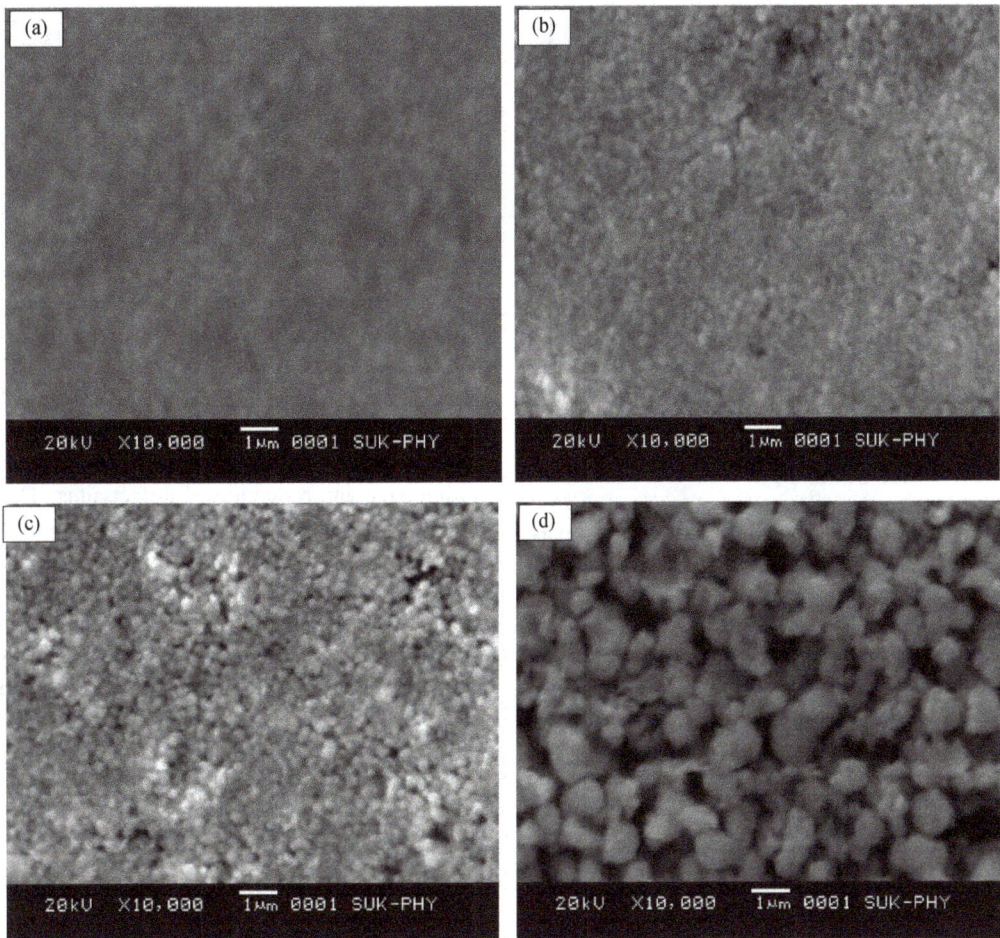

Figure 3. SEM of NiO thin films anneled at (a) 400°C; (b) 500°C; (c) 600°C; and (d) 700°C.

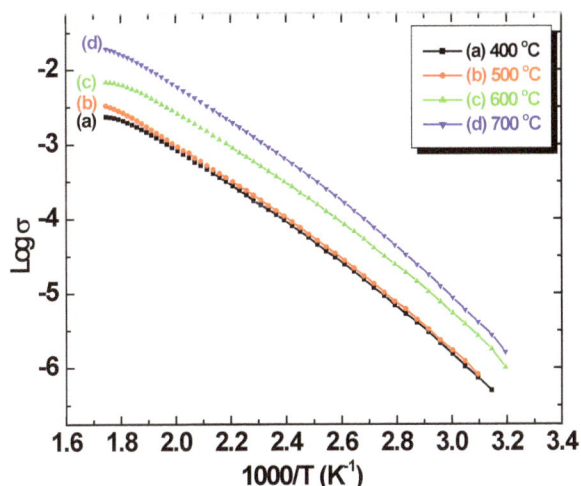

Figure 4. Arrhenius plot of log conductivity vs. 1000/T of NiO thin film annealed at different temperatures.

picted in **Figure 5**. The thermo-emf was measured as a function of temperature in the temperature range 300 – 500 K. The polarity of the thermo-emf was negative at the hot end with respect to the cold end which confirmed that nickel oxide thin films are of p-type semiconducting similar to earlier report [23]. The plot shows increase in thermo-emf with increase in temperature when film annealed from 400°C -700°C. This is attributed to the increase in hole concentration as the annealing temperature increases and also due to the increase in crystallite size as discussed in section 3.3. The thermoelectric power was found to be of the order of 10^{-3} V/K when film annealed from 400°C - 700°C

3.5. Optical Studies

The optical absorption spectra in the range of 200 - 1000 nm for NiO thin films annealed at 400 - 700°C were carried out at room temperature without taking in account of scattering and reflection. **Figure 6** shows the optical absorption spectra of NiO thin films annealed at 400°C - 700°C, it is observed that the absorption coefficient is very low for photon energy in the IR and visible region while the sudden increase in the absorption coefficient

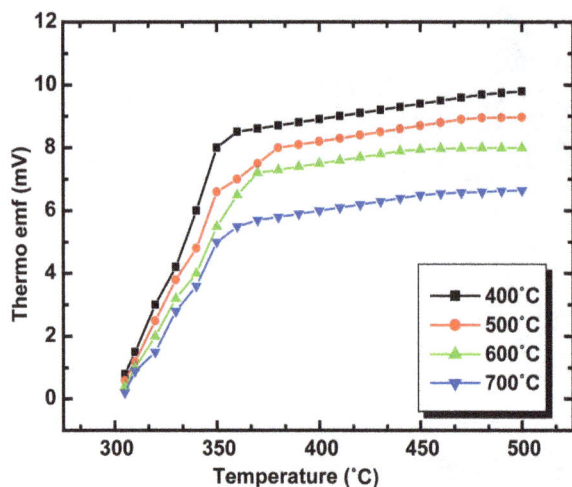

Figure 5. The variation of thermo-emf with temperature for of NiO thin film annealed at different temperatures.

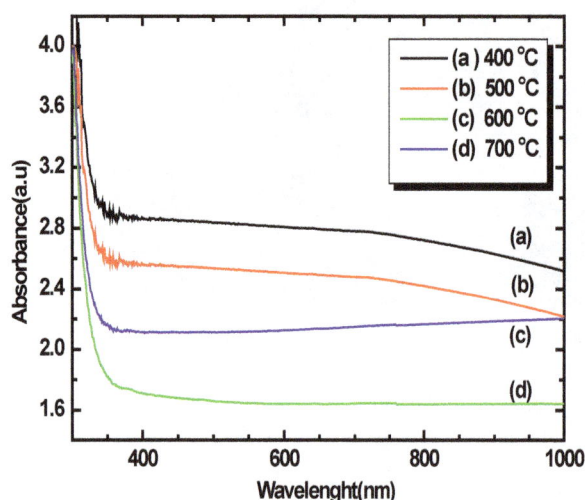

Figure 6. Variation of absorbance (αt) with wavelength (λ) of NiO thin film annealed at different temperatures.

occurs in the near UV region. It was found that, the absorption coefficient of films is increases with increase in annealing temperature. This could be because of increase in the density of states of holes with increase in annealing temperature, similar results are reported by Varkey *et al.* [24] and Pejova [25]. The optical band gap (E$_g$) of NiO thin films annealed at 400˚C - 700˚C is calculated on the basis of optical absorption spectra using the following equation:

$$\alpha = \frac{A\left(Eg - hv\right)^n}{hv} \qquad (4)$$

where '*A*' is a constant, '*E$_g$*' is the semiconductor band

gap and '*n*' is a number equal to 1/2 for direct gap and 2 for indirect gap compound.

The plots of $(\alpha \cdot hv)^2$ versus hv of films annealed at 400˚C - 700˚C are shown in **Figure 7**.

Figure 7 Plot of $(\alpha \cdot hv)^2$ versus (hv) of NiO thin film for different annealing temperatures.

Since the plots are almost linear, the direct nature of the optical transition in β-Ni(OH)$_2$ and NiO is confirmed. Extrapolation of these curves to photon energy axis reveals the band gaps. The band gap was found to be decreased from 3.86 to 3.47 eV for films annealed at 400˚C - 700˚C. Varkey and Fort [14] reported the slightly lower band gaps 3.75 and 3.25 eV for as-prepared NiOOH and annealed NiO thin films [24]. The decrease in E$_g$ with annealing temperature could be due to increase in crystalline size and reduction of defect sites. This is in good agreement with the experimental results of XRD analysis. According to XRD results, the mean grain size has increased with increased annealing temperature. As the grain size has increased, the grain boundary density of a film decreased, subsequently, the scattering of carriers at grain boundaries has decreased [25] .A continuous increase of optical constants and also the shift in absorption edge to a higher wavelength with increasing annealing temperature may be attributed to increase in the particle size of the crystallites along with reduction in porosity.

The decrease in optical band gap energy is generally observed in the annealed direct-transition-type semiconductor films. Hong *et al.* [26] observed a shift in optical band gap of ZnO thin films from 3·31 - 3·26 eV after annealing, and attributed this shift to the increase of the ZnO grain size. Chaparro *et al.* [27] ascribed this 'red shift' in the energy gap, E$_g$, to an increase in crystallite size for the annealed ZnSe films. Bao and Yao [28] also

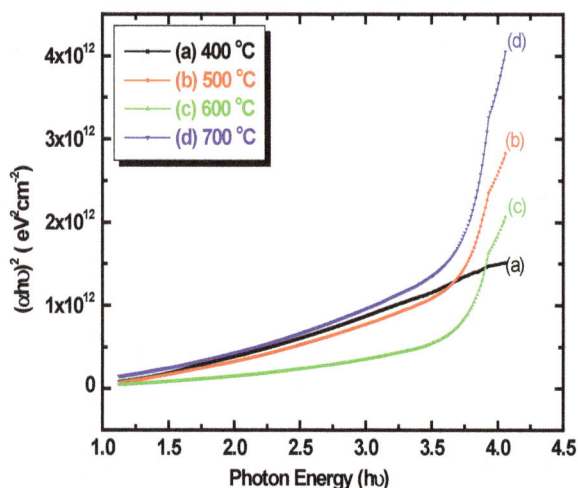

Figure 7. Plot of $(\alpha \cdot hv)^2$ versus (hv) of NiO thin film for different annealing temperatures.

reported a decrease in E_g with increasing annealing temperature for SrTiO$_3$ thin films, and suggested that a shift of the energy gap was mainly due to both the quantum-size effect and the existence of an amorphous phase in thin films. In present case, the mean crystallite size increases from 40 to 70 nm after annealing from 400°C - 700°C. Moreover, it is understood that the amorphous phase is reduced with increasing annealing temperature, since more energy is supplied for crystallite growth, thus resulting in an improvement in crystallinity of NiO films. Therefore, it is believed that both the increase in crystallite size and the reduction in amorphous phase cause are decreasing in band gap of annealed NiO films. The change in optical band gap energy, E_g, reveals the impact of annealing on optical properties of the NiO films.

4. Conclusions

Nanocrystalline nickel oxide thin films were prepared by low-cost sol gel spin coating technique. The NiO films were annealed for various temperatures between 400 to 700°C. The XRD results revealed that the NiO thin film has a good nanocrystalline cubic structure. The SEM results depict that a uniform surface morphology and the nanoparticles are fine with an average grain size of about 40 - 60 nm. The dc electrical conductivity is increased from 10^{-4} to 10^{-2} $(\Omega \cdot cm)^{-1}$ for films annealed at 400°C - 700°C. Optical absorption studies show low-absorbance in IR and visible region with band gap 3.86 eV (at 400°C) which was decreased to 3.47 eV (at 700°C). This has been attributed to the decrease in defect levels. The *p*-type electrical conductivity is confirmed from thermo-emf measurement with no appreciable change in thermoelectric power after annealing

5. Acknowledgments

Authors (VBP) are grateful to DAE-BRNS, for financial support through the scheme no.2010/37P/45/BRNS/1442. Thanks are also extended to Dr.P.S.Patil, Department of Physics, Shivaji University, Kolhapur for providing SEM facility.

REFERENCES

[1] C. L. Shao, X. H. Yang, H. Y. Guan, Y. C. Liu and J. Gong, "Electrospun Nanofibers of NiO/ZnO Composite," *Inorganic Chemistry Communications*, Vol. 7, No. 5, 2004, pp. 625-627.

[2] G.-J. Li, X.-X. Huang, Y. Shi and J.-K. Guo, "Preparation and Characteristics of Nanocrystalline NiO by Organic Solvent Method," *Materials Letters*, Vol. 51, No. 4, 2001, pp. 325-330.

[3] B. Sasi, K. Gopchandran, P. Manoj, P. Koshy, P. Rao and V. K. Vaidyan, "Preparation of Transparent and Semiconducting NiO Films," *Vacuum*, Vol. 68, No. 2, 2003, pp. 149-154.

[4] J. D. Desai, S. K. Min, K. D. Jung and O. S. Joo, "Spray Pyrolytic Synthesis of Large Area NiO$_x$ Thin Films from Aqueous Nickel Acetate Solutions," *Applied Surface Science*, Vol. 253, No. 4, 2006, pp.1781-1786.

[5] J.-K. Kang, S. W. Rhee, "Chemical Vapor Deposition of Nickel Oxide Films from Ni(C$_5$H$_5$)$_2$/O$_2$," *Thin Solid Films*, Vol. 391, No. 2, 2001, pp. 57-61.

[6] K. Nakaoka, J. Ueyama, K. Ogura, "Semiconductor and Electrochromic Properties of Electrochemically Deposited Nickel Oxide Films," *Journal of Electroanalytical Chemistry*, Vol. 571, No. 1, 2004, pp. 93-99.

[7] D. J. Taylor, P. F. Fleig, S. T. Schwab and R. A. Page, "Sol-Gel Derived Nanostructured Oxide Lubricant Coatings," *Surface and Coatings Technology*, Vol. 120, 1999, pp. 465-469.

[8] J. L. Garcia-Miquel, Q. Zhang, S. J. Allen, A. Rougier , A. Blyr, H. O. Davies, A. C. Jones, T. J. Leedham, P. A.William and S. A. Impey, "Nickel Oxide Sol-Gel Films from Nickel Diacetate for Electrochromic Applications," *Thin Solid Films*, Vol. 424, No. 2, 2003, pp.165-170.

[9] J. W. Park, J. W. Park, D. Y. Kim, J. K. Lee, "Reproducible Resistive Switching in Nonstoichiometric Nickel Oxide Films Grown by rf Reactive Sputtering for Resistive Random Access Memory Applications," J*ournal of Vacuum Science and Technology A*, Vol. 23, No. 5, 2005, pp.1309-1313.

[10] K. S. Ahn, Y. C. Nah and Y. E. Sung, "Surface Morphological, Microstructural, and Electrochromic Properties of Short-Range Ordered and Crystalline Nickel Oxide Thin Films", *Applied Surface Science*, Vol. 199, No. 1-4, 2002, pp. 259-269.

[11] H. L. Chen, Y. M. Lu and W. S. Hwang, "Thickness Dependence of Electrical and Optical Properties of Sputtered Nickel Oxide Films," *Thin Solid Films*, Vol. 514, No. 1-2, 2005, pp. 361-365.

[12] P. Pramanik, S. Bhattacharya, "A Chemical Method for the Deposition of Nickel Oxide Thin Films," *Journal of Electrochemical Society*, Vol. 137, No. 12, 1990, pp. 3869-3870.

[13] B. Pejova, T. Kocareva, M. Najdoski and I. Grozdanov, "A Solution Growth Route to Nanocrystalline Nickel Oxide Thin Films," *Applied Surface Science*, Vol. 165, No. 4, 2000, pp. 271-278.

[14] A. J. Varkey and A. F. Fort, "Solution Growth Technique for Deposition of Nickel Oxide Thin Films," *Thin Solid Films*, Vol. 235, No. 1-2, 1993, pp. 47-50.

[15] S. Y. Han, D. H. Lee, Y. J. Chang, S. O. Ryu and T. J. Lee, C. H. Chang, "The Growth Mechanism of Nickel Oxide Thin Films by Room-Temperature Chemical Bath Deposition," *Journal of Electrochemical Society*, Vol.

153, No. 6, 2006, pp. C382-C386.

[16] S. Banerjee, A. Santhanam, A. Dhathathrenyan and M. Rao, "Synthesis of Ordered Hexagonal Mesostructured Nickel Oxide," *Langmuir*, Vol. 19, No. 13, 2003, pp. 5522-5525.

[17] E. Comini, G. Faglia, G. Sberveglieri, Z. Pan and Z. L. Wang, "Stable and Highly Sensitive Gas Sensors Based On Semiconducting Oxide Nanobelts," *Applied Physics Letters*, Vol. 81, 2002, pp.1869-1871.

[18] A. Studenikin, N. Golego and M. Cocivera, "Fabrication of Green and Orange Photoluminescent, Undoped ZnO Films Using Spray Pyrolysis," *Journal of Applied Physics*, Vol. 84, No. 4, 1998, pp. 2287-2280.

[19] P. K. Ghosh, R. Maity, K. K. Chattopadhyay, "Electrical and Optical Properties of Highly Conducting CdO: F Thin Film Deposited by Sol-Gel Dip Coating Technique," *Solar Energy Materials and Solar Cells*, Vol. 81, No. 2, 2004, pp. 279-289.

[20] K. Gurumurugan, D. Mangalaraj, S. K. Narayandass and Y. Nakanishi, "DC Reactive Magnetron Sputtered CdO Thin Films," *Materials Letters*, Vol. 28, No. 4-6, 1996, pp. 307-312.

[21] C. N. R. Rao, S. R. C. Vivekchand, K. Biswas and A. Govindaraj, "Synthesis of Inorganic Nanomaterials," *Dalton Transactions*, Vol. 34, 2007, pp. 3728-3749.

[22] S. G. Pawar, S. L. Patil, M. A. Chougule and V. B. Patil, "Synthesis and Characterization of Nanocrystalline TiO_2 Thin Films," *Journal of Materials Science: Material in*

Electronics, Vol. 22, No. 3, 2011, pp.260-264.

[23] Y. K. Jeong and G. M. Choi, "Nonstoichiometry and Electrical Conduction of CuO," *Journal of Physics and Chemistry of Solids*, Vol. 57, No. 1, 1996, pp. 81-84.

[24] B. Pejova, T. Kocareva, M. Najdoski and I. Grozdanov, "A Solution Growth Route to Nanocrystalline Nickel Oxide Thin Films," *Applied Surface Science*, Vol. 165, No. 4, 2000, pp. 271-278.

[25] J. H .Lee, K. H. Ko and B. O. Park, "Electrical and Optical Properties of ZnO Transparent Conducting Films by the sol-gel method," *Journal of Crystal Growth*, Vol. 247, No.1-2, 2003, pp. 119-125.

[26] R. Hong, J. Huang, H. He, Z. Fan and Shao "Influence of Different Post-Treatments on the Structure and Optical Properties of Zinc Oxide Thin Films," *Applied Surface Science*, Vol. 242, No. 3-4, 2005, pp. 346-352.

[27] A. M. Chaparro, M. A. Martinez, C. Guillen, R. Bayon, M. T. Gutierrez and J. Herrero, "SnO_2 Substrate Effects on the Morphology and Composition of Chemical Bath Deposited ZnSe Thin Films," *Thin Solid Films*, Vol. 361, 2000, pp.177-182.

[28] D. H. Bao and X. Yao, Naoki Wakiya, Kazuo Shinozaki and Nobuyasu Mizutani, "Band-Gap Energies of Sol-Gel Derived $SrTiO_3$ thin films" *Applied Physics Letters*, Vol. 79, No. 23, 2001, pp. 3767-3772.

10

Development of the Biopolymeric Optical Planar Waveguide with Nanopattern

Seung H. Yoon[1], Won T. Jeong[1], Kyung C. Kim[1], Kyung J. Kim[2], Min C. Oh[2], Sang M. Lee[3]

[1]Mechanical Engineering, Pusan National University, Pusan, Korea; [2]Electrical Engineering, Pusan National University, Pusan, Korea; [3]Pusan National University, Pusan, Korea.

ABSTRACT

This paper demonstrates for fabricating the biopolymric optical planar waveguide. Gelatin and chitosan were mixed with ratio of 9 to 1 and stirred at 70 °C with 1300 rpm. The blended biopolymer was spincoated on silicon substrate with 500 rpm and then dried in the oven at 50 °C. The refractive indices of the prepared biopolymer clad and core layers of the waveguide were measured by the ellipsometry. The measured refractive indices of the two layers were obtained to be 1.516 and 1.52, respectively. The nanograting was successfully imprinted on surface of the biopolymeric waveguide.

Keywords: *Biopolymeric Optical Waveguide, Nanograting*

1. Introduction

This work aims to develop highly sensitive all-biopolymeric planar Bragg grating biosensor. Biopolymer [1-4] is biocompatible material which is developed by blending the chitosan and gelatin [5]. The optical planar waveguide is composed of a clad and core layers. The refractive index of the core layer is slightly higher than that of the clad layer so that light can be guided in the waveguide. The Bragg grating is printed on the developed optical planar waveguide by using the nanoimprint technique.

2. Theory

This section provides the basic mathematical foundation that will be used to design and perform data analysis of the proposed planar waveguide grating sensor. This section starts by providing the expressions needed to calculate the propagation constants for the slab waveguide in terms of the waveguide geometric and optical properties.

Then the coupled mode equations used to determine the grating properties are presented. The waveguide geometry of interest in this paper is illustrated in **Figure 1**, and consists of a biopolymer core bounded by air above and a biopolymer clad below. This waveguide is fabricated on a silicon substrate. The refractive indices of air, clad and core layers are denoted by n_1, n_2, and n_3

*Korean Government (NRF-2009).

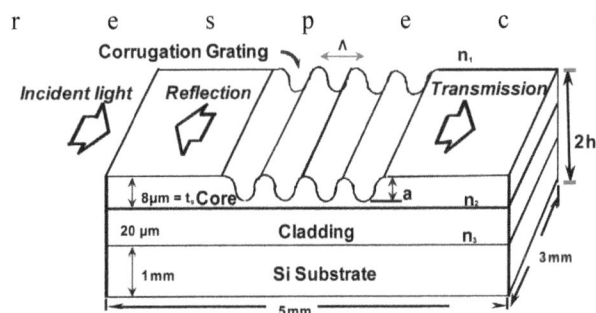

Figure 1. Schematic of the slab waveguide with surface corrugation grating.

tively and the core thickness is given by t_g. Also seen in **Figure 1** are the corrugations of pitch and depth a. The most important parameter in the design of Bragg grating in slab waveguide is the effective index. This effective index is found in the usual way [6] by solving the wave equation and applying the continuity boundary conditions at the respective core/cladding interfaces of the waveguide shown in **Figure 1**. The guided modes have a propagation constant β_s such that $k_0 n_3 < \beta_s < k_0 n_2$, where $n_1 < n_3$. This solution process leads to the following transcendental equation that yields the propagation constant:

$$\tanh t_g = \frac{p+q}{h\left(1 - \dfrac{pq}{h^2}\right)} \qquad (1)$$

where $h = \left(n_y^2 k_0^2 - \beta_s^2\right)^{1/2}$, $q = \left(\beta_s^2 - n_1^2 k_0^2\right)^{1/2}$, $p = \left(\beta_s^2 - n_1^2 k_0^2\right)^{1/2}$ and $k_0 \approx \omega/c = 2\pi/\lambda$. Given a set of refractive indices n_1, n_2, and n_3 and the waveguide thickness, t_g, of the planar waveguide, and the source wavelength, λ, Equation (1) in general yields a number of solutions for the propagation constant, β_s. However, the source wavelength and the waveguide thickness are restricted in the present study such that only one propagation mode is supported, and therefore Equation (1) has only one solution of interest. As a result, the effective index of the planar waveguide is given by $n_{eff} = \beta_s \lambda/2\pi$. The corrugated structure into the waveguide leads to a corresponding periodic perturbation of the refractive index distribution. Each groove of the grating acts like a weak mirror, and the cumulative effect of all of the weak reflectors results in a very strong combined reflection centered on what is known as the Bragg wavelength. The Bragg wavelength is related to the effective index calculated above and the grating period, Λ, by [6]

$$\lambda_b = 2n_{eff}\Lambda \qquad (2)$$

which when expressed in terms of the propagation constant is given by

$$\lambda_b = \frac{\beta_s \lambda \Lambda}{2} \qquad (3)$$

where λ_b is the Bragg wavelength and λ is the central wavelength of the optical source.

3. Experiment

3.1. The Production of Biopolymer through Blending of Chitosan and Gelatin

The biopolymer was developed in this study by blending chitosan and gelatin. The buffer solution which was composed of sodium acetate and acetic acid dissolved chitosan and gelatin in liquid state. In order to control the refractive index of the material, the ratios of chitosan to gelatin were 1:9, 2:8, 5:5, 8:2 and 9:1. **Figure 2** shows procedures for blending the biopolymer.

There are issues addressed in the process for dissolving the biopolymers. As the first issue, viscosity of chitosan in liquid state increases when the amount of chitosan in the buffer increases. When blending chitosan with gelatin in the liquid state, conglomeration of chitosan may occur and accordingly it causes surface quality. Another issue is that chitosan in acetic acid is not completely insoluble and remained to be particle state.

To measure the refractive index of the sample, surface of the sample must be kept clean and flat. In order to mitigate the problems mentioned above, dissolving the amount of chitosan was controlled to prevent increase in viscosity of chitosan solution, and chitosan particles existing in the solution have been removed using the centrifuge.

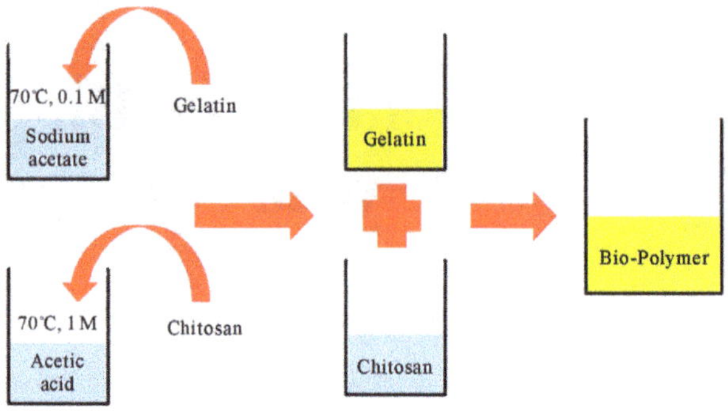

Figure 2. Schematic of blending process of a biopolymer.

sting in the solution have been removed using the centrifuge.

3.2. Fabrication of Biopolymeric Optical Waveguide

The ratio of chitosan and gelatin was varied to obtain the slightly different refractive indices of the biopolymers which consisted of the biopolymeric waveguide. The biopolymer to be used in core layer was mixed with the ratio of chitosan of 0.05 g and gelatin of 1.03 g. The clad layer was mixed with the ratio of chitosan of 0.03 g and gelatin of 1.05 g. The biopolymer was spincoted on the silicon wafer with 550 rpm for the 30 s and then baked for 5 hours at 50°C. The ellipsometer was used to measure the refractive index of the spincoted film of the biopolymer.

3.3. Optical Butt Coupling

The optical planar waveguide was butt-coupled to investigate the light propagation into the waveguide. The light is incident into the optical fiber which was butt coupled to the waveguide, and then the light transmitted through the waveguide was monitored by the CCD camera, as shown in **Figure 3**.

The Bragg grating which will be imprinted on the biopolymeric planar waveguide obtained in this study is designed by using the Equation (2). **Figure 4** shows schematic of the optical waveguide, indicating that the grating pitch was determined by the effective refractive index of and thickness of waveguide.

3.4. Imprint of Bragg Grating on the Biopolymric Waveguide

Holographic grating, as shown in **Figure 5**, is imprinted on the photoresist spincoated on the silicon substrate with grating period obtained by calculating the optical waveguide. PDMS grating was molded on the PDMS by using the photoresist grating on the silicon substrate. PDMS grating mold production process and a grating imprinting technique are shown in **Figures 6** and **7**.

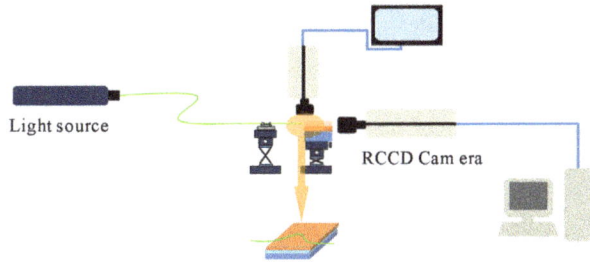

Figure 3. Schematic of optical butt coupling setup.

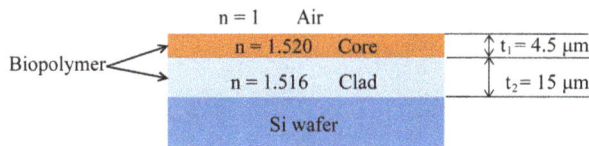

Figure 4. Schematic of the optical waveguide.

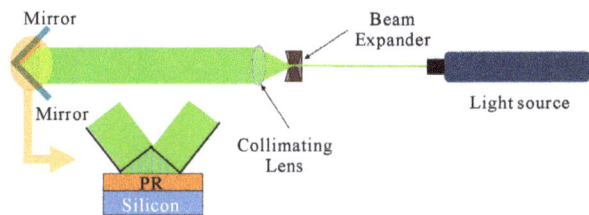

Figure 5. Schematic of holographic lithography of the grating.

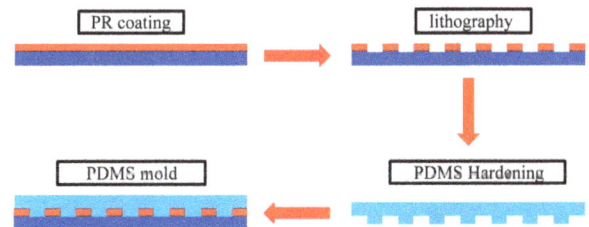

Figure 6. Schematic of PDMS grating mold.

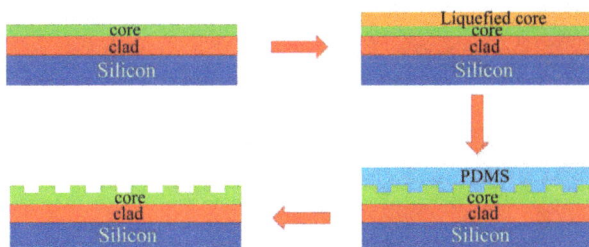

Figure 7. Schematic of grating imprinting process.

4. Results and Discussion

The ratio of blending of chitosan and gelatin makes the refractive index of the biopolymeric material changed. The blended biopolymer was coated on the substrate with the proper velocity of spincoating and time. The condi-tion for spincoating is shown in **Table 1**. The thickness of the spincoated layer of the biopolymer was measured to be 300 nm - 800 nm.

However, the biopolymer layer was spincoated more than once to obtain the thick layer of biopolymer as thick as 8 μm - 10 μm. The following method which may be able to increase the thickness of biopolymer layer was employed. The biopolymer was spincoated on the silicon wafer and then the buffer solution was evaporated in or-der to increase the viscosity of the spincoated biopoly-meric layer. The clad layer with the more than 8μm thickness of the sample was obtained in order to meet the requirement of optical coupling and then the core layer in the 4 μm thickness was spincoated on the clad layer to complete fabrication of the optical waveguide. **Figure 8** shows surface quality of the coated biopolymer. A cer-tain amount of conglomerated chitosan particles was ob-served. **Figure 9** shows magnification of the photo shown in **Figure 8**.

Table 1. Cladding layer thickness.

Spin-Coating Velocity (rpm)	Spin-Coating Time (s)	Thickness (μm)	Dry Temperature (°C)	Dry Time (hr)
500	30	0.640	50	5
500	30	4.8	50	5

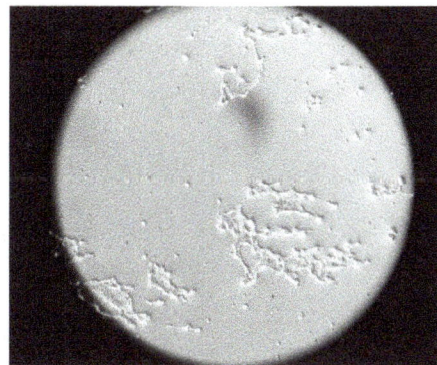

Figure 8. Photo of the surface of biopolymer.

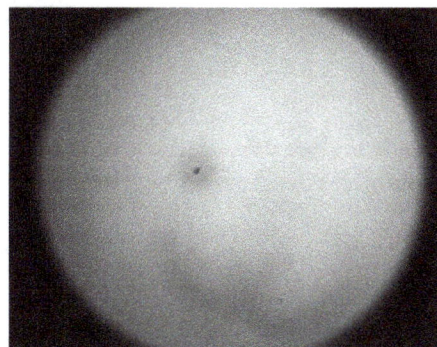

Figure 9. Photo of the sample is produced.

The ellipsometer was used to measure the refractive indices of biopolymer layers. As a result, the measured refractive indices are $n = 1.516$ for the clad layer and $n = 1.520$ for the core layer, respectively, as shown in **Figure 10**. Change in the ratio of chitosan and gelatin enables us adjust the refractive index.

The clad thickness of the sample waveguide, which was coupled with the light, was more than 8.5 μm, and the core thickness of the sample was more than 4.5 μm. The transmitted light through the waveguide was measured by using the CCD camera, as shown in **Figure 11**.

5. Conclusions

The biopolymric optical planar waveguide and Bragg grating were developed in this study. Gelatin and chitosan were blended with the proper ratio to develop the biopolymers with the different refractive index. The refractive indices of the spincoated biopolymer clad and core layers of the waveguide were obtained to be used in the planar waveguide. Then, the Bragg grating was successfully imprinted on the biopolymeric waveguide. The biopolymeric planar waveguide Bragg grating, which is biocompatible, implantable, and biodegradable, will have

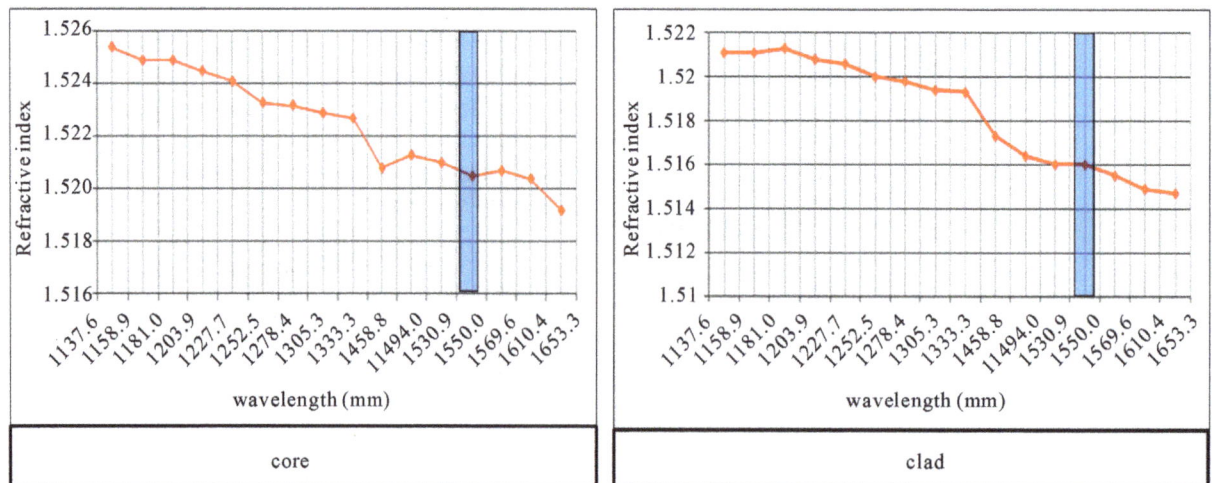

Figure 10. Meausred refractive index of biopolymer by using the ellipsometer.

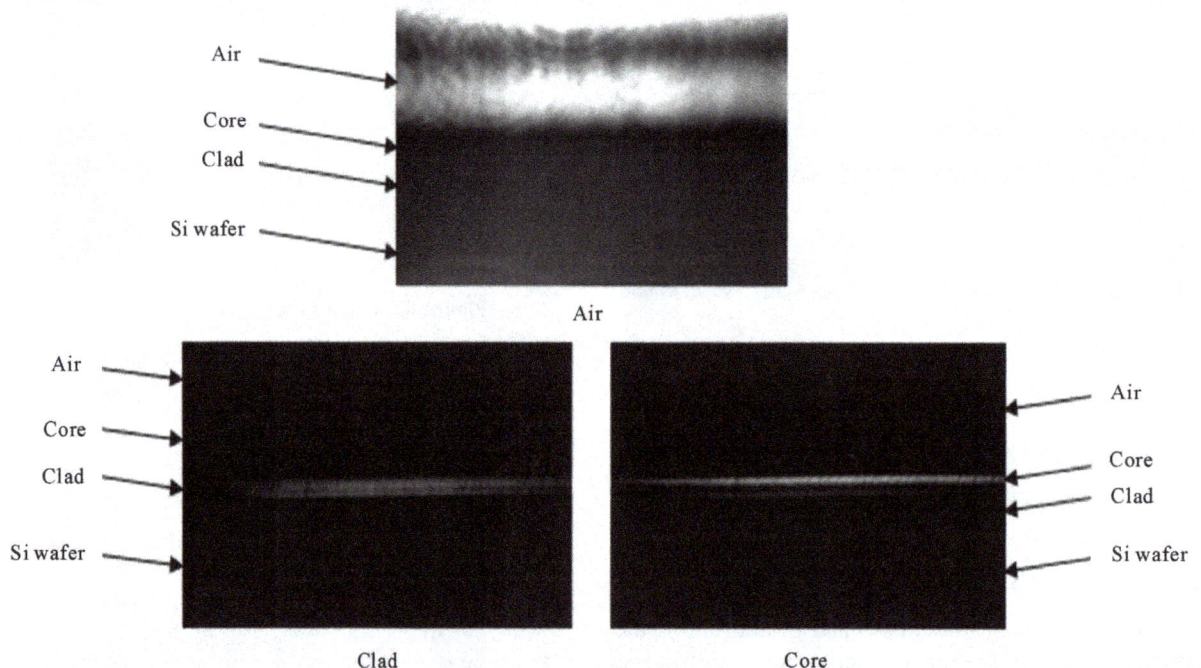

Figure 11. CCD image for light coupling measurements.

PR grating pattern

PDMS grating pattern

Biopolymer grating pattern

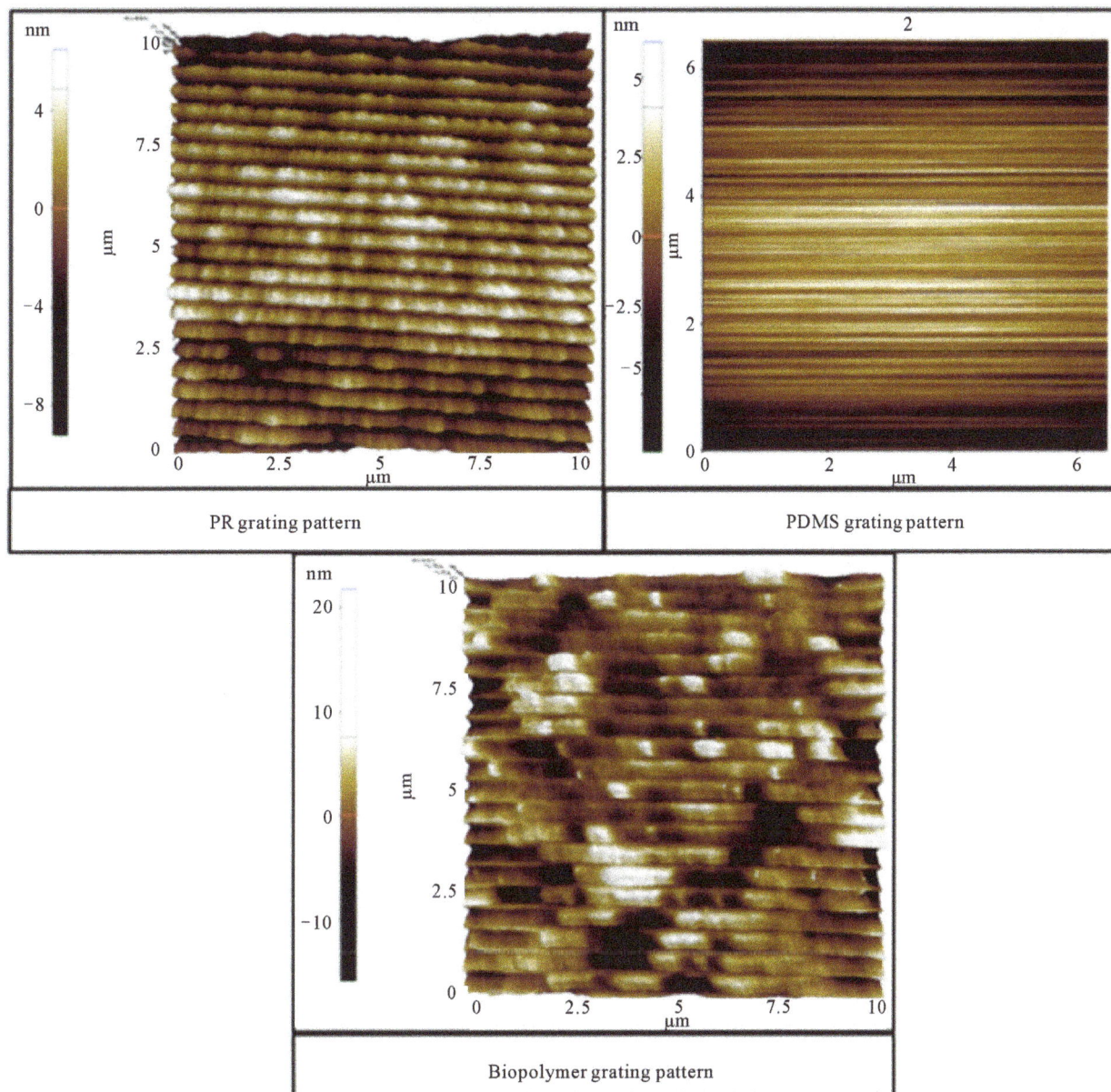

Figure 12. AFM images of the imprinted grating on the biopolymeric waveguide.

a great potential in application for biomedical diagnosis and monitoring as well as military and environmental

6. Acknowledgements

This work was supported by the National Research Foundation Korea Grant funded by the Korean Government (MEST) (NRF-2009-0076655).

REFERENCES

[1] K. C. Basavaraju, T. Damappa and S. K. Rai, "Preparation of Chitosan and Its Miscibility Studies with Gelatin Using Viscosity, Ultrasonic And Refractive Index," *Carbohydrate Polymers*, Vol. 66, No. 3, 2006, pp. 357-362.

[2] W. Ding, Q. Lian, R. J. Samuels and M. B. Polk, "Synthesis and Characterization of a Novel Derivative of Chitosan," *Polymer*, Vol. 44, No. 3, 2003, pp. 547-556.

[3] J. C. Martinez-Anton and E. Bernabeu, "Spectrogoniometry and the WANTED Method for Thickness and Refractive Index Determination," *Thin Solid Film*, Vol. 313, 1998, pp. 85-89.

[4] H. Jiang, W. Su, S. Caracci, T. I. Bunning, T. Cooper, and W. W. Adams, "Optical Waveguiding and Morphology of Chitosan Thin Films," *Journal of Applied Polymer Science*, Vol. 61, No. 7, 1996, pp. 1163-1171.

bohydrate Polymers, Vol. 66, No. 3, 2006, pp. 357-362.

[5] K.C. Basavaraju, T. Damappa and S. K. Rai,; "Preparation of Chitosan and Its Miscibility Studies with Gelatin Using Viscosity, Ultrasonic and Refractive Index," *Car-*

[6] A. Yariv, "Coupled-Mode Theory for Guided - Wave Optics," *IEEE Journal of Quantum Electronics*, Vol. 9, No. 9, 1973, pp. 919-933.

11

AZ91 Magnesium Alloys: Anodizing of Using Environmental Friendly Electrolytes

N. A. El Mahallawy[1], M. A. Shoeib[2], M. H. Abouelenain[3]

[1]The Design and Production Engineering Department, Faculty of Engineering, Ain Shams University, Cairo, Egypt; [2]Surface Coating Department, Central Metallurgical Research & Development Institute, Helwan, Cairo, Egypt; [3]Petroleum Marine Service, Cairo, Egypt.

ABSTRACT

An anodizing process, based on environmental friendly electrolyte solutions has been studied on AZ 91 magnesium alloys by using three types of electrolytes: the first is based on sodium silicate, the second on sodium hydroxide-boric acid-borax and the third on sodium silicate-potassium hydroxide-sodium carbonate-sodium tetra borate. A pretreatment including fluoride activation was applied before the anodizing process. It was found that the anodic film thickness increases as current density or anodizing voltage increases. It is also increased with deposition time until the deposition stops due to the formation of a thick anodic film. Optimization of the anodizing conditions - current density and deposition time - was made for each electrolyte. Characterization of anodizing layer was achieved by determination of surface morphology, microstructure, phase analysis, coat thickness, adhesion and corrosion resistance. In all cases, excellent adhesion and corrosion resistance was obtained. A corrosion efficiency ranging from 94% to 97% was reached; the highest value corresponding to the third electrolyte.

Keywords: *Anodizing, Magnesium AZ91, Corrosion, Characterization, Environmental Friendly Electrolytes*

1. Introduction

Magnesium and its alloys have received great attention because of their superior properties, such as low density, high specific strength/stiffness, excellent dimensional stability and electromagnetic shielding property, superior damping capacity, high creep strength, good machinability, weldability, high impact resistance, high recyclability, as well as thermal and electrical conductivities [1,2]. They are used in fields where weight reduction is critical or where particular technical requirements are required such as automotive, aeronautic and aerospace including space station, artificial satellite, space shuttle, nuclear energy, electronic and military industries, together with AVCC (Audio video-Computer-Communication) equipment, portable tools, supporting goods, *etc.* [3,4]. However, magnesium and its alloys are highly susceptible to corrosion especially in harsh environmental conditions.

Several techniques have been applied in order to improve the surface properties of magnesium alloys. Anodizing is among the promising techniques for surface protection of Mg alloys; however, most existing anodizing processes use toxic chromate, harmful phosphate or/and fluorides. Therefore, it is still worthy to develop new environmental friendly anodizing processes [5].

In this paper, the anodizing process, based on environmental friendly electrolyte solutions using DC current have been developed to enhance the corrosion resistance of magnesium AZ91alloy. The electrolytes contain none of the chromates, phosphates or fluorides solutions. Optimization of the process parameters was achieved and the anodizing layer was characterized by its adhesion strength, thickness, phase analysis, microstructure, as well as its corrosion resistance on AZ91 magnesium alloys.

2. Experimental Technique

The die-cast Mg alloy AZ91, whose chemical composition is shown in **Table 1**, was prepared as circular discs of 50 mm diameter and 10 mm thickness. The chemical composition was analyzed by Atomic Emission Spectroscopy (AES).

Anodizing experiments were done by using two groups of chemicals: chemicals for pretreatment and chemicals for anodizing processes. The pretreatment procedure used commonly in electroless processes [7], shown in **Table 2**, was applied.

Table 1. Chemical composition of AZ91 Mg alloy (wt%)

Al	Zn	Mn	Si	Cu	Fe	K	Mg
8.77	0.74	0.18	0.01	0.001	0.001	0.01	BAL

Table 2. Pretreatment procedure applied before anodizing process.

Stage no	Name		Symbol	Constituent or condition
1	Mechanical Preparation and Cleaning	Specimens are first polished by using finer grades of waterproof silicon carbide abrasive papers , rinsed in distilled water, supersonic degreasing in acetone and finally dried in air.		
2	Alkaline cleaning	Sodium hydroxide	NaOH	50g/L
		Sodium phosphate	Na_3PO_4	10g/L
		Temperature	Room temperature	
		Time	8 - 10 min	
	Specimens became brighter than in step one because Mg is passive in alkaline media so that dust, grease… etc were removed from the surface of magnesium alloy.			
3	Acid etching	Chromium trioxide	CrO_3	125 g/L
		Nitric Acid	HNO_3 (70% V/V)	110 ml/L
		Temperature	Room Temperature	
		Time	30 - 60 s	
	Gross surface scale or oxides are removed and replaced by preferred oxides to be removing later. Etching treatments also provide surface pits to act as sites for mechanical interlocking to improve adhesion.			
4	Fluoride activation	Hydrofluoric acid	HF (40%V/V)	385 ml/L
		Temperature	Room Temperature	
		Time	10 min	
	Removing residual oxides, that created in the above step and replacing it with a thin layer of MgF_2 .			

Three types of electrolytes were used in anodizing processes. The chemical composition and operating conditions of each type are shown in **Table 3**. The electrolyte cell was built with stainless steel hoop as cathode 10 cm height and 12 cm diameter wide enough to provide an even current distribution. The cell was connected to a DC power supply (MUNK- PSP - VARI-PULS, 10 A, 300 V, USA).The current density and voltage were measured using two digital ammeters (DT 9205 N, China).

In order to optimize the effect of deposition time and current density on the deposition layer thickness, the experiments were divided into two groups: in the first group, the deposition time was varied for the three electrolytes using a constant current density of 20 mA/cm^2; in the second group, the current density was varied while the deposition time was constant. The anodizing process was made at temperature of (30 - 40)°C .The observation of sparks, which indicates the occurrence of deposition showed that for a time less than 1.5 minutes almost no deposition occurs. The time was increased from 1.5 minutes until the deposition stopped-marked by no sparks.

After anodizing processes, the oxide layers obtained were sealed by immersion in boiling distilled water for a time equal to that of the deposition in each case. Sealing of the anodized film is necessary in order to achieve an abrasion and corrosion resistant film by precipitation of hydrated base metal species inside the pores [6,11].

After anodizing, the surface morphology was examined using a scanning electron microscope (SEM) model; JEOL JSM 5410. The grain size of the anodic film was determined by X-ray diffraction (XRD) model; PANalytical X'Pert. The XRD technique, using small-angle X-ray scattering (SAXS) was used to analyze the phases in the anodic coating. The corrosion resistance of anodized coat layer on AZ91D alloy was performed in 3.5 wt% NaCl solution to determine the polarization resistance at ambient temperature using AUTOLAB PGSTAT 30

The anodic film thickness was measured by the coating thickness meter (posi-tector 6000 FN). Hydraulic adhesion tester (Elcometer 108) measured the adhesion of the anodic film. The microhardness measurements

Table 3. Chemical composition and operating conditions of electrolytes.

Electrolyte (1) [8]	
sodium silicate (Na_2SiO_2)	122 g/l
Temp	$(30 - 40)°C$
Current density	16 mA/cm^2 to 31 mA/cm^2
Deposition time (minute)	1.5 to 11
Electrolyte (2)[9]	
Sodium hydroxide (NaOH)	50.0 g/L
Boric acid(H_3BO_3)	10.0 g/L
Borax or hydrous sodium borate ($Na_2B_4O_7.10H_2O$)	20.0 g/L
Temp	$(30 - 40)°C$
Current density	16 mA/cm^2 to 20 mA/cm^2
Deposition time (minute)	1.5 to 9
Electrolyte (3)[10]	
Sodium silicate (Na_2SiO_3)	50 g/l
Potassium hydroxide (KOH)	50 g/l
Sodium carbonate (Na_2CO_3)	50g/l
Sodium tetra borate (Borax $Na_2B_4O_7$)	30 g/l
Temp	$(30 - 40)°C$
Current density	16 mA/cm^2 to 31 mA/cm^2
Deposition time (minute)	1.5 to 9

were carried out by the hardness tester model Shimdsu type M, and the surface roughness by Elcometer 223. The microhardness measurements of coat layers were performed on specimens cross-sections using 50-gram load for 15 seconds.

3. Results and Discussion

3.1. Effect of Pretreatment on Surface Morphology

Study of the surface morphology after the pretreatment procedure, shown **Figure 1**, indicates the formation of pits, which is expected to improve the adherence of coat layer to the substrate. The use of fluoride activation in the pretreatment is also expected to add a layer of MgF_2 which minimizes the effect of local corrosion cells by creating an equipotentialized surface [7].

3.2. Effect of the Deposition Time on the Coating Thickness

Figure 2 shows that the coat thickness for the three electrolytes increases with the deposition time, rapidly in the first 3 minutes then with a slow rate after 3 to 9 minutes until it levels off after 9 minutes. This behavior is due to the increase in the thickness of the coat layer acting as a barrier to the flow of current, which decreases the rate of oxidation of magnesium. From the results, the optimum deposition time used to obtain maximum coat thickness was 9 minutes for the three electrolytes at the same operating conditions with 20 mA/cm^2 current density. In general, the coat thickness was the highest for electrolyte (1) - 42 μm, followed by electrolyte (2) - 32 μm, then electrolyte (3)-28 μm.

Comparison between the maximum coating thickness obtained in the present work and previous work [8-10] indicates a general agreement, however, difference in values are obtained due to the difference in base metal composition, current density and temperature. For example, for electrolyte 1, maximum coating layer thickness was 42 μm while in previous work [8] on AZ91, the anodic film was 25 μm at 60 C, 20mA/Cm2 and 9 minutes coating time. The large thickness in the present work is due to the lower anodizing temperature. This explanation is based on previous work [12,13], which

i   n   d   i   c   a   t   e   s   t   h   a   t

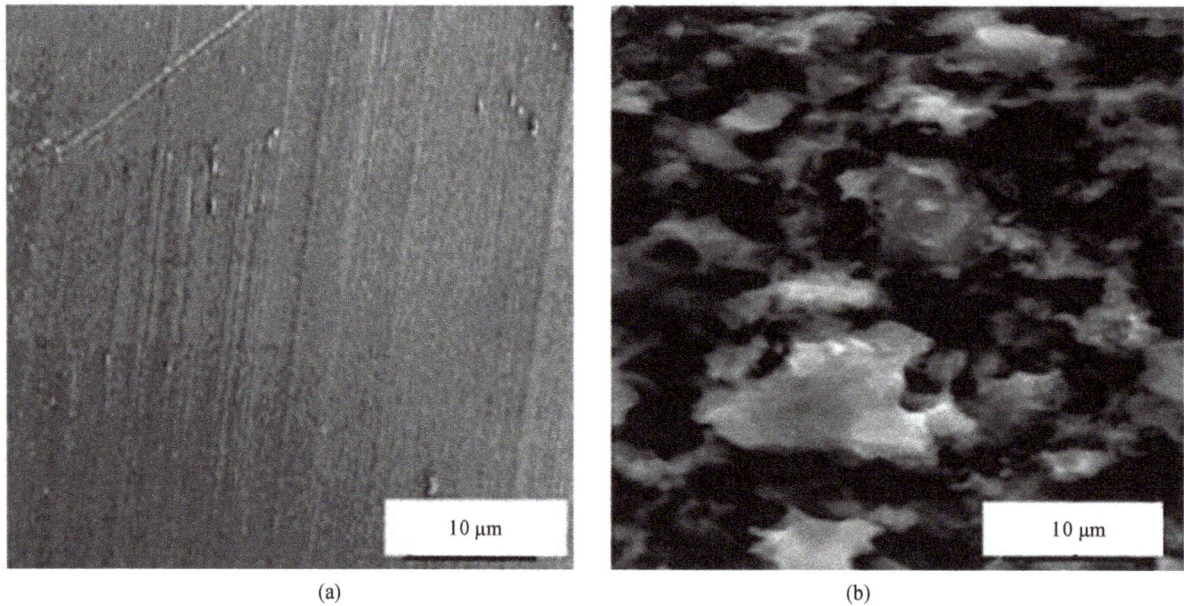

(a)   (b)

Figure 1. Surface morphology of the AZ91 magnesium alloy (a) before pre-treatment (b) after pre-treatment.

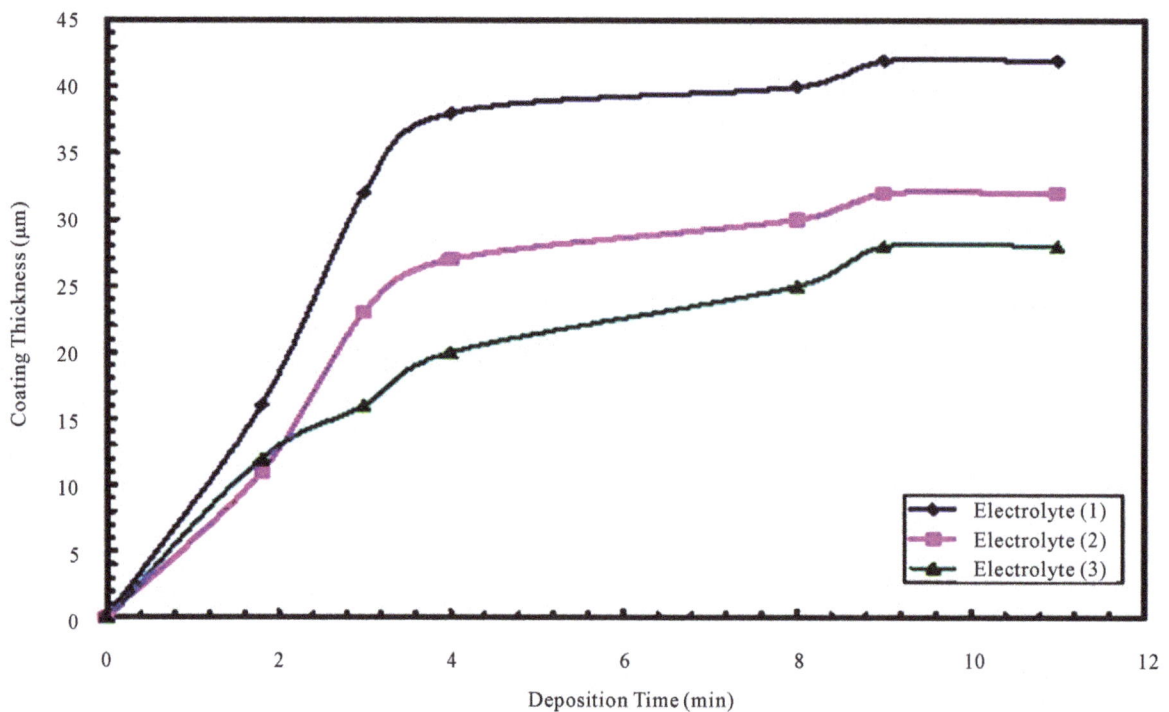

Figure 2. Effect of deposition time on coat thickness by using three electrolytes and pre-treatment procedure at temp. (30 - 40)°C and constant current density –DC-20 mA/cm².

the temperature and coat thickness are inversely proportional.

3.3. Effect of Current Density on Coating Thickness

The current density is a main electric parameter in anodizing. **Figure 3** shows the effect of current density on coat thickness for the three types of electrolytes, using a deposition time of 9 minutes (optimum) at a temperature (30 - 40)°C. With increasing current density, the driving force for anodizing increases, which enhances the coat development and formation. The thickness increased first

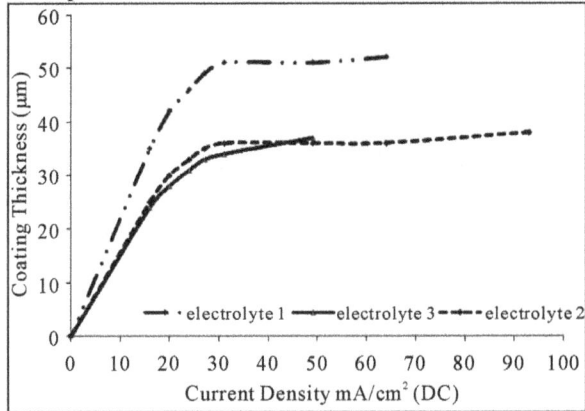

Figure 3. Effect of current density on coating thickness by using three electrolytes and constant deposition time at 9 minutes.

rapidly up to 20 mA/cm^2 then the rate slowed down until almost no increase in thickness of the deposited oxide layer was obtained. This threshold current density was 25mA/cm^2 for electrolyte (1) and (3) and 20 mA/cm^2 for electrolyte (2), resulting in coat thickness of 50 μm for electrolyte (1) and 30 μm for electrolytes (2) and(3).

For coat development, there are three ways by which anions move to the anode, mainly diffusion, convection and electric migration [14]. The latter being the main one during anodizing. It can be said that as the anodizing layer thickness increases beyond a certain value, the electric migration is suppressed due to the higher resistivity of the formed oxide layer marked by stop of sparking.

3.4. Oxide Formation Mechanism

The anodizing process is a dynamic equilibrium of partial processes of oxide formation, the others. The mechanism of anodizing process could be explained dissolution, and oxygen evolution. The dominance of one partial process will suppress according to the growth of anodic film, in which the pre-existing film was strengthened, modified or substituted repeatedly by sparking/break- down events. The sparking occurs by the action of the strong electric field created during anodization. Anions in the electrolyte first need to arrive at the anode/elec- trolyte interface and then enter into anodic coatings [15].

The general reactions, occurring in the anodizing process for Mg, are as follow:

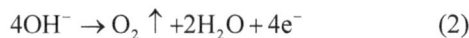

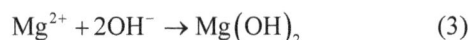

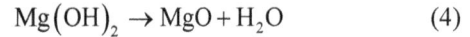

$$Mg \rightarrow Mg^{2+} + 2e^- \qquad (1)$$

$$4OH^- \rightarrow O_2 \uparrow +2H_2O + 4e^- \qquad (2)$$

$$Mg^{2+} + 2OH^- \rightarrow Mg(OH)_2 \qquad (3)$$

$$Mg(OH)_2 \rightarrow MgO + H_2O \qquad (4)$$

$$Mg(OH)_2 + SiO_2 \rightarrow Mg_2SiO_4 \qquad (5)$$

During the oxidation process, the Mg ions, produced by reaction (1), combine with the OH$^-$ in the electrolyte solution to form Mg(OH)$_2$ and Mg$_2$SiO$_4$ (reactions (3) and (5), respectively). Due to the thermal energy from the sparks in the PEO process, the hydroxides change to oxide compounds by the dehydration process, reaction (4). The coating formation processes, reactions (3), (4) and (5) may be promoted by a high concentration of the electrolyte, containing more SiO$_2$ and OH$^-$ ions.

Based on previous observations during anodizing of AZ 91 magnesium alloy [5], formation and dissolution of oxide/hydroxide films occur simultaneously. The present results indicate that the anodic film growth is dominant up to 20 - 25 mA/cm^2, beyond which the thickness of the oxide layer remains almost constant due to the breakdown and/or dissolution of this thin film as previously noticed.

3.5. Surface Morphology of Anodic Coat

The surface morphology of anodic coat, **Figures 4-6**, revealed the presence of pores with different shapes and sizes distributed all over the surface. However, the pores are very small and do not have full penetration to the substrate surface. Different pore size and density were observed depending on the electrolyte. Electrolyte (1) created many pores of approximately 7 μm diameter, **Figure 4**, while electrolyte (2) created a few pores of approximately 1 μm diameter, **Figure 5**, and electrolyte (3) created a few pores of (2 - 3) μm diameter, **Figure 6**. These differences in pore diameters result from the differences of spark behavior and evolution of gases, as in electrolyte (1) the spark was stronger and accompanied with higher evolution of gases compared to electrolytes (2) and (3).

Micro cracks are also visible on the coat surface. They are formed due to thermal stresses resulting from rapid cooling of the oxides by the electrolyte acting as a coolant.

Difference crack sizes are also observed in the coat layer due to the presence of different oxides, with different shrinkage rates [16,17] depending on the type of electrolyte (see next paragraph), **Figures 4-6**.

The cracks are probably formed in week regions, where the localized film layers were destroyed prior to other regions. However, new anodizing products formed in/around the broken regions rapidly and intensively so that the existing microdefects were filled up or mended by the fresh products. This is clear from the occurrence of cracks on the surface but not reaching the depth.

3.6. Phase Analysis in Coat Layer Using XRD and EDX

X-ray diffraction patterns of the specimen anodized by

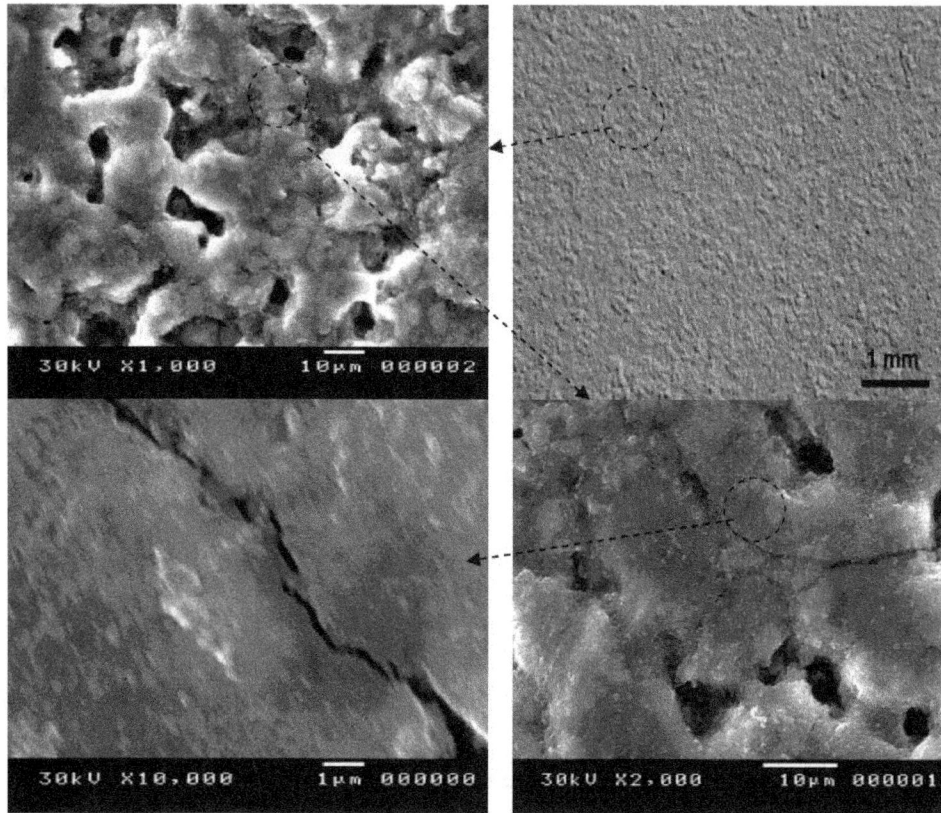

Figure 4. SEM morphology showing the coat surface using electrolyte (1).

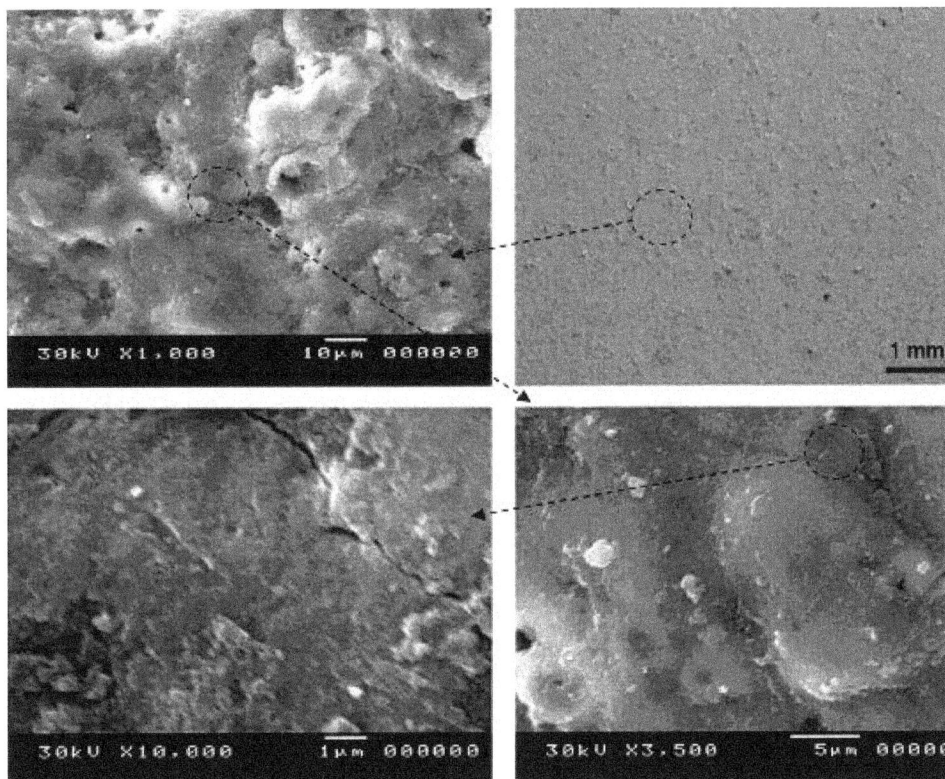

Figure 5. SEM morphology showing the coat thickness using electrolyte (2).

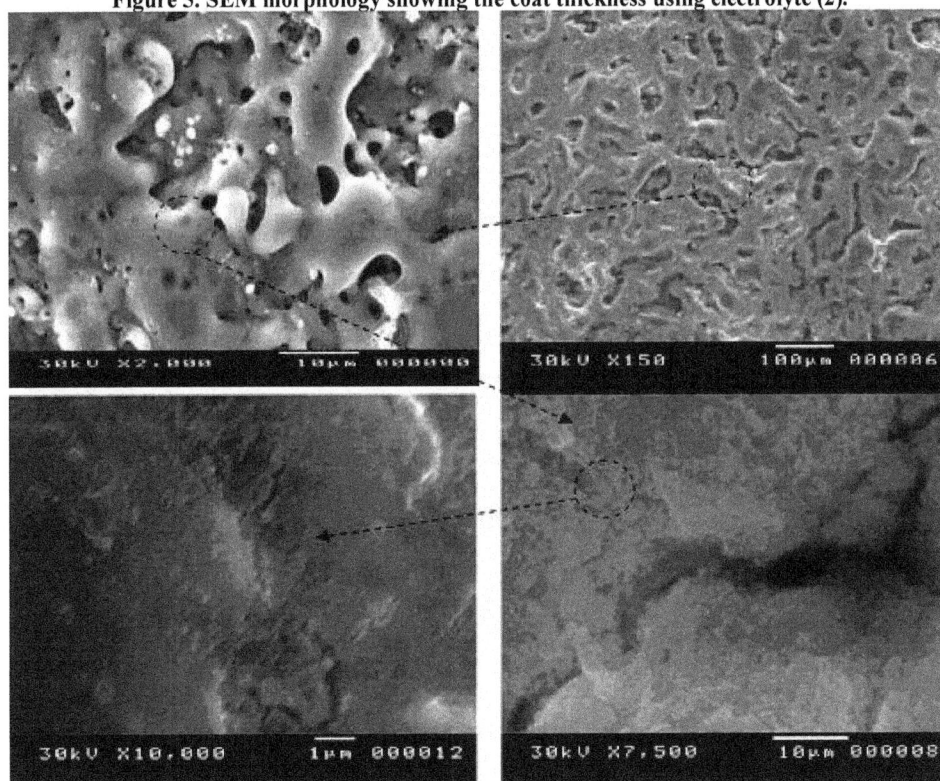

Figure 6. SEM morphology showing the coat surface using electrolyte (3).

using electrolyte (1), **Figure 7**, indicate that the dominating phase is MgO, followed by smaller amounts of SiO_2 and Mg as shown in **Table 4**, while in case of electrolyte(2), the dominating phase is MgO, Mg and B_2O_3. For electrolyte (3), the dominating phase is MgO, followed by SiO^2, Mg and Mg_2SiO_4. The occurrences of these phases indicate that the substrate and the electrolyte both contribute in forming the coat layer.

The EDX analysis across the anodic coat layer, **Table 5**, indicates that the percentage of oxygen and silicon decreases from the surface towards the substrate (zone III > zone II > zone I). This indicates that the coat layer acts as a barrier to decrease the rate of oxidation of magnesium. It was found previously [18] that when a relatively thick coating was established, an excessive oxygen evolution is observed. This could explain the increase in oxygen at the expense of the Mg on the outer layer.

3.7. Microstructure of the Coat Layers and Crystal Size

The cross-section view of the coat layers for all three electrolytes, **Figure 8**, shows that anodized coating films are relatively uniform in thickness and that the pores are present but do not completely extend to the base metal. Pores result from the generation of sparks and evolution of gases from the hydrolysis of water [16]. These pores are relatively uniform in distribution.

The crystal size of the anodic coat determined by X-ray diffraction (XRD), indicate that all crystal phases are in the nano scale with sizes 4 to 5 nm, **Table 6**. This means that the three electrolytes result in almost similar crystal size, as the MgO is the dominating phase in all coats.

3.8. Coat Adhesion

An excellent adhesion exceeding 18 MPa between the anodic film and the substrate was found for all specimens. This value is larger than for electroless nickel plating on anodized magnesium alloy (about 11 MPa) [19]. This adhesion is enhanced by the formation of pits formed by the pretreatment procedure, which increased the surface roughness and contributed to increased adhesion of the coat layer as reported by Adachi *et al.* [20]. In additions, the analysis of the obtained anodic film layers showed that they contain (Mg - MgO- Mg_2SiO_4). This indicates that the metallic substrate has diffused into coating layer, resulting in further enhancement of the adhesion.

3.9. Surface Roughness

Ten surface roughness measurements were taken on the surface of the specimens and the results in **Figure 9** indicate that the surface roughness ranges from 13 μm to

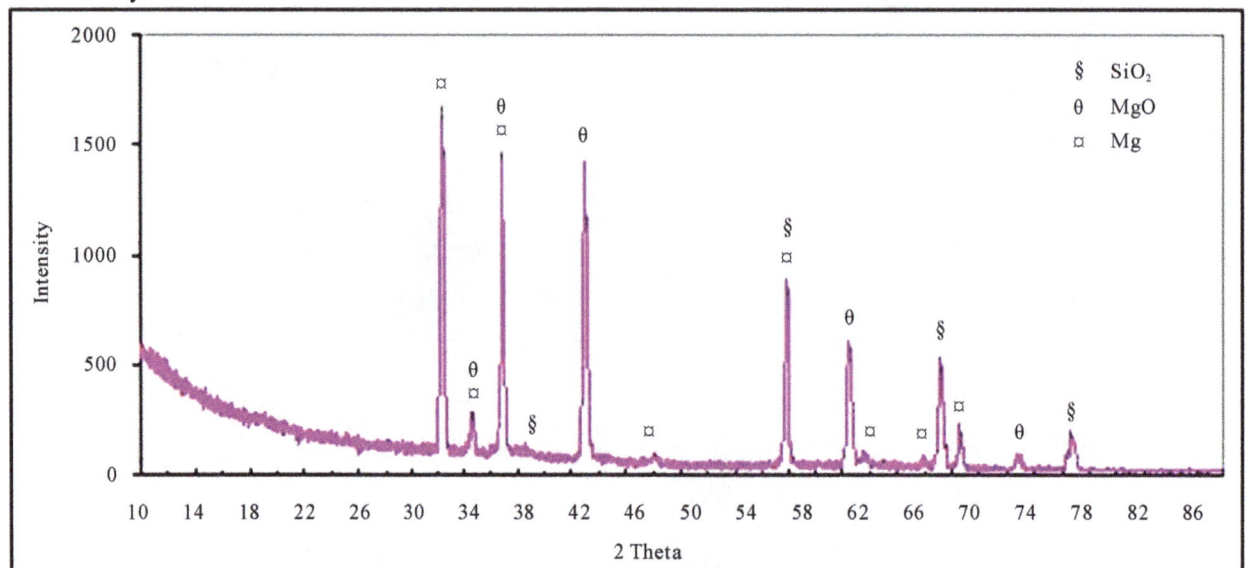

Figure 7. XRD spectra of the anodic coatings by using electrolyte (1), [temp. (30 - 40)°C, deposition time 9 min and current density 31 mA/cm²].

Table 4 Phases present in coating layer using the different electrolytes.

Compound name	Chemical formula	Score	Percentage (wt%)
	Electrolyte (1)		

Magnesium	Mg	31	0.25
Periclase	MgO	68	0.55
Silicon Oxide	SiO_2	25	0.20
	Electrolyte (2)		
Magnesium	Mg	53	0.3
Periclase	MgO	75	0.42
Ceramic Oxide Or (Boric oxide)	B_2O_3	51	0.28
	Electrolyte (3)		
Magnesium	Mg	52	0.28
Silicon Oxide	SiO_2	47	0.25
Periclase	MgO	69	0.37
Forsterite	Mg_2SiO_4	16	0.10

Table 5. EDX analysis of anodic coat layer (wt%).

Zone No.	O%	Mg%	Al%	Si%
Zone I (close to the Mg)	27.7	33.4	0.9	37.8
Zone II (central region)	29.2	27.6	0.8	42.2
Zone III (close to surface)	31.1	24.3	0.7	43.7

*The EDX analysis of the anodic coat layer by using electrolyte (3) after the pretreatment, deposition time 9 min and current density 31 mA/cm^2 at temp. (30 - 40)°C.

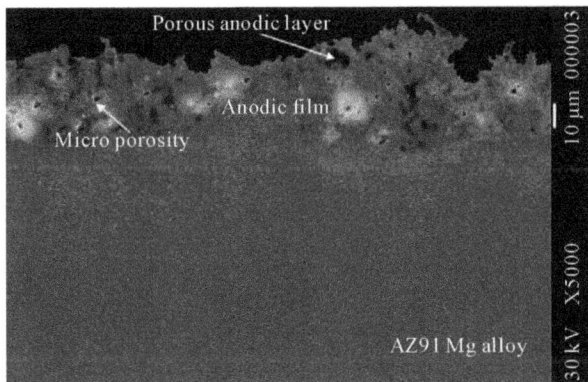

(a)

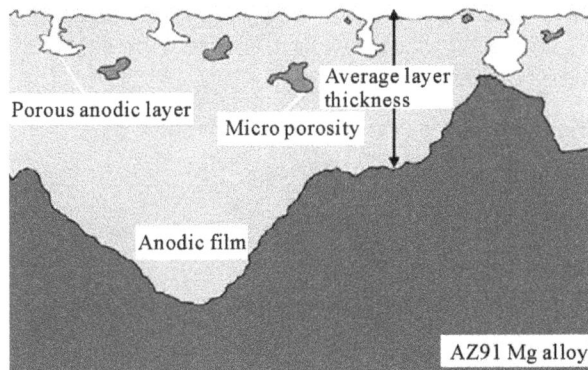

(b)

Figure 8. (a) Cross section morphology (b) sketch of a typical cross section, for anodic coating of anodic coating film on magnesium alloy AZ91D at 31 mA/cm^2 current density, 9 minutes deposition time and temperature (30 - 40)°C.

Table 6. Crystalline size for the coat layers for the three electrolytes.

Type of bath	Crystal size (nm)
Electrolyte (1)	5.1
Electrolyte (2)	3.95
Electrolyte (3)	4.12

24 µm. It increases by increasing the current density and is lower for electrolyte (2). In addition, **Figure 10** shows that the surface roughness increases by increasing the coating thickness. For electrolyte (2), the surface roughness was found to be constant for current density ranging from 24 mA/cm^2 to 31 mA/cm^2, which corresponds to an almost constant coat thickness, as indicated in **Figure 3**.

The surface roughness is partly related to the non-homogeneity of coat layer and the formation of pores and cracks on the anodic film layers. These surface defects were greater in case of electrolyte (1) followed by (3) and (2).

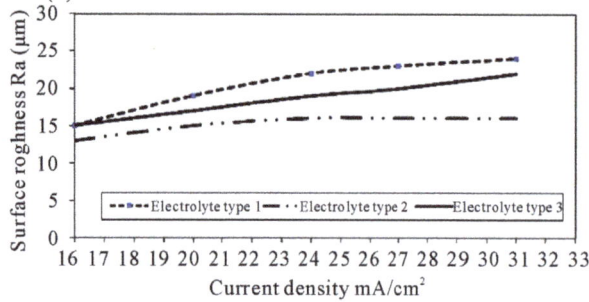

Figure 9. Effect of change of current density on surface roughness of coating layers by using three electrolytes at temperature (30 - 40)°C and constant time 9 minutes.

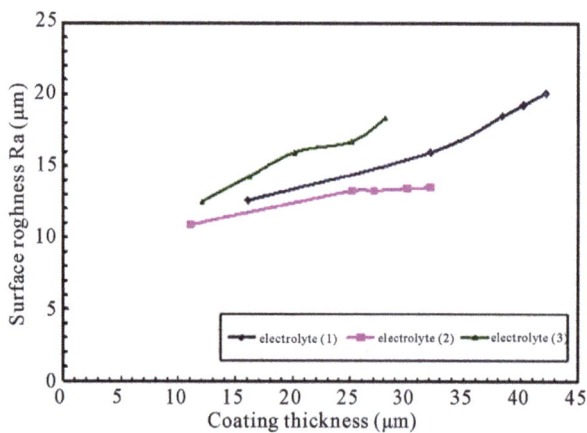

Figure 10. Relation between coating thickness and surface roughness for anodizing process by using three electrolytes, pre-treatment procedure 9 minute deposition time, and current density (16 to 31) Am/cm².

3.10. Microhardness

The results indicate that the microhardness increases with increasing the current density, **Figure 11**, and coat thickness, **Figure 12** reaching value of 570 VHN for electrolyte (3) with 28 μm thickness. This is expected as the coat phases include MgO and SiO₂ hard phases. **Table 5** indicates that the Si increases towards the coat surface which could results in higher concentration of the SiO₂ hard phase. In general, the hardness of the coat phases is higher than that of the magnesium substrate (VHN = 360 HV).

3.11. Corrosion Tests

Figure 13 shows potentiodynamic curves obtained for AZ91 magnesium alloys before and after anodizing processes by using the three electrolytes. Potentiodynamic curves for the anodized specimens are shifted in the noble direction indicating a higher corrosion resistance.

The electrolytes can be ranked to give the best anti-corrosion properties as electrolyte (3) followed by (2),

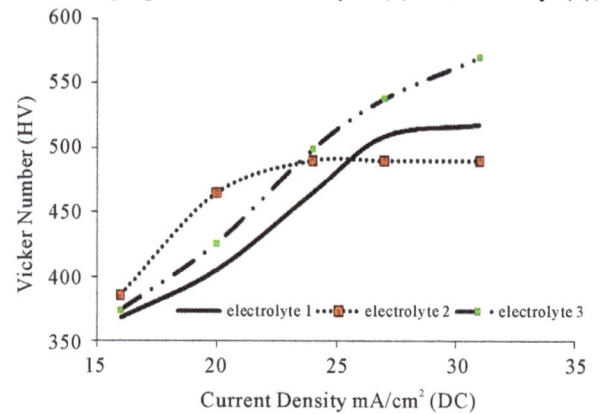

Figure 11. Effect of change current density in microhardness of coating layers by using three electrolytes at temperature (30 - 40)°C and time at 9 minute.

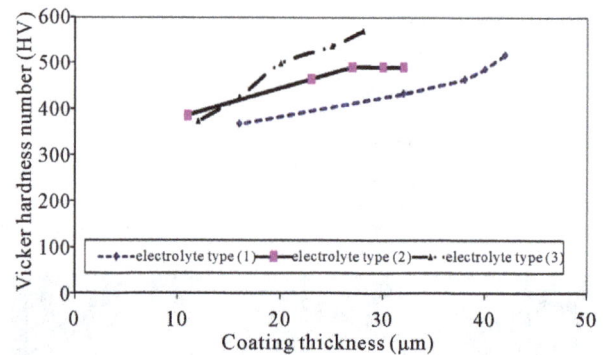

Figure 12. Relation between coating thickness and micro-hardness of coating layers for anodizing process by using three electrolytes.

then by (1). The results of linear polarization experiments summarized in **Table 7** shows that the corrosion resistance of AZ91 magnesium alloys is enhanced significantly by using the three electrolytes in anodizing processes. This is marked by the increase of R_p, decrease in Icorr and shift of E_{corr} in the noble direction (more positive values). The corrosion rate drops from 57.48 mpy to 3.39, 2.64 and 1.70 mpy using electrolytes (1), (2) and (3) reaching excellent efficiency of 94%, 95% and 97% respectively.

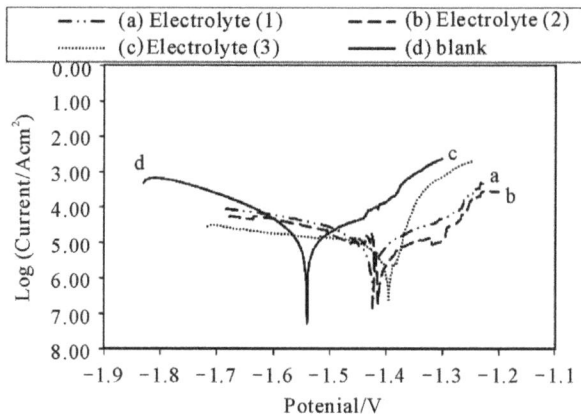

Figure 13. Potentiodynamic polarization curves for the anodized specimens by using three-electrolyte to gather with the AZ91 Mg Substrate (Blank).

4. Conclusions

1) The anodizing process of AZ91 based on environmental friendly electrolytes was successful to form a smooth anodic film, with high corrosion resistance and excellent bonding strength to substrate. The solutions used were sodium silicate for electrolyte (1), sodium hydroxide - boric acid - borax for electrolyte (2), sodium silicate-potassium hydroxide- odium carbonate-sodium tetra borate for electrolyte (3),

2) The anodized layer included the MgO phase as the dominating phase followed by smaller amounts of SiO_2, MgO, B_2O_3 and Mg_2SiO_4 indicating that the substrate and the electrolyte both contribute in forming the coat layer.

3) High improvement in corrosion is due to the formation of anodic film with layer thickness between 30 micron (electrolyte (2) and (3)) and 50 micron (electrolyte (1)) with excellent adhesion to the magnesium substrate. The higher improvement in corrosion resistance is associated with less microcracks and micropores in the coat which in all cases do not reach the Mg substrate.

4) The best anticorrosion performance was obtained for electrolyte (3) where the corrosion resistance efficiency 97% was reached, while it was 94% and 95% for electrolyte (1) and (2) respectively.

Table 7. Results of linear polarization experiments.

Specimen for	E_{corr}, V	R_p, $\Omega.cm^2$	I_{corr}, $\mu A/cm^2$	Corrosion Rate, mpy	Efficiency
AZ91 Mg substrate (Blank)	−1.585	4.168 E+2	5.027×10^{-5}	57.481	-----------
Electrolyte (1)	−1.427	1.863E+3	4.171E-6	3.39	94.10%
Electrolyte (2)	−1.394	2.71E+3	3.193 E-6	2.64	95.40%
Electrolyte (3)	−1.371	3.67E+2	2.093 E-6	1.70	97.04%

*Efficiency = (Corrosion Rate for Mg blank - Corrosion Rate using Electrolyte)/Corrosion Rate for Mg blank

REFERENCES

[1] G. L. Song, A. Atrens, D. Stjohn, J. Nairn and Y. Li, "The Electrochemical Corrosion of Pure Magnesium in 1N NaCl," *Corrossion Science*, Vol. 39, No. 5, 1997, pp.855-856.

[2] A. J. Zozulin and D. E. Bartak, "Anodized Coatings for Magnesium Alloys," *Metal Finishing*, Vol. 92 No. 3, 1994, pp.39-42.

[3] J. E. Gray and B. Luan, "Protective Coatings on Magnesium and Its Alloys — A Critical Review," *Journal of Alloys Compounds*, Vol. 336, No. 1-2, 2002, pp. 88-13.

[4] E. Ghali and W. Dietzel, "General and Localized Corrosion of Magnesium Alloys: A Critical Review," *Journal of Materials Engnerring and Performance*, Vol. 13, No. 1, 2004, pp. 7-23.

[5] Y. Zhang, C. Yan, F. Wang, H. Lou and C. Cao, "Study on the. Environmentally Friendly Anodizing of AZ91D Magnesium Alloy," *Surface and Coating Technology*, Vol. 161, No. 1, 2002, pp. 36-43.

[6] C. K. Mittal, "Chemical Conversion and Anodized Coatings," *Transactions of the Metal Finishers Association of India*, Vol. 4, 1995, pp. 227-231.

[7] A. K. Ehmeda "Anodizing of Magnesium Alloy (AZ91D) in Extreme Alkaline and Acidic Media," MSC Thesis, Tabbin Institute for Metallurgical Studies, Cairo, 2008

[8] W. Li , L. Zhu and H. Liu , "Preparation of Hydrophobic Anodic Film on AZ91D Magnesium Alloy in Silicate Solution Containing Silica Sol," *Surface and Coatings Technology*, Vol. 201, No. 6, 2006, pp. 2573-2577.

[9] C. S. Wu, Z. Zhang, F. H. Cao, L. J. Zhang, J. Q. Zhang and C. N. Cao, "Study on the Anodizing of AZ31 Magnesium Alloys in Alkaline Borate Solutions," *Applied Surface Science*, Vol. 253, No. 8, 2007, pp. 3893-3898.

[10] M. Hara, K. Jimatsuda, W. Yamauchi, M. Sakaguchi and T. Yoshikata. "Optimization of Environmentally Friendly

Anodic Oxide Film for Magnesium Alloys," *Materials Transactions*, Vol. 47, No. 4, 2006, pp. 1013-1019.

[11] O. Khaselev, J. Yahalom and J. Electrochem. "Constant Voltage Anodizing of Mg-Al Alloys in KOH-Al (OH) (3) Solutions," *Journal of the Electrochemical Society*, Vol. 145, No. 1, 1998, pp. 190-193.

[12] A. K. Sharma, R. Uma Rani and K. Giri, "Studies on Anodization of Magnesium Alloy for Thermal Control Applications," *Metal Finishing*, Vol. 95, No. 3, 1997, pp.43-54.

[13] J. Yahalom and J. Zahavi, "Experimental Evaluation of Some Electrolytic Breakdown Hypotheses," *Electrochimica Acta*, Vol. 16, No. 5, 1971, pp. 603.

[14] R. F. Zhang, D. Y. Shan, R. S. Chen and E. H. Han, "Effects of Electric Parameters on Properties of Anodic Coatings Formed on Magnesium Alloys," *Materials Chemistry and Physics*, Vol. 107, No. 2-3, 2008, pp. 356-363.

[15] Y. Zhang and C. Yan, "Development of Anodic Film on Mg Alloy AZ91D," *Surface and Coatings Technology*, Vol. 201, No. 6, 2006, pp. 2381-2386.

[16] X. Zhou, G. E. Thompson, P. Skeldon, G. C. Wood, K. Shimizu and H. Habazaki, "Film Formation and Detachment During Anodizing of Al-Mg Alloys", *Corrosion Science*, Vol. 41, No. 8, 1999, pp. 1599-1605.

[17] H. F. Guo, M. Z. An, H. B. Huo and S. Xu, "Microstructure Characteristic of Ceramic Coatings Fabricated on Magnesium Alloys by Micro-Arc Oxidation in Alkaline Silicate Solutions," *Applied Surface Science*, Vol. 252, No. 22, 2005, pp. 2187-2191.

[18] P. Zhang, X. Nie and D. O. Northwood, "Influence of Coating Thickness on the Galvanic Corrosion Properties of Mg Oxide in an Engine Coolant," *Surface and Coating Technology*, Vol. 203, No. 20-21, 2009, pp. 3271-3277.

[19] S. Sun, J. Liu, C. Yan and F. Wang, "A Novel Process for Electroless Nickel Plating on Anodized Magnesium Alloy," *Applied Surface Science*, Vol. 254, No. 16, 2008, pp. 5016-5022.

[20] S. Adachi and K.Nakata, "A Novel Process for Electroless Nickel Plating on Anodized Magnesium Alloy," *Plasma Processes and Polymers*, Vol. 4, No. S1, 2007, pp. 512-515.

12

H$_2$ and CH$_4$ Sorption on Cu-BTC Metal Organic Frameworks at Pressures up to 15 MPa and Temperatures between 273 and 318 K

Yves Gensterblum

Institute of Geology and Geochemistry of Petroleum and Coal, RWTH Aachen University, Lochnerstr Aachen, Germany.

ABSTRACT

Sorption isotherms of methane and hydrogen on Cu$_3$(BTC)$_2$ have been measured in the temperature range from 273 to 318 K and at pressures up to 15 MPa. H$_2$ excess sorption capacities of the Cu$_3$(BTC)$_2$ amounted to 3.9 mg/g at 14 MPa. Promising maximum CH$_4$ excess sorption capacities on the same sample were reached at approximately 5 MPa. They amounted to 101, 100, 92 and 80 mg/g at 273, 278, 293 and 318 K, respectively. The sorbed phase density was essentially the same for all temperatures and amounted to ~600 kg/m^3. Structural changes of the Cu$_3$(BTC)$_2$ samples after thermal activation and treatment with high pressure H$_2$ and CH$_4$ were tested. It was found that the initial micropore structure has virtually disappeared as evidenced by a decrease of the Langmuir specific surface area by a factor ~3 and CO$_2$ micropore volume by a factor of ~4 for H$_2$ and ~3 for CH$_4$. This is in line with an increase in the average pore diameter from initially 9.2 to 15.7 for H$_2$ and 12.8 for CH$_4$.

Keywords: *Metal Organic Framework (MOF), Sorption, Methane, Hydrogen, Pore Structure*

1. Introduction

The concept of reticular design in the synthesis of metal organic frameworks (MOF) permits custom-tailoring of regular pore structures on the nanometer scale [1]. This approach opens new perspectives for the development of gas and energy storage systems [2]. So far, the gas sorption capacity of metal-organic frameworks (MOF) of various chemical and structural compositions has mainly been determined at low pressures (< 0.1 MPa) and temperatures (77 to 87 K). Only a few studies [3,4] reported H$_2$ and CO$_2$ measurements at ambient temperatures (ideal operating temperature).

Several authors reported the internal structure of different MOFs [5,6]; however, significant amounts of side products were detected using the synthesis described by Chui *et al.* (1999)[5]. One major problem concerning the storage of gases on these materials at high pressures (> 1 MPa) and temperatures (especially at the activation temperature of 458 K) is their structural instability and hence, a decrease in specific surface area associated with struc- tural rearrangements for Cu-(BTC) [6].

In this study the sorption capacities of CH$_4$ and H$_2$ on Cu$_3$(BTC)$_2$ (Copper benzene-1,3,5-tricarboxylate,C$_{18}$H$_6$ Cu$_3$O$_{12}$) have been investigated at temperatures between 273 and 318 K and pressures up to 15 MPa. Specific surface areas (SSA) and pore size distributions (PSD) have been determined on untreated samples before and after high-pressure sorption experiments in order to detect any structural changes of the sample due to the interaction with the gases. A similar approach has been used by [7], who reported changes in SSA and PSD on MOF-5 during hydrogen sorption.

Two slightly different MOFs both consist of the same organic ligands but with different basic metals (Al^{3+} and Cr^{3+}) [8]. At 77K the Al^{3+}–MOF shows higher sorption capacities than the corresponding Cr^{3+} sample. However, both samples nearly show the same specific surface area of ~1000 m^2/g (**Figure 1**). This finding confirms the importance of micropore volume and the electrostatical potential of the different metals.

Data provided in **Figure 1** reflects the general correlation between SSA (determined by low pressure N$_2$-isotherms at 77 K) and H$_2$ sorption capacity at 77 K. It is known that this trend is ambiguous for isotherms recorded at higher temperatures [9]. One suggestion is that

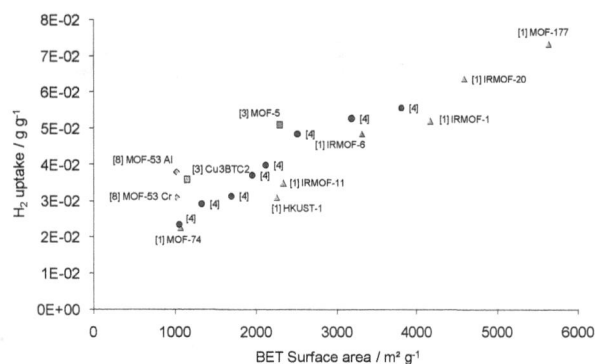

Figure 1. Comparison of high-pressure (~5 MPa) H₂ sorption capacities at 77 K vs. N₂-BET surface area for different MOFs and carbon materials.

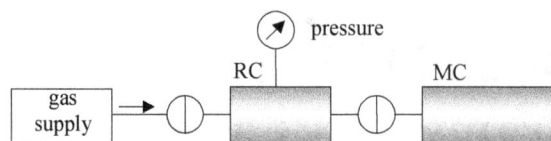

Figure 2. Schematic diagram of sorption apparatus used for high pressure/high temperature H2 and CH4 sorption experiments on Cu-BTC MOFs. (RC = calibrated reference volume ; MC = measuring cell).

a better correlation can be obtained by separating the hydrogen adsorbed in the micropores from that on the surface of the mesopores. Furthermore, Nijkamp et al. (2001) concluded that a higher storage capacity could be achieved with adsorbents containing a large micropore volume with suitable diameter [9].

In a study on hydrogen storage in chemically activated carbons and carbon nanomaterials, Beneyto (2007) [10] proposes that at 77 K the H₂ adsorption capacity depends on the SSA and the total micropore volume of the activated carbon. At 298 K it depends on both the micropore volume and the micropore size distribution. To date, very few high pressure/temperature isotherms and corresponding micropore volumes have been determined on MOFs.

In comparison to hydrogen, it was shown for methane in different studies that $Cu_3(BTC)_2$ has a high CH_4 storage capacity at room temperature [11,12] (295 to 298 K, Wang et al. 2002; Lin et al. 2006). Sorption values on Cu-BTC for CH_4 of up to 72 mg/g (at 295 K and 0.1 MPa) with nearly linear sorption isotherms up to 0.1 MPa [11]. Maximum CH_4 sorption capacities of 31.4 mg/g at 0.9 MPa and 298 K on Zn–MOF and a hysteresis between the sorption and desorption curves [12]. Senkovska and Kaskel (2007) [13] reported high pressure CH_4 adsorption on $Cu_3(BTC)_2$, $Zn_2(bdc)_2dabco$, $Cr_3F(H_2O)_2$, $O(bdc)_3$. Among the three materials, $Cu_3(BTC)_2$ shows the highest excess adsorption at 303 K (15.7 mg/g).

2. Experimental

2.1. High-Pressure Sorption Experiments

High pressure single-gas adsorption experiments have been performed using a manometric experimental set-up (Figure 2) consisting of a stainless-steel measuring cell (MC) and a calibrated reference volume (RC) connected with a set of actuator-driven valves and a high-precision

pressure transducer (max. pressure 25 MPa, with a precision of 0.05% of the full scale value). The entire device is placed in a temperature-controlled oven (variations in temperature < 0.1 K). Calculation of the excess sorption was performed based on equations of state by Setzman and Wagner (1991) [14] and McCarty et al. (1981) [15] for methane and hydrogen, respectively. Quality control is ensured by participation in laboratory comparison studies (e.g. Gensterblum et al. 2009 [16], 2010 [17] and Goodman et al. 2004 [18]). For further details on the experimental approach, please refer to e.g. Busch et al. (2004) [19].

As pointed out by Férey (2003) [8] for the MIL-53 and Schlichte et al. (2004) [6] for the MOFs - which were also investigated in this study - the activation of the metal-organic framework compounds is an important issue affecting the quality and reproducibility of sorption capacity measurements. The activation procedure was carried out in three stages: 12 h at 105°C, 12 h at 155°C, and finally 18 h at 185°C. This activation procedure led to well-reproducible starting conditions for the isotherm measurements.

Error bars shown in the diagrams and error margins listed in the **Table 2** were determined based on the Gauss error propagation method considering the individual errors of the individual experimental parameters.

2.2. Characterisation of Pore Structure and Surface Area

A Quantachrome Autosorb 1 instrument was used to characterise the pore structure and surface properties of the samples. The latter has been performed using the equation proposed by Brunauer, Emmet and Teller (1938, BET) [20]. The pore size distribution for 1.2 up to 40 nm pore diameters were derived by N₂-isotherms measured at 77 K. Pore size distributions (PSD) were analysed [21,22] and the nonlinear density functional theory (NLDFT, Thommes et al. 2006) [23]. The submicro- and micropores, in the range of 0.3 to 1.5 nm, were charac- terised by CO_2 isotherms measured at 273 K. These iso- therms were also evaluated according to the Dubinin and NLDFT methods. For the NLDFT evaluation, slit and/or cylindrical pores were assumed which represents a com- promise between availability of mathematical PSD-al- gorithms with spherical pore structure and real pore structures.

3. Results and Discussion

3.1. Sorption Capacities

3.1.1. Hydrogen

Figure 3 shows the H_2 excess sorption isotherms for $Cu_3(BTC)_2$ (HKUST-1), measured at 318 K, in compare-son with similar literature data reported for $Cu_3(BTC)_2$ at 298 K [7]. The higher sorption capacities reported by Panella *et al.* (2005) [7] can be explained by lower experimental temperatures used. The rather linear trend (Freundlich or Langmuir-Freundlich type) of the sorption isotherms is comparable to other isotherm data (see **Figure 1**), but it is not a classical Langmuir type as would be expected for porous materials.

Structural changes (pore structure, specific surface area) of the $Cu_3(BTC)_2$ after H_2 sorption have been determined; the results are provided in **Table 1**. It is obvious that specific surface areas (SSA) decrease by a factor of 3-4, N_2 pore volumes by a factor of 2-3, and CO_2 micropore volumes by a factor of 4-5, while an increase in the Dubinin-Astakhov (DA) average pore diameter from 9.2 to 15.7 Å (10^{-10} m) can be observed.

Compared to the N_2 micropore volumes for $Cu_3(BTC)_2$ determined in this study 0.63 cm^3g^{-1} (**Table 1**), several authors obtained values between 0.41 and 0.76 cm^3g^{-1} on virgin samples [13,24,25]. Other studies report BET surface areas for this material of 1781 m^2/g (single-point BET) [26], 1154 m^2/g [7] and 1239 m^2/g [11] which is similar to the BET surface area of 1246 m^2/g ascertained in this study (all measured with N_2 at 77 K).

3.1.2. Methane

Methane sorption isotherms on $Cu_3(BTC)_2$ have been measured at temperatures between 273 and 318 K and pressures up to 15 MPa and are provided in **Figure 4**. Maximum excess sorption values obtained in these measurements are listed in **Table 2**. They show a generally decreasing trend with temperature from 101 to 79 mg CH_4/g MOF. After passing a maximum value, the excess sorption isotherms in **Figure 4** decrease slightly until the final pressure value is reached. This decline above ~6 MPa can be attributed to a volumetric effect (non-negligible volume of the sorbed phase) that needs to be considered when calculating absolute sorption capacities. Among others, Humayun and Tomasko (2000) [27] proposed a method to determine the sorbed phase density, which has been applied in this study. This procedure resulted in nearly identical values of 597 to 601 kg/m^3 for

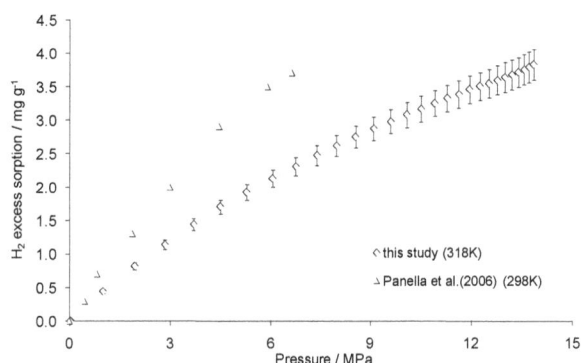

Figure 3. H_2 excess sorption capacity measured at 318K on $Cu_3(BTC)_2$. For comparison the H_2 excess sorption data at 298 K on $Cu_3(BTC)_2$ reported by Panella *et al.*[7] has been included.

Table 1. Structural changes of $Cu_3(BTC)_2$ after high-pressure H_2 sorption experiment at 318 K. DR = Dubinin Radushkewic; NLDFT = non-local density functional theory.

$Cu_3(BTC)_2$	method	N_2 surface area (m^2/g)	CO_2 micropore surface area (m^2/g)	N_2 pore volume (cm^3/g)	CO_2 micropore volume (cm^3/g)	DA pore diameter (Å)
before H_2 sorption test	DR	1731	2708	0.63	0.94	9.2
	NLDFT	1994	2340	0.64	0.89	
after H_2 sorption test	DR	651	752	0.23	0.26	15.7
	NLDFT	576	641	0.33	0.20	

Table 2. CH₄ Sorption parameters of $Cu_3(BTC)_2$ at 273 to 318 K considering helium densities.

T (K)	max. excess CH_4 sorption capacity (mg/g)	absolute CH_4 sorption capacity (mg/g)	CH_4 sorbed phase density (kg/m^3)
273	101 ± 7	113 ± 3	599 ± 10
278	100 ± 7	112 ± 3	601 ± 8
293	92 ± 2	103 ± 2	597 ± 5
318	79 ± 2	90 ± 3	n.d.

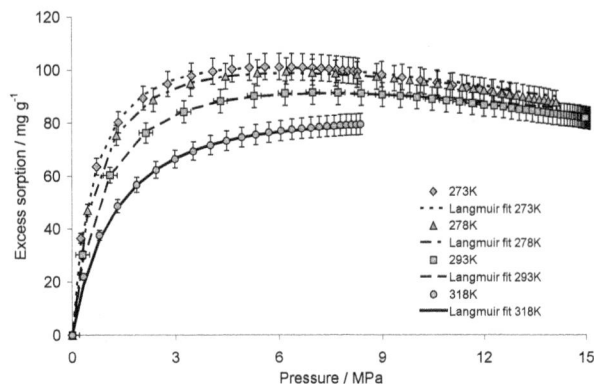

Figure 4. CH$_4$ sorption isotherms measured on Cu$_3$(BTC)$_2$ at 273, 278, 293, and 318 K.

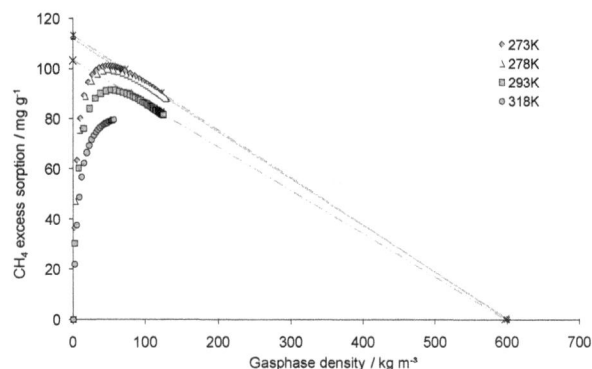

Figure 5. CH$_4$ excess sorption capacities vs. CH$_4$ gas phase density at different temperatures (Cu$_3$(BTC)$_2$.

the CH$_4$ sorbed phase density at three of the four experimental temperatures (**Figure 5**, **Table 2**). For the isotherm measured at 318 K, the pressure is not high enough for a graphical determination. This suggests that the density of the sorbed CH$_4$ phase is independent of the experimental temperature. Maximum absolute sorption capacities, taking into consideration the volume of the sorbed phase, have been determined to be as high as 113, 112, 103 and 90 mg/g at 273, 278, 293 and 318 K, respectively (**Table 2**), by applying the following relationship:

$$n_{abs} = \left(\frac{1+ap}{ap}\right)\frac{n_{ex}}{1-\left(\dfrac{\rho_{gas}}{\rho_{sorbed}}\right)} \qquad (1)$$

where n_{ex} denotes the excess sorption capacity, ρ_{gas} the gas phase density, "a" is the Langmuir parameter (1/MPa) and ρ_{sorbed} the sorbed phase density as determined graphically in **Figure 5**.

The Langmuir curves shown in **Figure 4** were calculated with explicit consideration of the sorbed phase density and therefore, reproduce the decline of the isotherms at high pressures.

In comparison to the study by Senkovska and Kaskel (2007) [13], the CH$_4$ sorption values in this study are lower by a factor of ~1.4. One explanation might be the reference to the sample mass (or sample density). Senkovska and Kaskel (2007) [13] used a crystallographic density of 0.88 g/cm^3. In this study, sorption values are related to initial weight, while the sample density after activation is 1.77 ± 0.03 g/cm³ as determined by helium pycnometry. This lead to a slightly lower sorption amounts for this study in comparison to Senkovska and Kaskel (2007) [13].

3.2. Isosteric Heats of Adsorption

Based on the sorption isotherms measured between 273 and 318 K, isosteric heats of adsorption for CH$_4$ have been calculated using the Clausius-Clapeyron relation (Equation (2)). In this first approximation, the dependence of the adsorption enthalpy on surface coverage was neglected.

$$\Delta h_{ads} = \frac{R \cdot T_1 T_2}{T_2 - T_1}\ln\left(\frac{p_2}{p_1}\right)_\theta = \frac{R \cdot T_1 T_2}{T_2 - T_1}\ln\left(\frac{a_1}{a_2}\right) \qquad (2)$$

where T_1 and T_2 are 273 and 318 K, respectively, p_1 and p_2 are the CH$_4$ pressures for each isotherm corresponding to an equal fractional coverage Θ and R denotes the universal gas constant. The parameters a_1 and a_2 are constants derived from the linearised Langmuir curve for the two temperatures. For further details, please refer to Panella et al. (2006) [3].

The isosteric heat of adsorption was calculated based on a Langmuir fit of the experimental data. The decrease in CH$_4$ excess sorption at pressures above 10 MPa is reproduced by introducing a term that accounts for the increase in sorbed phase volume with increasing loading (Equation (1)). For this purpose, the density of the sorbed phase was determined graphically (see **Figure 5**). The procedure resulted in a very good fit (**Figure 4**) of the experimentally determined excess sorption values. The Langmuir parameters a_1 and a_2 obtained with this method for isotherms at two different temperatures were used in Equation (2) for the calculation of the isosteric heat of adsorption.

Isosteric heats of adsorption for CH$_4$ show an average value of −11.2 ± 0.7 kJ/mol. In this case, a constant density of the sorbate phase was assumed. Using the parameters for the best least square Langmuir fit, an average value of −12.2 ± 0.7 kJ/mol was obtained. Evaluation of the isosteric heat of adsorption in the low pressure range (0 - 4 MPa) with the linearised Langmuir fit (providing the same values of absolute adsorption as the graphical evaluation) results in a significantly higher value for the heat of adsorption (-17.9 ± 1.3 kJ/mol).

3.3. Impact of CH₄ Sorption on Pore-Size Distribution

The pore-size distributions of the initial $Cu_3(BTC)_2$ and the samples after successive CH_4 and H_2 sorption measurements were determined by N_2 adsorption at 77 K. In order to describe adsorption on a wider range of microporous materials the Dubinin-Astakhov (DA) equation was used (Dubinin and Astakhov, 1971) [21]. It is a generalized form of the Dubinin-Radushkevich (DR; Dubinin, 1975 [22]) equation and was found to reasonably fit adsorption data for heterogeneous micropores. The results of the evaluations according to the DA method are displayed in **Figure 6**; the pertaining fitting parameters are listed in **Table 3**. The heterogeneity value (n) is higher for homogeneous distributions and is typically between 1 and 3. For instance, zeolite has an n-value higher than 3, demonstrating that these materials are very homogeneous. In this study, a value of $n = 1$ was determined for $Cu_3(BTC)_2$. The characteristic energy E after Dubinin (1975) was determined to be 20.97 kJ/mol for the initial (unspoiled) sample and ranged between 8 and 18 kJ/mol after high-pressure CH_4 and H_2 sorption tests. The interaction constant is given as $k = 2.96$ kJ·nm³/mol for nitrogen.

From **Figure 6** it is evident that CH_4 affects the stability of the MOF to a lesser extent than H_2. Although a significant decrease in total pore volume is shown after two successive CH_4 sorption experiments, accompanied by a shift to larger mean pore radii (4.7 Å for the original sample; 6.4 Å after six CH_4 sorption isotherms; **Table 3**), a reduction comparable to the H_2 treatment is only observed after 6 CH_4 sorption isotherms. As observed for the H_2 treatment, the pore volume in the 5 to 15 Å range is most heavily affected. In this study it is likely that we had chosen a slightly too high activation temperature. However the reason for the different influence of CH_4 and H_2 on the sample structure is not clear will be investigated in further detail in a follow-up study.

Results of the evaluation of the micropore structure of the $Cu_3(BTC)_2$ before and after high-pressure CH_4 sorption isotherms based on the DR method and the nonlinear density functional theory (NLDFT) are summarised in **Table 4**. These results also support a significant decrease in micropore volume and subsequent reduction in specific surface area of the sample after high-pressure CH_4 sorption tests.

Figure 7 compares the high-pressure CH_4 excess sorp-

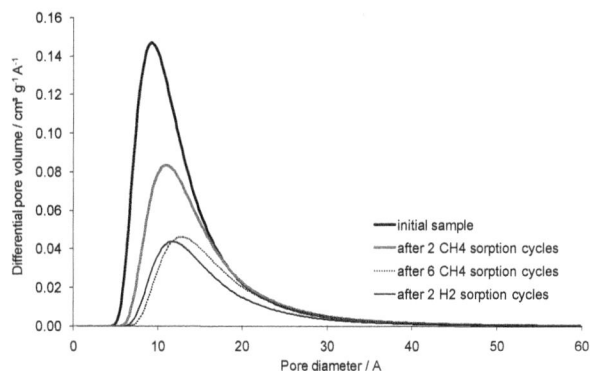

Figure 6. Dubinin Astakov differential pore volume distributions of $Cu_3(BTC)_2$ before and after high-pressure CH_4 and H_2 sorption isotherms. Evaluation based on low-pressure N_2-sorption at 77 K.

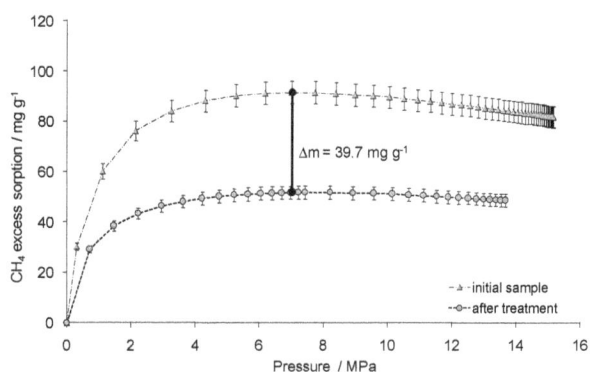

Figure 7. Comparison between initial and CH_4 treated $Cu_3(BTC)_2$ at 293 K.

tion isotherms measured on the initial Cu-BTC MOF with a decomposed sample of the same material at 293 K. The treated sample shows significantly lower sorption values (decrease of up to 40 mg/g) compared to the initial samples, while the shape of the isotherms is similar between the two isotherms.

4. Conclusions

At pressures up to 7 MPa, Cu-BTC MOF is one of the most promising MOF materials for methane storage with a large uptake capacity and an excess sorption of 9.2 wt% at 293 K. The repeated thermal activation and high-pressure sorption tests with CH_4 on HKUST-1 at temperatures between 273 and 318 K resulted in a reduction of specific micropore volumes and surface area. The processes behind the reduction of the specific micropore volumes and surface area during each CH_4 sorption test (**Table 4**) are ambiguous and further research on this topic is needed.

Recent studies indicate thermal deterioration effects and

Table 3. Fitting parameters of the DA equation for micropore characterisation of Cu$_3$(BTC)$_2$ before and after CH$_4$ and H$_2$ high-pressure sorption isotherms.

Cu$_3$(BTC)$_2$	N$_2$ characteristic energy E [kJ/mol]	DA heterogeneous value n	average pore radius [Å]
initial sample	20.97	1	4.7
after 2 Ch$_4$ isotherms	18.66	1	4.9
after H$_2$ isotherms	11.17	1	5.8
after 6 CH$_4$ isotherms	8.66	1	6.4

Table 4. Results of pore structure analysis of Cu$_3$(BTC)$_2$ before and after high-pressure sorption isotherms with CH$_4$.

	method	N$_2$ surface area (m^2/g)	N$_2$ pore volume (cm^3/g)	CO$_2$ surface area (m^2/g)	CO$_2$ micropore volume (cm^3/g)	N$_2$ DA pore diameter (Å)
initial sample	DR	1731	0.62	2708	0.94	9.2
	NLDFT	1994	0.64	2340	0.89	
after 2 CH$_4$ isotherms	DR	1174	0.42	1226	0.43	12.4
	NLDFT	1186	0.55	1069	0.35	
after 6 CH$_4$ isotherms	DR	737	0.26	997	0.35	12.8
	NLDFT	591	0.48	887	0.28	

by interaction with atmospheric air could cause the observed degradation of the sample.

Table 4 shows a stepwise deterioration of the sample after successive CH$_4$ sorption isotherms and the thermal activation process on the identical sample. Further changes caused by exposure to the atmosphere (e.g. oxidation of the sample) can be neglected, since samples did not come in contact with atmospheric air from the beginning of the thermal activation process (only H$_2$, CH$_4$ and He). However, it is unclear as of yet why micropores volumes decrease drastically and larger pores remain preserved.

Finally, this study has shown that the CH$_4$ sorbed phase density is, within reasonable errors, independent of temperature.

5. Acknowledgments

I am greatly indebted to Stefan Kaskel for providing 2007 the samples and Joachim Borchardt for technical support. The fruitful discussion and remarks of Bernd Krooß, Andreas Busch, Susan Giffin and Dirk Prinz are greatly appreciated.

REFERENCES

[1] M. Eddaoudi, J. Kim, N. Rosi, D. Vodak, J. Wachter, M. O'Keeffe and O. M. Yaghi, "Systematic Design of Pore Size and Functionality in Isoreticular MOFs and Their Application in Methane Storage," *Science Magazine*, Vol. 295, No. 5554, 2002, pp. 469-472.

[2] N. Amaroli and V. Balzani, "The Future of Energy Supply: Challenges and Opportunities," *General and Introductory Chemistry*, Vol. 46, No. 1-2, 2007, pp. 52-66.

[3] B. Panella, M. Hirscher, H. Pütter and U. Müller, "Hydrogen Adsorption in Metal-Organic Frameworks: Cu-MOFs and Zn-MOFs Compared," *Advanced Functional Materials*, Vol. 16, No. 4, 2006, pp. 520-524.

[4] H. Li, M. Eddaoudi, M. O'Keeffe and O. M. Yaghi, "Design and Synthesis of an Exceptionally Stable and Highly Porous Metal-Organic Framework," *Nature*, Vol. 402, pp. 276-279.

[5] S. Y. Chui, *et al.*, "A Chemically Functionalizable Nanoporous Material [Cu$_3$(TMA)$_2$(H$_2$O)$_3$]$_n$," *Science Magazine*, Vol. 283, No. 5405, 1999, pp. 1148-1150.

[6] K. Schlichte, T. Kratzke and S. Kaskel, "Improved synthesis, Thermal Stability and Catalytic Properties of The Metal-Organic Framework Compound Cu$_3$(BTC)$_2$," *Microporous and Mesoporous Materials*, Vol. 73, No. 1-2, 2004, pp. 81-88.

[7] B. Panella and M. Hirscher. "Hydrogen Physisorption in Metal-Organic Porous Systems," *Advanced Material*, Vol. 17, No. 5, 2005, pp. 538-541.

[8] G. Férey, M. Latroche, C. Serre, F. Millange, T. Loiseau and A. Percheron-Guégan, "Hydrogen Adsorption in the Nanoporous Metal-Benzenedicarboxylate M(OH)(O$_2$C–C$_6$H$_4$–CO$_2$)(M = Al$_3$+, Cr$_3$+), MIL-53," *Chemical Communications*, No. 24, 2003, pp. 2976-2977.

[9] M. G. Nijkamp, J. E. M. J. Raaymakers, A. J. van Dillen and K. P. de Jong "Hydrogen Storage Using Physisorption – Materials Demands," *Applied Physics A Materials Science & Processing*, Vol. 72, No. 5, 2001, pp.

619-623.

[10] J. Beneyto, F. Suárez-García, D. Lozano-Castelló, D. Cazorla-Amorós and A. Linares-Solano, "Hydrogen Storage on Chemically Activated Carbons and Carbon Nanomaterials at High Pressures," *Carbon*, Vol. 45, No. 2, 2007, pp. 293-303.

[11] Q. M. Wang, D. Shen, M. Bulow, M. L. Lau, S. Deng, F. R. Fitch and N. O. Lemcoff, J. Semanscin, "Metallo-Organic Molecular Sieve for Gas Separation and Purification," *Microporous and Mesoporous Materials*, Vol. 55, No. 2, 2002, pp. 217-230.

[12] X. Lin, A. J. Blake, C. Wilson, X. Z. Sun, N. R. Champness, M. W. George, P. Hubberstey, R. Mokaya and M. Schroder, "A Porous Framework Polymer Based on a Zinc(II) 4,4'-Bipyridine-2,6,2', 6'-Tetracarboxylate: Synthesis, Structure, and 'Zeolite-Like' Behaviors," *Journal of American Chemical Society*, Vol. 128, No. 33, 2006, pp. 10745-10753.

[13] J. Senkovska and S. Kaskel, "High Pressure Methane Adsorption in the Metal-Organic Frameworks $Cu_3(BTC)_2$, $Zn_2(bdc)_2dabco$, and $Cr_3F(H_2O)_2O(bdc)_3$," *Microporous and Mesoporous Materials*, Vol. 112, No. 1-3, 2008, pp. 108-115.

[14] U. Setzmann and W. Wagner, "A New Equation of State and Tables of Thermodynamic Properties for Methane Covering the Range From the Melting Line to 625 K at pressures up to 1000 MPa," *Journal of Physical and Chemical Reference Data* , Vol. 20, No. 6, 1991, pp. 1061-1155.

[15] R. D. McCarty and V. D Arp, "A New Wide Range Equation of State for Helium," *Advances in Cryogenic Engineering*, Vol. 35, 1990, pp. 1465-1475.

[16] Y. Gensterblum, P. van Hemert, P. Billemont, A. Busch, D. Charriere, D. Li, B. M. Krooss, G. de Weireld, D. Prinz and K.-H. A. A. Wolf, "European Inter-Laboratory Comparison of High Pressure CO_2 Sorption Isotherms. I: Activated Carbon," *Carbon*, Vol. 47, No. 13, 2009, pp. 2958-2969.

[17] A. L. Goodman, A. Busch, G. Duffy, J. E. Fitzgerald, *et al.*, "An Inter-laboratory Comparison of CO_2 Isotherms Measured on Argonne Premium Coal Samples," *Energy and Fuels*, Vol. 18, No. 4, 2004, pp. 1175-1182.

[18] Y. Gensterblum, P. Van Hemert, P. Billemont, *et al.*,

"European inter-laboratory comparison of high pressure CO_2 sorption isotherms II: Natural coals," *International Journal of Coal Geology*, Vol. 84, No. 2, 2010, pp. 115-124.

[19] A. Busch, Y. Gensterblum, B. M. Krooss and R. Littke, "Methane and Carbon Dioxide Adsorption-Diffusion Experiments on Coal: An upscaling and Modelling," *International Journal of Coal Geology*, Vol. 60, No. 2-4, 2004, pp. 151-168.

[20] S. Brunauer, P. H. Emmett and E. Teller, "Adsorption of Gases in Multimolecular Layers," *Journal of American Chemical Society*, Vol. 60, No. 2, 1938, pp. 309-319.

[21] M. M. Dubinin, "Physical Adsorption of Gases and Vapors in Micropores," Academic Press, New York, 1975, p. 1-70.

[22] M. M. Dubinin and V. A. Astakhov, "Description of Adsorption Equilibria of Vapors on Zeolites over Wide Ranges of Temperature And Pressure," *Advances in Chemistry*, Vol. 102, No. 69, 1971, pp. 65-69.

[23] M. Thommes, B. Smarsly, M. Groenevolt, P. I. Ravikovitch and A.V. Neimark, "Adsorption Hysteresis of Nitrogen and Argon in Pore Networks and Characterization of Novel Micro- and Mesoporous Silicas," *Langmuir*, Vol. 22, Vol. 2, 2006, pp. 756-764.

[24] M. Kramer, U. Schwarzer, S. Kaskel, "Synthesis and properties of the metal-organic framework $Mo_3(BTC)_2$ (TUDMOF-1)" *Journal of Material Chemistry,* Vol. 16, 2006, pp. 2245-2248.

[25] P. Krawiec, M. Kramer, M. Sabo, R. Kunschke, H. Fröde and S. Kaskel, "Improved Hydrogen Storage in the Metal-Organic Framework $Cu_3(BTC)_2$" *Advanced Engineering Material*, Vol. 8, No. 4, 2006, pp. 293-296.

[26] A. R. Millward and O. M. Yaghi, "Metal-Organic Frameworks with Exceptionally High Capacity for Storage of Carbon Dioxide at Room Temperature," *Journal of American Chemical Society*, Vol. 127, No. 51, 2005, pp 17998-17999.

[27] R. Humayun, D. L. Tomasko, "High-Resolution Adsorp-Tion Isotherms of Supercritical Carbon Dioxide on Activated Carbon," *AICHE Journal*, Vol. 46, No. 10, 2000,pp. 2065-2075.

13

High Pressure Water-Jet Technology for the Surface Treatment of Al-Si Alloys and Repercussion on Tribological Properties

Md. Aminul Islam[1*], Zoheir Farhat[1], Jonathon Bonnell[2]

[1]Department of Process Engineering and Applied Science, Dalhousie University, Halifax, Canada; [2]Vector Aerospace, Engine Services-Atlantic, Summerside, Canada.

ABSTRACT

Recent developments in high pressure water-jet technology have brought the process to the forefront as a means of surface treatment. Water jet technology offers cleaning, cutting, processing as well as potential refinement of surface properties. By adapting the process parameters the surface characteristics can be changed while the profile remains the same. In the present study, water-jet technology was used for the surface treatment of Al-Si alloy to investigate its effect on tribological properties. Dry sliding wear behavior was investigated against AISI 52100 bearing steel ball using a reciprocating ball-on-flat configuration. Optical microscopy examination reveals that ploughing of grains, transgranular and intergranular propagation of cracks; are the mechanisms by which material is removed during water jet treatment. While, on the other hand, SEM observation of the wear track reveals that plastic deformation and delamination are the dominant wear mechanism during the wear process. Water jet treatment was compared to hot isostatic pressing in terms of its effects on wear resistance and surface porosity of Al-Si alloy. It was found that, hot isostatic pressing reduces the total amount of porosity at the expanse of hardness while water jet treatment produces a compressed surface having higher hardness and compressive residual stress, which ultimately increases wear resistance.

Keywords: *Tribology, High Pressure Water Jet Treatment, Porosity, Surface Treatment, XRD, Reciprocating Wear*

1. Introduction

In recent years, the use of high-pressure water jet technique became widely accepted practice in order to meet the requirements of production and maintenance. Water-jet technologies offer cleaning, machining, surface treatment and cutting of materials. Both roughening as well as polishing is possible with this procedure. It is therefore excellent for pre-treatment of engineering surfaces. For the last few decades, water-jet technology covered the following applications: removal of paint, grease, dirt from aircraft in the aviation industry; prevent pump cavitation; removal of cement lips, incrustations, lime, solidified dust from autoclave vessel; removal of worn protective coating, incrustations, solidified materials in the chemical industry; semiconductor frame cleaning in the electronic industry; removal of weld slag, water scale, mill scale and rust in steel mills; water jet cutting; vibration free demolition by abrasive water jets and so on [1-11].

The conventional water jet technique has been used for cutting and polishing by mixing abrasive materials into the water which transport the abrasive materials to strike the substrate. On the other hand, high pressure water jet can produce a compressed surface layer without abrasive materials, while the core of the component remains unchanged. The new surface layer 'tribolayer' is expected to have enhanced tribological properties due to possible work hardening, residual stress and reduction of surface porosity. Hence, water jet treatment may be an effective technique for improved surface properties.

Aluminum-silicon based alloys have received considerable attention due to their high strength to weight ratio. The reduction in weight of components leads to significant impact on fuel economy in dynamic systems. In addition, these alloys are reported to have a reasonably high wear resistance. The wear resistance of Al-Si alloys depends on a number of material related parameters, *i.e.*, size, shape, composition and distribution of micro-constituents, in addition to service conditions [12-16]. Porosity is a common feature of sintered Al-Si alloys and

High Pressure Water-Jet Technology for the Surface Treatment of Al-Si Alloys and Repercussion on Tribological
Properties

89

strongly influences their properties and applications. In general the presence of pores is accompanied by a drop in strength, ductility and wear resistance of materials [17]. Not only the total volume percentage of porosity influences the degradation of properties but also size, shape and interconnectivity of pores play an important role [18-20]. Meanwhile, initial attempts on the mechanical characterization of these materials were focused on static tests. However, in order to employ these materials in new applications, their dynamic properties have to be carefully assessed. In recent years, fatigue behavior has been studied in numerous investigations. The influence of pores on crack initiation and propagation under cyclic loading was examined as well [21,22]. The effect of water jet peening toward improving fatigue resistance was examined by various experimental tests and proved as an effective techniques for preventing fatigue in structural components [23-25].

The objective of this study is to assess the potential of employing high pressure water-jet technology for the surface treatment of Al-Si alloys to improve wear resistance for automobile and aerospace applications.

2. Experimental Details

In order to obtain a homogeneous distribution of porosity, samples were prepared using powder metallurgy method. Two powders (*i.e.*, Al-Si master alloy and Al-Mg master alloy), were mixed to produce the following alloy composition; 88.8 wt% Al, 6.0 wt% Si, 4.5 wt% Cu, 0.5 wt% Zn, 0.2 wt% Fe. A Lico wax C was used as a pressing lubricant. Specimens were pressed at 100, 200 and 600 MPa and were sintered in a tube furnace in the presence of nitrogen gas at 5600 C for 20 minutes and then slow cool to 4800 C. The green and sintered densities of samples were determined in accordance with MPIF Standard 42 [26] and are listed in **Table 1**. For micro-structural experiments, specimens were cut, mounted and ground using 240, 320, 400 and 600 grit SiC abrasive papers and then polished using 1µm, 0.3µm and 0.05µm gama alumina suspension. Olympus BX51 research microscope, equipped with bright-field objectives was used to analyze the microstructure at high resolution.

Table 1. Basic properties of sintered Al - 6 wt% Si alloy.

Specimens	Green Density (g/cc)	Sintered Density (g/cc)
Pressed at 100 MPa	2.11	2.33
Pressed at 200 MPa	2.29	2.46
Pressed at 600 MPa	2.58	2.6

Water-jet treatment was performed using a highly pressurized water-jet. The system is capable of employing ultrasonic pulses of water-jet for more intense treatment. Several tests were performed using the ultrasonic system. During experiments, specimens were individually mounted on a turntable inside the water jet enclosure. Specimens containing different amount of porosity (compacted at 100, 200 and 600 MPa) were subjected to water jet at four different pressures, 34, 48, 55 and 70 MPa, employing a nozzle with an orifice size of 1.6 mm. The standoff distances between the nozzle and the specimen were 25, 64, 76 mm and the translation speeds were 25, 50, 150 and 250 mm/s.

Surface porosities of samples were calculated using image analysis software. A series of images were taken to cover the whole surface area of the sample. Porosities were identified based on their gray-level intensity differences compared to the matrix. Gray-level threshold settings were selected to permit independent detection of porosity, using the 'flicker method' of switching back and forth between porosity and the matrix. Second phase particles and dendrite, may be counted as porosity because their gray level range is similar to that of porosity. The grey-level thresholds as well as boundary conditions, (*i.e.*, aspect ratio, min radius and area) were set to avoid second phase particles and dendrite. A counting protocol was chosen to correct for edge effects so that a porosity lying across a field boundary is counted only once [27]. For each field the area fraction of the detected area of porosity was measured by dividing the detected area of porosity by the area of the measurement field. Surface porosity values are given in **Table 2**.

Table 2. Surface porosity of sintered Al - Si alloy.

Specimens	Untreated		After HIP		After water jet		
	% Surface Porosity	Average Pore Size (µm)	%Surface Porosity	Average Pore Size (µm)	% Surface Porosity	% Sub surface Porosity	Average Pore Size (µm)
Pressed at 100 MPa	6.71 ± 0.01	82.4	2.13 ± 0.01	34.6	9.14± 0.01	5.33 ± 0.01	86.5
Pressed at 200 MPa	4.22 ± 0.01	70.2	1.34 ± 0.01	23.5	4.37 ± 0.01	3.65 ± 0.01	73.2
Pressed at 600 MPa	2.35 ± 0.01	46.9	0.75 ± 0.01	20.4	3.80 ± 0.01	1.97 ± 0.01	48.8

Micromet micro-hardness tester was used to measure the micro-hardness of the specimens. This was conducted to assess the effect of water jet treatment on the sub-surface properties of the Al-Si alloy. In this test method, Vickers hardness measurements were performed using a diamond indenter and 15 g load. The size of the indentation was measured using a light microscope equipped with a filar type eyepiece. HV measurements were taken as a function of depth from the treated surface.

In order to measure residual stresses induced as a result of the high pressure water jet treatment, X-ray peak broadening diffraction techniques was employed. X-ray diffraction (XRD) experiment was carried out on a specimen treated ultrasonically at 48 MPa from 76 mm standoff distance and 150 mm/s translation speed, employing a high-speed Bruker D8 Advance system using Cu-Kα radiation having a wave length (λ) of 1.54 Å, tube voltage of 40 KV, and tube current of 40 mA. The strongest four peaks (i.e., (111), (200), (220) and (311)) were selected for slow scan of 0.02°/sec. The integral breadth of XRD peaks were analysed using EVA software package. Peaks were corrected for the effects of Kα2 radiation and the background was removed. A standard stress-free Al from the same alloy was used to measure instrumental broadening. The data was treated according to Williamson-Hall method [28]. The integral breadth (i.e., total area under the peak/peak height) of the diffraction peak can be expressed as,

$$B_{obs} = B_{ins} + B_{size} + B_{\varepsilon} \quad (1)$$

Here B_{size} and B_{ε} are the grain size and micro-strain contributions to the observed peak broadening B_{obs} respectively. B_{ins} is the peak broadening at a stress-free state, referring to the instrument contribution. The crystalline size 't' is related to B_{size} by the Scherrer equation [29],

$$t = \frac{0.9\lambda}{B_{size} \cos \theta_B} \quad (2)$$

Here, θ_B is Bragg's angle. While the micro-strain can be calculated from,

$$B_{\varepsilon} = 2 \left(\frac{\Delta d}{d} \right) \tan \theta_B \quad (3)$$

Where, d is the interplaner spacing. Substituting equation (2) and (3) into (1) gives,

$$\left(B_{obs} - B_{ins} \right) \cos \theta_B = \frac{0.9\lambda}{t} + 2 \left(\frac{\Delta d}{d} \right) \sin \theta_B \quad (4)$$

For grain size larger than 10 μm (as in the present case) X-ray peak broadening is due to micro-strain effect alone, hence, the first term in equation (4) can be eliminated.

Dry reciprocating wear tests were performed using a Universal Micro-Tribometer. This test method utilizes a ball upper specimen that slides against a flat lower specimen in a linear, back and forth, sliding motion having a stroke length of 5.03 mm. All tests were conducted at room temperature and a relative humidity of 40 - 55 %. The load is applied downward through the ball specimen against the flat specimen mounted on a reciprocating drive. The tester allows for monitoring the dynamic normal load, friction force and depth of the wear track during the test. A 6.3 mm diameter AISI 52100 bearing steel ball having a hardness of HRA 83 was used as a counter-face material. The ball was mounted inside a ball holder which was attached directly to a suspension system, which, in turn, is attached to a load sensor that controls and records forces during the test. The weight of the specimen was measured before and after each wear test to determine individual weight loss at selected time intervals. 10 N normal load and 15 Hz frequency were employed for three different time intervals (10 min, 45 min and 90 min). After wear tests, worn surfaces and cross section of the wear track were examined using optical and scanning electron microscopy to determine possible wear mechanisms.

3. Results and Discussion

The size, shape and amount of pores are largely dependent on processing parameters. As all the specimens were sintered under the same sintering conditions, the compaction pressure plays the most significant role in determining pore size, shape and amount. Increasing the compaction pressure decreases the amount of porosity and reduces pore size, while the pore shape changes from large irregular to small round shape. The percent surface porosity ranges from 6.71% to 2.35%, while the average pore size ranges from 82.4 μm to 46.9 μm respectively as the compaction pressure of the Al-Si alloy powder is raised (**Table 2**). The low standard deviations of surface porosities indicate a uniform distribution of porosity throughout the structures. To mitigate the detrimental effect of porosity, specimens were subjected to hot isostatic pressing (HIP) treatment. HIP gave rise to a decrease of approximately 68% in surface porosity and 60% in pore size. On the other hand, when the sintered Al-Si alloy was subjected to high pressure water-jet (using 70 MPa pressure, 64 mm distance and 25 mm/s) it is found that there is an increase of 36%, 4% and 61% in surface porosity at compact pressure of 100, 200 and 600 MPa, respectively. Moreover, there is approximately 4.3% increase in the average pore size after water jet treatment. While there is around 16% reduction in porosity in the subsurface region (22 ± 3μm from the surface)

High Pressure Water-Jet Technology for the Surface Treatment of Al-Si Alloys and Repercussion on Tribological
Properties

91

after the water jet treatment. It has also found that treated surfaces exhibit high roughness. The discrepancy in the amount of porosities and size of pores between the surface and near surface (**Table 2**) is attributed to the difficulty of measuring surface porosity when surface roughness is high. Therefore, near surface porosity is a better indicator of the amount and size of porosity.

In the water jet process, jet pressure, standoff distance and translation speed have a significant impact on final surface properties.

During the water jet treatment, water jet strikes the surface at a relatively high velocity which causes a significant amount of material loss. The amount of material loss increases with water jet pressure. **Figure 1** shows a representative image showing the surface damage after water jet treatment. In this particular situation, the specimen compacted at 100 MPa was subjected to 55 MPa jet pressure operated from 25 mm standoff distance at 50 mm/s translation speed reveal a depth of the material removal of about 900 µm. The insert on the upper right corner of the image (**Figure 1**) shows the position where the depth measurement was taken.

Optical microscopy examination reveals that ploughing of grains, transgranular, intergranular propagation of cracks and as a consequence, break away and pulling out of grains are the mechanisms for material removal during water jet treatment. **Figure 2(a)-(e)** shows a cross-section of a specimen compacted at 600 MPa and treated ultrasonically at 34 MPa at a translation speed of 250 mm/s and standoff distance of 64 mm. In **Figure 2(a)**, shows transgranular crack propagating through a grain, while **Figure 2(b)** reveals a portion of a grain breaking away. It was also observed that the water jet penetrates along the grain boundaries promoting intergranular cracks to develop (**Figure 2(c)**). This leads to pulling out of grains as shown in **Figure 2(d)**. **Figure 2(e)** shows ploughing of grains along the water jet path. SEM observation

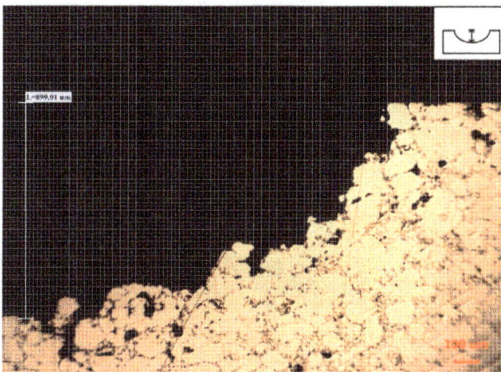

Figure 1. Surface damage due to 55 MPa water jet pressure, 25 mm standoff distance and 50 mm/s translation speed.

of the surface on the other hand, reveal that water jet pressure causes collapse of porosity underneath the surface. **Figure 2(f)** shows an SEM image of the specimen compacted at 100 MPa and treated ultrasonically at 48 MPa from 76 mm standoff distance and at 150 mm/s translation speed. It is observed from the figure what appears to be a sub-surface porosity collapse due to the pressure created by the water jet, hence, producing low porosity subsurface 'tribolayer'.

In order to investigate the effect of water-jet treatment on hardness, a series of micro-hardness measurements were conducted according to ASTM standard [30]. Hardness was measured at three different depths, 17 ± 3, 122 ± 5 and 1000 ± 5 µm below the treated surface. **Figure 3** shows the variation of hardness at different distance below the surface of the specimen compacted at 100 MPa and treated at 55 MPa water jet pressure from 25 mm standoff distance at 50 mm/s translation speed. Each hardness value in the figure represent an average of 10 measurements. It is evident from the figure that the hardness drops gradually away from the treated surface.

There is about 15% increase in hardness in the vicinity of the surface as compared to hardness of the base metal about 1 mm from the surface. It is believed that the observed rise in hardness closer to the surface is due to work hardening effects as a result of the high pressure water jet treatment.

Figure 4 shows the superimposed XRD diffraction peaks of the standard and water jet treated specimens as representative peaks. It should be noted that neither broadening nor sifting of peaks was observed for the untreated sample. XRD of the all scanned peaks were only slightly broadened but significantly shifted to the right after water jet treatment. According to peak broadening calculations the microstrain of the treated specimen was found to be 10×10^{-5}. It can be shown that the residual stress resulting from microstraiin analysis is 7 MPa (*i.e.*, $\sigma_{res} = \varepsilon E$, where, E = 70 GPa [31]). On the other hand, the macrostrain (uniform strain) calculated from the shifts in peak positions, as $(\Delta d/d)$, is 6×10^{-4} and the residual stress was calculated to be 42 MPa. The measured macrostrain and, hence, the associated residual stress are compressive as peak shifts are to the right (toward higher Bragg angle). As in the XRD diffraction method, only planes parallel to the surface contribute to the measured intensity, crystal planes parallel to the surface are under compression due to the water jet treatment. The measured compressive residual stress promotes the closing of surface and near surface pores, hence, raising the hardness in the 'tribolayer' and retarding crack initiation and propagation during wear.

In order to investigate wear performance of the sintered

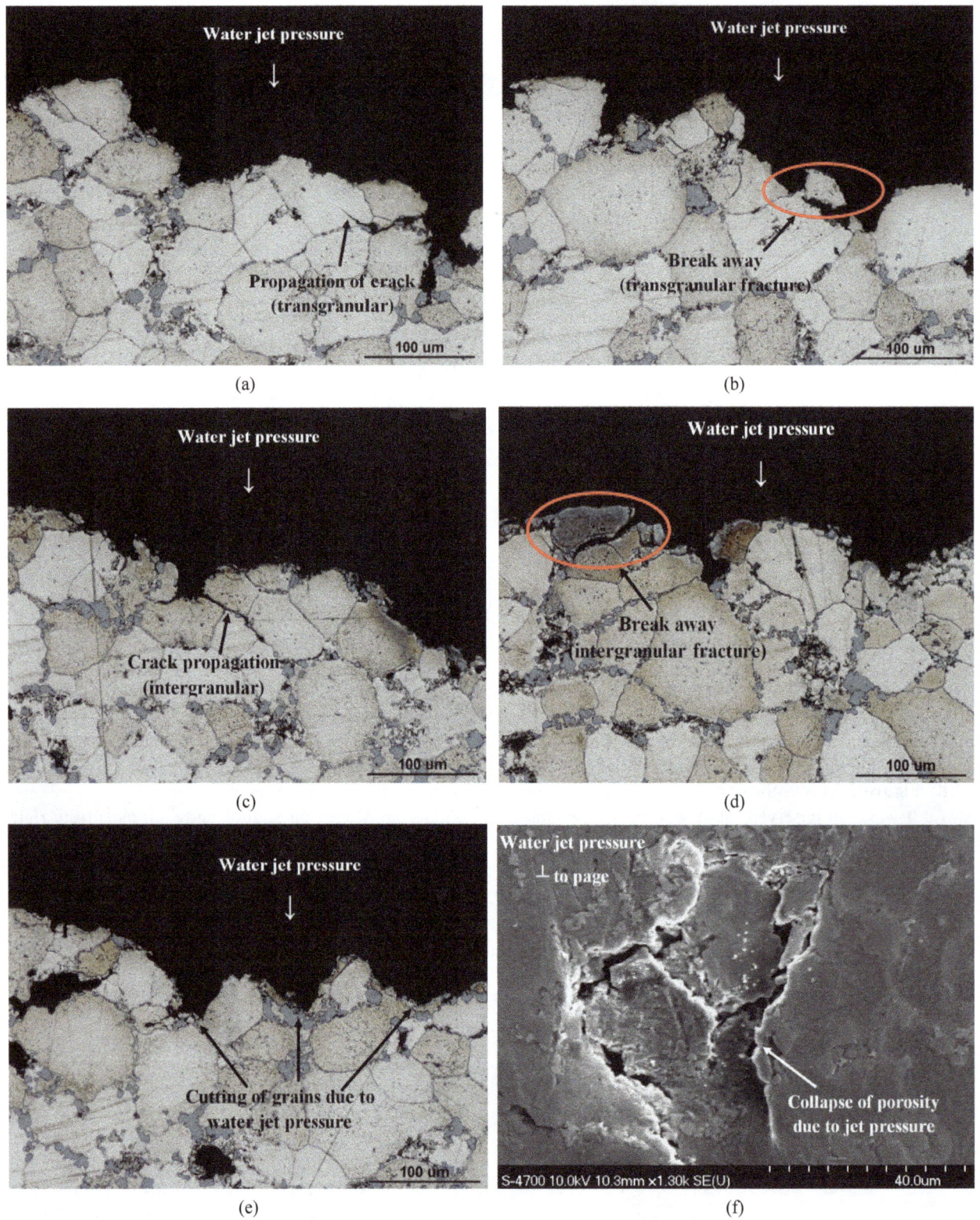

Figure 2. X-section of a specimen compacted at 600 MPa, (a) transgranular crack propagation; (b) break away of grains; (c) intergranular crack propagation, (d) pulling out of grains, (e) ploughing of grains, (f) collapse of porosity due to water jet pressure.

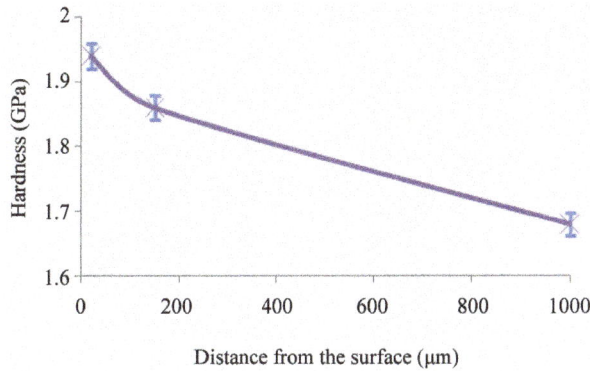

Figure 3. Variation of hardness at different depth below the surface.

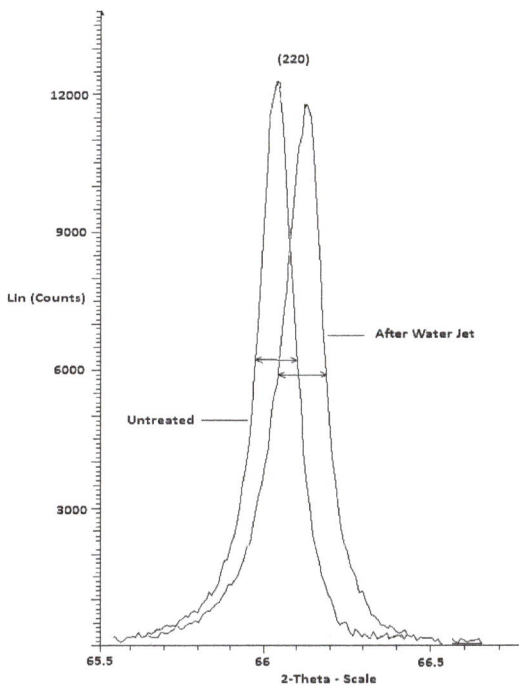

Figure 4. (220) X-ray diffraction peak of water jet treated and standard Al specimens.

Al-Si alloy after water jet treatment and compare to untreated and HIPed conditions, reciprocating wear tests were performed under 10 N normal load and 15 Hz frequency at three time intervals (10 min, 45 min and 90 min). For the water-jet treated specimens, the operating condition are as follows: 70 MPa from 64 mm standoff distance and 25 mm/s translation speed. **Figure 5** depicts the data collected during wear tests. The figure shows a somewhat linear increase (with no transition) in weight loss with sliding distance. The fact that there are no wear transitions may indicate that the same wear mechanism(s) is operative throughout the duration of the tests. As we

observed from the figure, 100 MPa specimens (under all conditions) exhibit higher weight loss compared to 600 MPa specimens (under corresponding conditions) due to lower porosity content in the specimens. On the other hand, for any given compaction pressure (*i.e.*, percent porosity), water jet treated specimens exhibit lower weight loss compared to HIPed and untreated specimens.

Wear rates (*i.e.*, slope of best fit weight loss versus sliding distance lines) are calculated from **Figure 5** and their variation under various compaction pressures are plotted in **Figure 6**. It is clear that the specimens subjected to water jet treatment exhibit superior wear resistance compared to both untreated and HIPed specimens. For the specimens compacted at 100, 200 and 600 MPa, wear resistance increases by 4.3, 3.8 and 14.7% respectively after HIP, while after water jet treatment, there is 34.7, 36.4 and 57.3% increase in wear resistance, respectively.

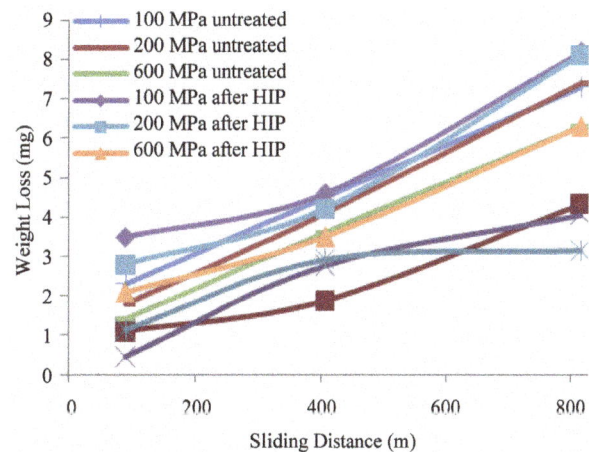

Figure 5. Weight loss vs sliding distance.

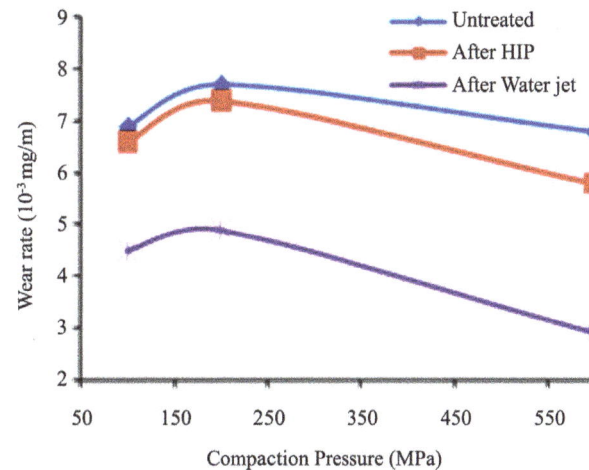

Figure 6. Wear rate at different operating condition.

The improvement in wear resistance of water jet treated specimens as opposed to hipped and untreated specimen can be explained in terms of variation in porosity and hardness levels post treatments. In the HIPed specimens, simultaneous application of heat and pressure reduces near surface internal voids and micro-porosity through a combination of plastic deformation, creep, and diffusion bonding. However, it has been found that, the final structure exhibit lower hardness due to softening at higher temperature due to HIP [34]. Water jet treatment was found to produce a compress layer and simultaneously increases hardness of the material.

To explain the initial increase and subsequent drop in wear rate with compaction pressure, detailed examination of the effect of porosity on wear mechanism is necessary. Scanning electron microscopy examinations of wear tracks reveal that delamination, plastic deformation and surface fracture are the dominant mechanisms during the wear process. Based on experimental observations in this study, several factors can be identified that affect the wear resistance of the water jet treated Al-Si alloys, namely, amount of porosity, pore size and shape and hardness of the subsurface region. The wear tracks of the samples were characterized by plastic deformation and damage in the form of plastic flow extending parallel to the sliding direction as shown in **Figure 7(a)**. Surface pores act as stress concentration sites, the area around pores tend to fracture as a result of the stress exerted by the slider. This is evident from the cracks initiating from pore edge shown in **Figure 7(b)**. Furthermore, Hertzian type cracks develop as a result of surface tensile stresses that evolve during Hertzian contact. These types of cracks normally extend perpendicular to the sliding direction. **Figure 7(c)** shows cracks perpendicular to the sliding direction which represents a clear evidence of Hertzian cracks developing during the wear process.

The other mechanism contributing to the observed wear is delamination. Initially, plastic deformation of surface layers due to cyclic normal and tangential loads take place, followed by crack or void nucleation in the deformed layers at pore sites (as pores act as stress concentration zones). Cracks nucleate at depth where shear stress is maximum as predicted by the Hertzian theory [35]. Then cracks grow nearly parallel to the surface and eventually become unstable and extend to the surface leading to the formation of thin, plate-like wear debris. Furthermore, the cracks extend and create a network of cracks by connecting different subsurface pores (**Figure 7(d)**). Pores serve as the origin and end of crack propagation, hence reducing the required length of crack propagation. This represents clear evidence that porosity in the subsurface layer accelerates delamination wear.

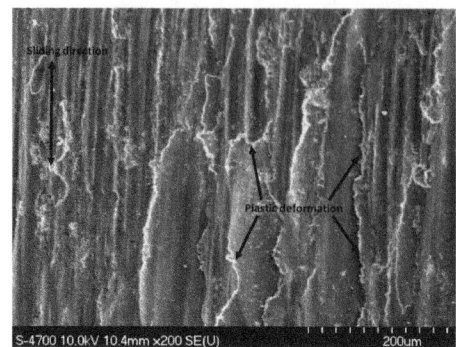

(a)

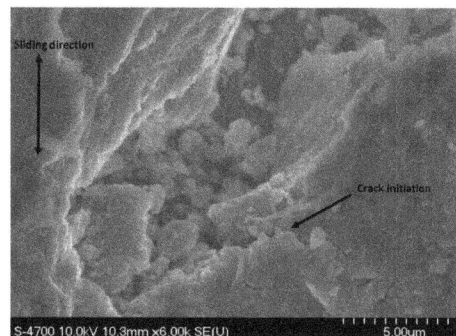

(b)

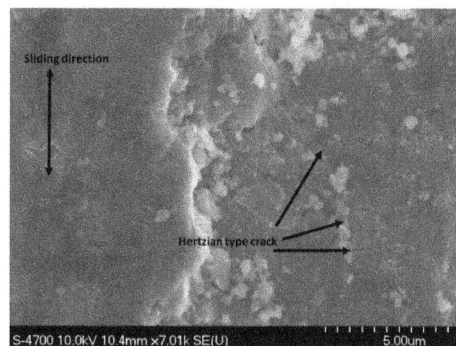

(c)

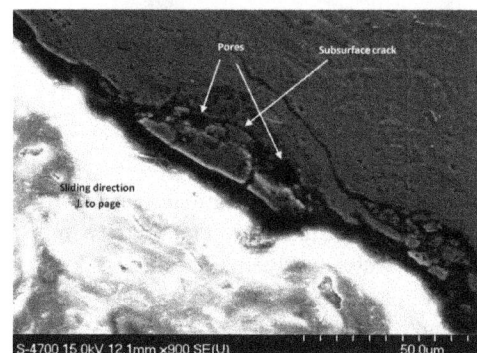

(d)

Figure 7. SEM observation of wear tracks. (a) plastic deformation, (b) crack nucleation from pore edge, (c) Hertzian crack perpendicular to the sliding direction, (d) cross section of wear track, pore connects in the sub surface region.

Based on SEM examination and Hertzian analysis (discussed in details in a previous work [13]), as expected, wear increase as porosity increases. However, at a critical pore size, where the contact area between the slider and sample surface is in the same order of magnitude as pore size, wear drops as the counterface slides into the pore, hence, not contributing to wear loss.

4. Conclusions

In the present work, the potential of using high pressure water-jet as a surface treatment for Al-Si alloys to mitigate the effect of surface porosity was assessed and the following main conclusions have emerged.

a. The amount of subsurface porosity and average pore size drops after water jet treatment.

b. Subsurface work hardening and compressive residual microstrain develop as a result of water jet pressure.

c. Ploughing of grains, transgranular and intergranular propagation of cracks are the mechanisms for material removal during water jet treatment.

d. High pressure water-jet treatment is more effective than HIP in promoting superior wear resistance. Prolonged heating at high temperature during HIP decreases the total amount of porosity at the expanse of decreased hardness. While water jet produces hard compressed subsurface 'tribolayer' leading to increased wear resistance of the Al-Si alloy.

5. Acknowledgements

The author would like to thank Auto-21 for financial support, Bodycote International for performing HIP experiments and Vector Aerospace for performing water jet treatment.

REFERENCES

[1] S. A. Hofacker, "The Large Aircraft Robotic Paint Stripping System," Proc. 7th American Water Jet Conference," Water Jet Technology Association, St Louis, 1993, pp. 613-628.

[2] M. Wen, L. Zhang, H. Han, Y. Dong, Z. Che, "Water-Jet-Technology and Its Application for Preventing Pump-Cavitation," Renewable Energy Resources, Vol. 28, No. 2, 2010, pp. 145-147.

[3] B. Wood, "A Water-Cooled, Hydraulically Positioned 20,000 Psi Lance for Waterblasting Inside a Hot Klin," Jetting Technology, Mechanical Engineering Publication Limited, London, 1996, pp. 379-392.

[4] A. W. Momber and A. G. Nielsen, "Pipeline Rehabilitation by Water Jetting," Material Evaluation, Vol. 37, 1998, pp. 97-101.

[5] R. Yasui, A. Yanari and F. M. Carletti, "The Removal of

Excessive Resin from Semiconductor Leadframes with Spot-Shot Waterjets," Proceeding 7th American Water Jet Conference, Water Jet Technology Assoiation, St Louis, 1993, pp. 813-827.

[6] M. Raudensky, J. Horsky and L. Telecky, "Thermal and Mechanical Effect of High-Pressure Spraying of Hot Surfaces-Descaling," Proceeding 3rd International Metallurgic Conference, Continental Casting Dillets, Trinec, 1999, pp. 217-221.

[7] A. W. Momber and R. Kovacevic, "Principals of Abrasive Water Jet Machining." Springer-Verlag Ltd., London, 1998.

[8] G. Anirban, M. Ronald and R. Balachandar, "An Experimental and Numerical Study of Water Jet Cleaning Processes," Journal of Materials Processing Technology, Vol. 211, No. 4, 2011, pp. 610-618.

[9] A. W. Momber, "Water Jet Applications in Construction Engineering," A A Balkema, Rotterdam, 1998.

[10] S. Srinivas and N. Ramesh, "An Analytical Model for Predicting Depth of Cut in Abrasive Water Jet Cutting of Ductile Materials Considering the Deflection of Jet in Lateral Direction," International Journal of Abrasive Technology Vol. 2, No. 3, 2009, pp. 259-278.

[11] H. Orbanic and M. Junkara, "Analysis of Striation Formation Mechanism in Abrasive Water Jet Cutting," Wear Vol. 265, No. 5-6, 2008, pp. 821-830.

[12] H. Torabiana, J. P. Pathaka and S. N. Tiwaria, "Wear Characteristics of Al-Si Alloys," Wear, Vol. 172, 1994, pp. 49-58.

[13] M. Elmadagli, T. Perry and A. T. Alpas, "A Parametric Study of the Relationship between Microstructure and Wear Resistance of Al-Si Alloys," Wear, Vol. 262, No. 1-2, 2007, pp. 79-92.

[14] N. Saka, A. M. Eleiche and N. P. Suh, "Wear of Metals at High Sliding Speeds," Wear, Vol. 44, No. 1, 1977, pp. 109-125.

[15] T. M. Chandrashekharaiah and S. A. Kori "Effect of Grain Refinement and Modification on the Dry Sliding Wear Behaviour of Eutectic Al-Si Alloys," Tribology International, Vol. 42, No. 1, 2009, pp. 59-65.

[16] S. Hegde and K. N. Prabhu, "Modification of Eutectic Silicon in Al-Si Alloys," Journal of Material Science, Vol. 43, 2008, pp. 3009-3027.

[17] G. F. Bocchini, "The Influence of Porosity on the Characteristics of Sintered Materials," International Journal of Powder Metallurgy, Vol. 22, No. 3, 1986, pp. 185-202.

[18] F. Akhlaghi and A. A. Zare-Bidaki, "Influence of Graphite Content on the Dry Sliding and Oil Impregnated Sliding Wear Behavior of Al 2024-Graphite Composites Produced by in Situ Powder Metallurgy Method," Wear, Vol. 266, No. 1-2, 2009, pp. 37-45.

[19] H. Danninger, G. Jangg, B. Weiss and R. Stickler, "Microstructure and Mechanical Properties of Sintered Iron, Part I: Basic Considerations and Review of Literature," *Powder Metallurgy International*, Vol. 25, No. 3, 1993, pp. 111-117.

[20] M. A. Islam and Z. N. Farhat, "Effect of Porosity on Dry Sliding Wear of Al-Si Alloys," *Tribology International*, Vol. 44, No. 4, 2011, pp. 498-504.

[21] D. A. Gerard and D. A.Koss., "Low Cycle Fatigue Crack Initiation: Modeling the Effect of Porosity," *International Journal of Powder Metallurgy*, Vol. 26, No. 4, 1990, pp. 337-343.

[22] Y. Mai, B. Cotterell, S. Q. He and Y. B. Ke, "Handbook of Fatigue Crack Propagation in Metallic Structures," *Amsterdam*: *Elsevier*, Vol. 1, 1991, pp. 221-246.

[23] A. Shimamoto, R. Kubota and S. Yang, "Effects of Water Jet Peening and Hardening Treatment on Fatigue Properties of SCr420," *ASME Conference Proceeding*, Vol. 3, 2009, pp. 425-432.

[24] K. Hirano, K. Enomoto, M. Mochizuki, M. Hayashi, E. Hayashi and S. Shimizu, "Improvement of Residual Stress on Material Surface by Water Jet Peening," Translation of the 14th International Conference on Structural Mechanicsin Reactor Technology, Lyon, France, 1997, pp. 17-22.

[25] B. Han and D. Y. Ju, "Compressive Residual Stress Induced by Water Cavitation Peening: A Finite Element Analyses," *Materials & Design*, Vol. 30, No. 8, 2009, pp. 3325-3332.

[26] "Method for Determination of Density of Compacted or Sintered Powder Metallurgy Products-Standard 42," Standard Test Methods of Metal Powders and Powder Metallurgy Products, Metal Powder Industries Federation, New Jersey, Princeton, 2002, pp. 59-61.

[27] "Standard Practice for Determining the Inclusion or Second-Phase Constituent Content of Metals by Automatic Image Analysis," Annual Book of ASTM Standards: E 1245-03.

[28] G. K. Williamson and W. H. Hall, "X-ray Line Broadeningfromjiled Aluminium and Wolfram," *Acta Metallurgica*, Vol. 1, No. 1, 1953, pp. 22-31.

[29] A. L. Patterson, "The Scherrer Formula for X-Ray Particle Size Determination," *Physical Review*, Vol. 56, No. 10, 1939, pp. 978-982.

[30] "Standard Test Method for Knoop and Vickers Hardness of Materials," *Annual book of ASTM Standards*: *E* 384-10^{e2}

[31] S. P. Nikanorov, M. P. Volkov, V. N. Gurin, Y. A. Burenkov, L. I. Derkachenko, B. K. Kardashev, L. L. Regel and W. R. Wilcox, "Structural and Mechanical Properties of Al-Si Alloys Obtained by Fast Cooling of a Levitated Melt," *Materials Science and Engineering A*, Vol. 390, No. 1-2, 2005, pp. 63-69.

[32] H. G. Jiang, M. Ruhle and E. J. Lavernia, "On the Applicability of the X-Ray Diffraction Line Profile Analysis in Extracting Grain Size and Microstrain in Nanocrystalline Materials," *Journal of Materials Research*, Vol. 14, No. 2, 1999, pp. 549-559.

[33] G. K. Williamson and W. H. Hall, "X-ray Line Broadening from Filed Aluminium and Wolfram," *Acta Metallurgica*, Vol. 1, No. 1, 1953, pp. 22-31.

[34] M. A. Islam and Z. N. Farhat, "The Influence of Porosity and Hot Isostatic Pressure Treatment on Wear Characteristics of Cast and P/M Aluminum Alloys," *Wear*, 2011.

[35] K. Johnson, "Contact Mechanics," 1st Edition, Cambridge University Press, United Kingdom, 1987.

14

Microstructure, Corrosion, and Fatigue Properties of Alumina-Titania Nanostructured Coatings

Ahmed Ibrahim[1*], Abdel Salam Hamdy[2]

[1]Department of Mechanical Engineering, Farmingdale State College, Farmingdale, New York, USA; [2]Max Planck Institute of Colloids and Interfaces, Am Mühlenberg, Germany.

ABSTRACT

Air Plasma spray process was used to deposit a conventional and nanostructured Al_2O_3-13 wt% TiO_2 coatings on a stainless steel substrates. Morphology of the powder particles, microstructure and phase composition of the coatings were characterized by XRD and SEM. Potentiodynamic polarization tests and Electrochemical Impedance Spectroscopy (EIS) were used to analyze the corrosion of the coated substrate in 3.5% NaCl solutions to determine the optimum conditions for corrosion protection. The fatigue strength and hardness of the coatings were investigated. The experimental data indicated that the nanostructured coated samples exhibited higher hardness and fatigue strength compared to the conventional coated samples. On the other hand, the conventional coatings showed a better localized corrosion resistance than the nanostructured coatings.

Keywords: *Nanostructured Coatings, Alumina-Tiania, Fatigue Strength, Corrosion Resistance*

1. Introduction

The plasma-sprayed Al_2O_3 coatings have been extensively used in many applications due to their thermal, chemical and mechanical stability. The Al_2O_3 phase is characterised by the highest chemical resistance among all oxides, good heat and electric insulations, high hardness and wear resistance, etc [1].

Plasma sprayed Al_2O_3–TiO_2 (AT-13) coating is one of the most important coatings for many industrial applications [1-6]. It provides a dense and hard surface coating which is resistant to abrasion, corrosion, cavitation, oxidation and erosion. AT-13 has been used for wear resistance, electrical insulation, thermal barrier applications etc. several researchers reported that the Al_2O_3–TiO_2 coating containing 13 wt.% of TiO_2 showed the most excellent wear resistance among the Al_2O_3–TiO_2 ones [2-6].

Nanostructured materials are one of the highest profile classes of materials in science and engineering today, and will continue to be well into the future. Development of nanostructured ceramic coatings has become an important research area mainly due their interesting chemical, physical, and mechanical properties. For example, nanostructured AT-13 ceramic coatings show much higher wear resistance than conventional AT-13 coatings [3-7].

Fatigue and corrosion resistance are important properties for many coatings selected for critical applications. Plasma-sprayed Al_2O_3 coatings are often used in corrosion-resistant applications [8,9]. Because of their lamellar structure, ceramic coatings usually are characterized by a relatively high open porosity and incomplete bonding between lamellae, which are detrimental when the coatings have to perform in an aggressive environment. The porosity allows a path for electrolytes from the outer surface to the substrate [10,11].

It is widely recognized that thermal spray coatings can significantly influence the fatigue strength of coated components [12-14].

This paper presents the findings of a research on the microstructure, fatigue and corrosion behavior of thermally sprayed nanostructured and conventional AT-13 titania coatings.

2. Experimental procedure

2.1. Feedstock powders

Nanostructured (2613S) and conventional (ALO187) AT-13 feedstock powders employed in this study obtained from Inframat Corp. (Farmington, CT, USA) & Praxair, Indiana, USA. The morphologies of both AT-13 powders are shown in **Figure 1**. The nanostructured was agglomerated, spherical nanoparticles with high flowability and an average diameter of 30 μm. The conventional powder exhibited an angular and irregular morphology with size between 10 and 45 μm. Coatings were deposited using a Sulzer-Metco 9MB plasma torch under atmospheric conditions. Stainless steel cylindrical coupons were used as substrates. The typical spraying parameters for both conventional and nanostructured coatings are summarized in **Table 1**.

(a)

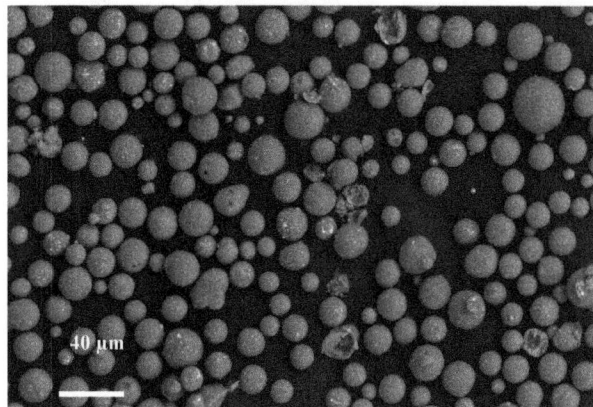

(b)

Figure 1. Morphologies of the AT-13 powders: (a) conventional; (b) nanostructured.

Table 1. Summary of the plasma spraying parameters.

Parameters	Conventional Coating	NanoCoating
Torch Metco	9MB	9MB
Current	470 A	500 A
Voltage	75 V	70 V
Argon Flow Rate Argon	80 SCFH	80 SCFH
Pressure Hydrogen	100 PSI	100 PSI
Pressure Powder Carrier	50 PSI	50 PSI
Flow	15-18 SCFH	15-18 SCFH
Spray Distance	4.5 inch	4 inch

2.2. Characterization of Coatings

The phase compositions of the as sprayed coatings were determined by X-ray diffraction (XRD) using a Philips X-ray diffractometer (Philips APD 3520). The microstructure of the as-sprayed coatings was examined by a LEO field emission scanning electron microscopy (SEM).

The microhardness measurements were conducted on the cross section of the as-sprayed coatings using Vickers Indentor. Microhardness values of the coatings were measured by digital hardness tester with load of 300 g on the cross-section of the polished samples.

2.3. Electrochemical Testing

2.3.1. Electrochemical Impedance

EIS technique was used to evaluate the electrochemical behaviour of the coated samples in 3.5% NaCl solution open to air and at room temperature for up to three weeks. A three-electrode set-up was used with impedance spectra being recorded at the corrosion potential Ecorr. A saturated calomel electrode (SCE) was used as the reference electrode. It was coupled capacitively to a Pt wire to reduce the phase shift at high frequencies. EIS was performed between 0.01 Hz and 65 kHz frequency range using a frequency response analyzer, FRA, (Autolab PGSTAT30, Eco-Chemie, The Netherlands). The amplitude of the sinusoidal voltage signal was 10mV.

2.3.2. Polarization Measurements

Linear polarization measurements were performed for the samples previously immersed for 30 minutes in 3.5% NaCl solution using Autolab PGSTAT30, Eco-Chemie, system. The scan rate was 0.05 mV/sec and the scan range was +/–20 mV with respect to the open circuit potential. The exposed surface area was 4 cm^2. All curves were normalized to 1 cm^2

2.4. Fatigue Testing

Rotating-beam fatigue testing was conducted on coated AISI low-carbon steel. The fatigue-testing machine is an RBF-200, rotating beam fatigue machine (Fatigue Dynamics Inc., Walled Lake, MI). The test specimen used in the fatigue testing was a 12.7 mm (1/2-inch) hourglass bar prepared according to ASTM E466 [15]. The nominal coating thickness was 100 μm (0.004 inches). The fatigue experiments were conducted at room temperature under a rotating beam fatigue and stress ratio of R = –1 configuration at a load frequency of 50 Hz. The surface of the specimen was prepared for coating by grit blasting with # 24 alumina, and no grinding was performed so as to not alter the surface roughness of the coatings. Fatigue life data generated in the fatigue tests were analyzed to determine the relationship between number of cycles to failure, N and probability of failure, P$_f$, for the samples tested.

3. Results and Discussion

3.1. Phase Composition of the Coatings

The XRD analysis of the AT-13 powders was confirmed in several previous studies [2-6] that both powders are prominently α-Al$_2$O$_3$ Rutile phase of TiO$_2$. The XRD analysis of the conventional and nanostructured coatings are shown in **Figure 2(a)** & **(b)**. The XRD patterns of the coatings show that most of α-alumina in the nanostructured powder converted into γ-Al$_2$O$_3$ after plasma spraying process, which was similar to that in the conventionally commercial powder. It is well established that γ-Al$_2$O$_3$ tends to be nucleated from the melt in preference to α-Al$_2$O$_3$ due to the higher cooling rate [4]. It can be seen from **Figure 2** that both nanostructured and conventional coatings mostly contained the γ-Al$_2$O$_3$ phase.

3.2. Microstructure of Coatings

The cross-sectional morphologies of the plasma sprayed coatings are shown in **Figure 3**. From the cross-sectional microstructures, it can be seen that both coatings consist of the lamella built up from the molten droplets impinging on the substrate. In case of the nanostructured coating, the interface between the steel substrate and the coating appears much stronger than that of the conventional coating. The "conventional" coating (**Figure 3(a)**) has a layered microstructure, typical of plasma sprayed coatings, which is the result of full melting (FM) of the ceramic feedstock powder and its solidification as "splats" on the substrate. The FM regions in the conventional coating consist of nanocrystalline γ-Al$_2$O$_3$ [1,17-20]. In all the coatings, some pores are observed, and splat boundaries are not clearly visible. A considerable amount (~16%) of partially melted regions (PM) is observed in the nanostructured coating (**Figure 3(b)**).

An SEM micrograph of the nanostructured coating is shown in **Figure 3(b)**. This coating shows a bimodal microstructure composed of the two regions where nanopowders were fully (FM) and partially melted (PM). The partially melted region was formed when TiO$_2$ was selectively melted because the temperature of nanopowders was not high enough during spray coating [2]. The partially melted (PM) rounded feature appears to consist of grains surrounded by a matrix phase, something similar to micro- structures of liquid-phase sintered materials [6].

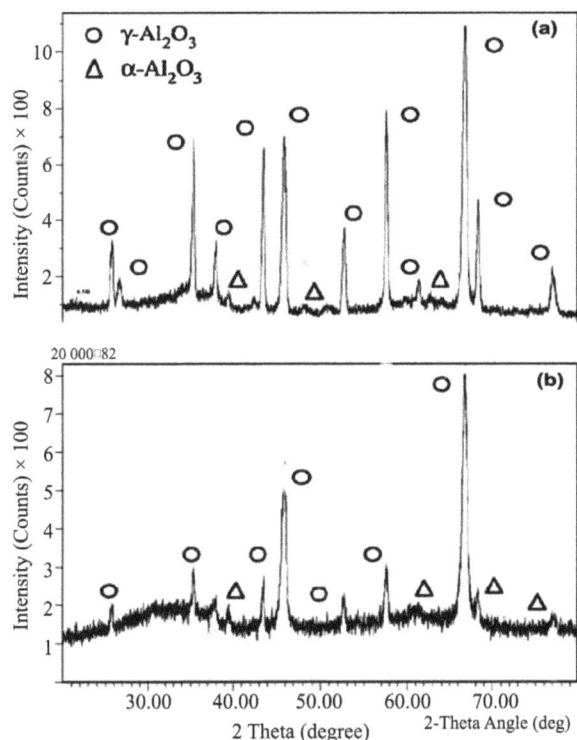

(a)

(b)

Figure 3. Cross-sectional morphologies of the as-sprayed AT-13 coatings: (a) conventional; (b) nanostructured.

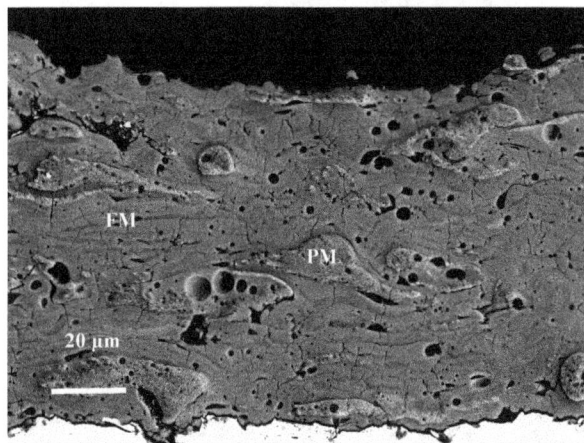

Figure 2. X-ray diffraction of the AT-13 sprayed coatings: (a) conventional coating; (b) nanostructured coating.

Figure 4 shows a high-magnification SEM image of the partially-melted (PM) region in the nanostructured coating. The partially-molten (PM) microstructural features consists mainly of submicron α-Al$_2$O$_3$ fine equiaxed grains surrounded by a TiO$_2$-rich amorphous phase.

3.3. Electrochemical Impedance Spectroscopy

EIS has been successfully applied to the study of corrosion systems for over thirty years and has been proven to be a powerful and accurate method for measuring corrosion rates especially for coatings and thin films. An important advantage of EIS over other laboratory techniques is the possibility of using very small amplitude signals without significantly disturbing the properties being measured. To make an EIS measurement, a small amplitude signal is applied to a specimen over a range of frequencies.

The expression for impedance is composed of a real and an imaginary part. If the real part is plotted on the Z axis and the imaginary part on the Y axis of a chart, we get a "Nyquist plot". However, Nyquist plots have one major shortcoming. When you look at any data point on the plot, you cannot tell what frequency was used to record that point. Therefore other impedance plots such as "Bode plots" are important to make a correct interpretation. In Bode plots, the impedance is plotted with log frequency on the x-axis and both the absolute value of the impedance ($|Z| = Z0$) and phase-shift on the y-axis. Unlike the Nyquist plot, the Bode plot explicitly shows frequency information.

In this work, coated samples were immersed in NaCl solution for 21 days. Nyquist and Bode plots have been used to evaluate the corrosion resistance of the nanostructured and conventional sprayed coatings.

The EIS results shown in **Figure 5** and **6** as well as the visual inspection of the tested samples suggested that the conventional sprayed samples have a strong surface resistance to chloride ion corrosion after three weeks of immersion in NaCl solution. **Figure 5** (Nyquist plot) showed that the surface resistance of the conventional coated samples is 2.0×10^4 Ω.cm^2 which is almost 10 times of that obtained from the nanostructured coated samples, Al-Ti N, (2.5×10^3 Ω.cm^2).

Figure 6 indicated that the pitting resistance of the conventional coated samples was generally improved which was confirmed by the relaxation of the impedance spectra. It seems that the presence of partially melted regions (PM) in the nanostructured coating which contain many micro- and/or nano-pores affect negatively the the pitting corrosion resistance. Pitting corrosion can easily be done through the micro- or nano-pores in the nanocoatings. Therefore, the number of pitting areas increased sharply in the nanocoated samples compared with the conventional.

3.4. Linear Polarization Measurements

Polarization measurements (**Figure 7**) were performed for the conventional and naocoatings after 30 minutes of

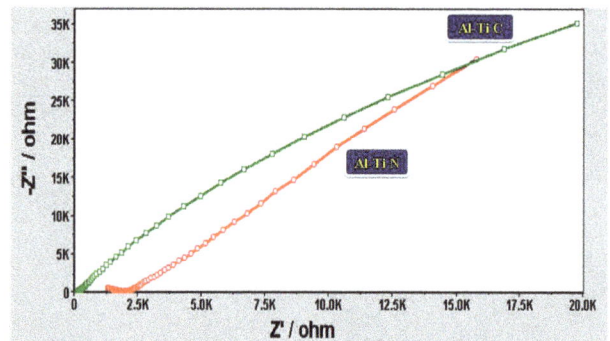

Figure 5. Electrochemical impedance spectroscopy (nyquist Plots) of alumina_titania coated samples: Al-Ti C (Conventional), Al-Ti N (Nanostructured).

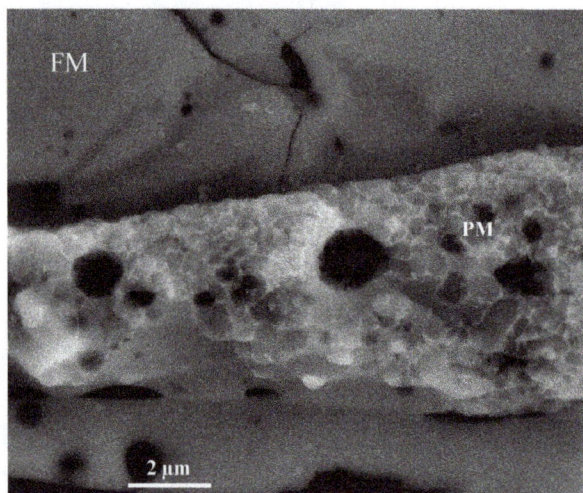

Figure 4. The partially melted region 'PM' consists of α-Al$_2$O$_3$ (dark) embedded in TiO$_2$-rich amorphous phase.

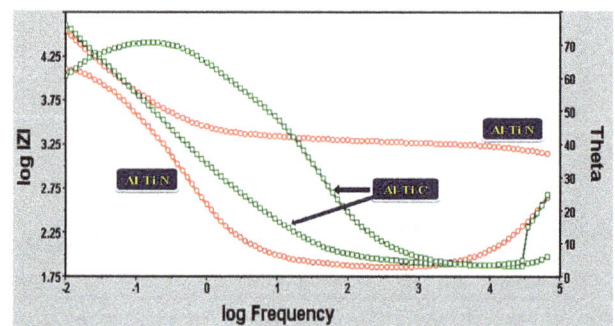

Figure 6. Electrochemical impedance spectroscopy (bode plots) of alumina-titania coated samples: Al-Ti C (Conventional), Al-Ti N (Nanostructured).

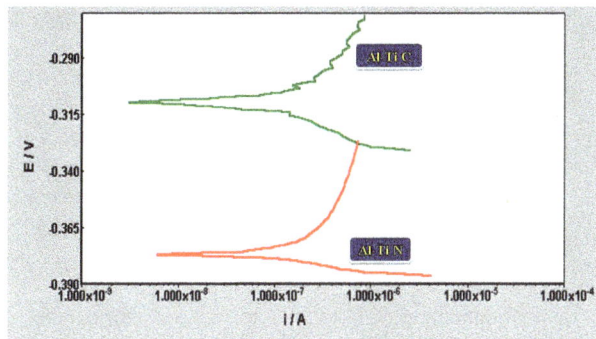

Figure 7. Polarization resistance measurments in NaCl solution.

immersion in NaCl solution. Conventional coatings samples showed a better polarization resistance than the nanocoated ones The corrosion potentioal of the conventional coated samples was shifted about 50 mV toward the passive direction compared with the nanocoated samples.

3.5. Hardness of the Coatings

Vickers microhardness measurements were performed under a 300 gf load for 15 s on the cross-sections of the coatings. A total of 10 microhardness measurements were carried out for each coating. The mean microhardness of conventional coating is about 840, which is lower than the mean microhardness of 965 achieved by the nanostructured coating. The observed hardness difference is believed to result partially from different coating microstructures and phase compositions, even though all these coatings have the same nominal chemical composition. Specifically, the distribution of TiO_2 in these Al_2O_3/TiO_2 coatings affects the coating hardness. It is interesting to notice that the nanostructured coating contains partially melted regions (PM) contain many micro- or nano-pores and thus these regions should have a relatively low microhardness.

3.6. Results of the Fatigue Tests

Fatigue life data generated in the fatigue tests were analyzed to determine the relationship between stress level, S, number of cycles to failure, N and probability of failure, P_f, for the samples tested. The failure probability [12], P_f, corresponding to the order number i is given by:

$$P_f = \frac{i}{n+1} \qquad (1)$$

Figure 8 shows the fatigue life (cycles) at 50% failure probability for the coated and uncoated specimens. The results indicate that the nanostructured alumina-titania coating exhibited higher fatigue lives compared to the conventional alumina-titania coating. The increase in the

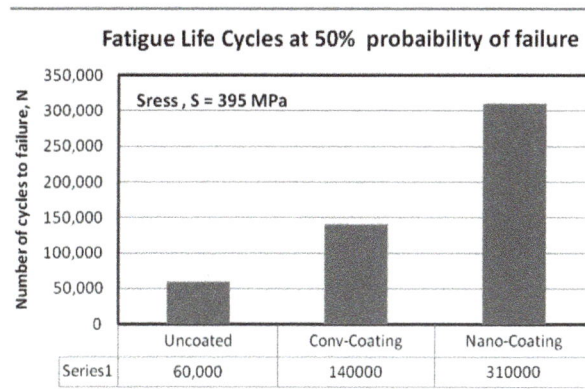

Figure 8. Fatigue life (cycles) for coated and uncoated specimens.

fatigue strength of nanostructured coatings can be related to the crack propagation resistance of these coatings. Several investigators have shown that the crack propagation resistance of nanostructured coatings is superior to that of the conventional coatings [2,19,20]. It is important to notice that the partially melted (PM) regions in the coating act as a crack arrest as seen in **Figure 3**. Another important factor that can play a major role in increasing the fatigue resistance is the interfacial toughness. Bansal *et al.* [20] have measured the interfacial toughness of the conventional and the nanostructured AT-13 plasma prayed coatings on steel substrates; the values were 22 and 45 $J \cdot m^{-2}$, respectively.

4. Conclusions

- The microstructure of the conventional coating consists primarily of fully-molten (FM) regions. The microstructure of nanostructured coating is a bimodal in nature and it consists of regions of (FM) mixed with partially-molten (PM) microstructural features.
- The presence of the partially melted (PM) region in the nanostructured coating act as a crack arrest and play a major rule in strengthening the crack propagation resistance.
- The significant increase in the fatigue strength of the nanostructured coatings compared to the conventional coating is attributed to the improvement in the crack propagation resistance resulted from the presence of the partially melted (PM) region.
- The presence of the partially melted regions (PM) in the nanostructured coating which contain many micro- and nano-pores affect negatively the pitting corrosion resistance. Therefore, conventional coatings showed a much better localized corrosion resistance than the nanocoatings.

REFERENCES

[1]   J. Iwaszko, "Surface Remelting Treatment of Plasma-Sprayed Al_2O_3 + 13 wt% TiO_2 Coatings," *Surface and Coatings Technology*, Vol. 201, No. 6, 2006, pp. 3443-3451.

[2]   E. H. Jordan, M. Gell and Y. H. Sohn, *et al.*, "Fabrication and Evaluation of Plasma Sprayed Nanostructured Alumina-Titania Coatings with Superior Properties," *Materials Science and Engineering A*, Vol. 301, No. 1, 2001, pp. 80-89.

[3]   E. Song, J. Ahn, S. Lee and N. Kim, "Effects of Critical Plasma Spray Parameter and Spray Distance on Wear Resistance of Al_2O_3 – 8 wt% TiO_2 Coating Plasma-Sprayed with Nanopowders," *Surface & Coatings Technology*, Vol. 202, No. 2, 2008, pp. 3625-3632.

[4]   M. Gell, E. H. Jordan, Y. H. Sohn, D. Goberman, L. Shaw and T. D. Xiao, "Development and Implementation of Plasma Sprayed Nanostructured Ceramic Coatings," *Surface and Coating Technology*, Vol. 146-147, No. 1, 2001, pp. 48-54.

[5]   A. Ibrahim, H. Salem and S. Sedky, "Excimer Laser Surface Treatment of Plasma Sprayed Al_2O_3 + 13 wt% TiO_2 Coatings," *Surface and Coating Technology*, Vol. 203, No. 2, 2009, pp. 3579-3589.

[6]   D. Goberman, Y. H. Sohn, L. Shaw, E. Jordan and M. Gell, "Microstructure Development of Al_2O_3 – 13 wt% TiO_2 Plasma Sprayed Coatings Derived from Nanocrystalline Powders," *Acta Materialia*, Vol. 50, No. 5, 2002, pp. 1141-1152.

[7]   H. Gleiter, "Nanostuctured Materials: Basic Concepts and Microstructure," *Acta Materialia*, Vol. 48, No. 1, 2000, pp. 1-29.

[8]   M. Wang and L. Shaw, "Effects of Powder Manufacturing Methods on Microstructure and WearPerformance of Plasma Sprayed Alumina-Titania Coaings," *Surface and Coating Technology*, Vol. 202, No. 1, 2007, pp. 34-44.

[9]   J. Zhang, Z. Wang, P. Lin, W. LU, Z. Zhou and S. Jiang, "Effect of Sealing Treatment on the corrosion resistance of Plasma-Sprayed $NiCrAl/Cr_2O_3$ – 8 wt% TiO_2 Coating," *Journal of Thermal Spray Technolog*, Vol. 20, No. 3, 2010, pp. 508-513.

[10] S. Liscano, L. Gil and M. H. Staia, "Effect of Sealing Treatment on the Corrosion Resistance of Thermal-Sprayed Ceramic Coatings," *Surface and Coatings Technology*, Vol. 188-189, No. 1, 2004, pp. 135-139.

[11] J. Creus, H. Idrissi, H. Mazille, F. Sanchette and P. Jacquot, "Corrosion Behaviour of Al/Ti Coating Elaborated by Cathodic Arc PVD Process onto Mild Steel Substrate," *Thin Solid Films*, Vol. 346, No. 1-2, 1999, pp. 150-154.

[12] A. Ibrahim and C. Berndt, "The Effect of High-Velocity Oxygen Fuel, Thermally Sprayed WC-Co on the High Cycle Fatigue (HCF) of Aluminum and Steel Thermally Sprayed with WC-Co," *Journal of Material Science*, Vol. 33, No. 2, 1998, pp. 3095-3100.s Science, Vol. 33, 12, 199

[13] A. Ibrahim and C. Berndt, "Fatigue and Deformations of HVOF Sprayed WC-Co Coatings vs. Hard Chrome Plating," *International Thermal Spray Conference*, Vol. 1, No. 1, 2003, pp. 377-380.

[14] H. J. C. Voorwald, R. C. Souza, W. L. Pigatin and M. O. H. Cioffi, "Evaluation of WC-17Co and WC-10Co-4Cr Thermal Spray Coatings by HVOF on the Fatigue and Corrosion Strength of AISI 4340 Steel," *Surface and Coatings Technology*, Vol. 190, No. 2, 2005, pp. 154-164.

[15] R. Ahmed and M. Hadfield, "Mechanisms of Fatigue Failure in Thermal Spray Coatings," *Journal of Thermal Spray Technology*, Vol. 11, No. 3, 2002, pp. 333-349.

[16] Y. Wang, S. Jiang, M. Wang, S. Wang, T. D. Xiao and P. R. Strutt, "Abrasive Wear Characteristics of Plasma Sprayed Nanostructured Alumina/Titania Coatings," *Wear*, Vol. 237, No. 1, 1999, pp. 176-185.

[17] R. S. Lima and B. R. Marple, "Process-Property-Performance Relationships for Titanium Dioxide Coatings Engineered from Nanostructured and Conventional Powders," *Materials & Design*, Vol. 29, No. 1, 2008, pp. 1845-1855.

[18] R. S. Lima, S. Dimitrievska , M. N. Bureau, B. R. Marple, A. Petit, F. Mwale and J. Antoniou, "HVOF-Sprayed Nano TiO_2-HA Coatings Exhibiting Enhanced Biocompatibility," *Journal of Thermal Spray Technology*, Vol. 19, No. 1-2, 2010, pp. 336-343.

[19] R.S. Lima and B. R. Marple, "Thermal Spray Coatings Engineered from Nanostructured Ceramic Powders for Structural, Thermal Barrier and Biomedical Applications: A Review," *Journal of Thermal Spray Technology*, Vol. 16, No. 1, 2007, pp. 40-63.

[20] P. Bansal, N. P. Padture, A. Vasiliev, "Improved Interfacial Mechanical Properties of Al_2O_3 – 13 wt% TiO_2 Plasmasprayed Coatings Derived from Nanocrystalline Powders," *Acta Materials*, Vol. 51, No. 10, 2003, pp. 2959-2970.

15

Electrochemical Characterization of Plasma Sprayed Alumina Coatings

Magdi F. Morks[1*], Ivan Cole[1], Penny Corrigan[1], Akira Kobayashi[2]

[1]CSIRO Division of Materials Science and Engineering, Clayton South, Victoria, Australia; [2]Joining and Welding Research Institute, Osaka University, Osaka, Japan.

ABSTRACT

Open circuit potential (OCP), potentiodynamic polarization, and electrochemical impedance spectroscopy (EIS) were employed to characterize the corrosion behavior of plasma-sprayed alumina-coated mild steel in 3.5 wt% NaCl solution. Alumina-coated steel showed higher OCP and lower corrosion current ($i_{corr.}$) compared with the steel substrate. However, localized corrosion probably occurs at the coat/steel interface when immersed in the corrosive media. The reason for that is the penetration of corrosive solution into the steel surface through the pores of accumulated alumina layers. The corrosion products (mainly iron oxides) accumulate inside the pores and on the coating surface. The presence of iron oxide slightly improved the corrosion resistance.

Keywords: *Plasma Spraying, Alumina Coating, Mild Steel, Corrosion, Polarization, EIS*

1. Introduction

Ceramic coatings generated by plasma spraying have been extensively used as anti-corrosion, abrasion and wear resistant layers for metallic structural components [1-3]. Such coatings are used by many industries such as automotive, mining and aerospace to protect internal components from corrosion and friction under severe condition such as temperature and corrosive materials. Pla-sma-sprayed (PS) alumina and alumina/titania ceramic coatings have been studied extensively because they are electrical and thermal insulators and improve the corrosion, wear, and erosion resistance of steel that has been configured [4-7].

The disadvantages associated with plasma-spraying of ceramic coatings are the rapid solidification of the flight particles and thermal stresses which propagate microcracks within the coatings and limit the coating thickness to a few hundred microns. Moreover, the pores initiated from unmelted and semi-melted particles cause defects in the thermal spray coatings. In addition, exposed metallic oxide powders can undergo phase changes, as in the case of the phase transition of plasma-sprayed ~Al_2O_3 to Al_2O_3 [8,9] due to the difficulty of nucleating stable -phase [10] under the rapid solidification rate associated with the plasma spraying process.

Plasma sprayed alumina coatings are corrosion-resistant; however, the porous structure of PS alumina coatings enables localized corrosion by penetration to the surface/coating interface of corrosive agents. In the case of an alumina/mild steel system, localized corrosion occurs and iron oxide is released by immersion in 3.5 wt% NaCl solution. The corrosion products at the interface could cause a failure of the whole coatings due to the decrease in adhesive strength.

In mining industries, PS alumina coatings provide excellent protection for steel tools and components against wear, abrasive and corrosion. However, the coatings defects such as cracks and pores usually cause localized corrosion in wet environment.

Pore sealing is one method to improve the corrosion resistance of the coatings. This can be performed by applying inhibitors or a phosphate conversion treatment. Heung *et al.* [11] found that the electrochemical impedance spectroscopy (EIS) is advanced technique to investigate the thermal barrier coatings contain metallic and ceramic layers.

In this work, the corrosion behavior of PS alumina coatings on mild steel was evaluated by means of OCP, polarization and EIS. The work will focus on the problem of corrosion in porous PS ceramic coatings and investigate it with electrochemical.

2. Experimental Method

2.1. Materials

Mild steel samples (50 × 50 × 3 mm) were grit-blasted

on one side to clean and roughen the surface and followed by ultrasonic cleaning using acetone to remove grease and dust. Alumina coating was applied on the prepared mild steel surface by plasma spray system (Bay-tate-Coaken Techno-Osaka-Japan) in air. The spray parameters are listed in **Table 1**. Argon was employed as the working gas as well as a carrier for the alumina particles into the plasma flame. Under plasma power of 25 kW, the alumina was sprayed continuously by moving the plasma gun, attached to a programmable robot, in front of the substrate. For all coatings, the spray distance was 100 mm.

2.2. Characterization

Cross-sectioned alumina coated steel was mounted in resin and polished with diamond paste (15, 9, 3, 1 m). Scanning electron microscopy (SEM) imaging and mapping, and energy-dispersive spectroscopy (EDS) was carried out using a Philips XL30 FESEM with a Link-ISIS X-ray analysis system (Oxford Instruments). EDS was carried out with a beam current of 5 kV. All samples were mounted using conductive carbon tape.

A JEOL JDX-3530 M X-ray diffractometer system was employed to analyze the phase structure of both the alumina feedstock and coatings. In the phase analysis, the radiation source was CuK; the operating voltage was 40 kV and current 40 mA.

A three-electrode corrosion cell, interfaced with a potentiostat (model HSV-100, Hokuto Denko Co., Japan), measured Tafel polarization plots. Electrochemical measurements were conducted at 25°C. Cathodic and anodic polarizations were recorded from -1.5 to -0.2V with a sweep rate of 0.5 mV/s in a 3.5 wt% NaCl solution. The electrochemical cell consisted of Calomel electrode (Hg/ Hg_2Cl_2) as the reference electrode, a Ti mesh (10 $\times$ 20 mm^2) as counter electrode, with the samples mounted in the substrate holder as working electrode. The area of the sample exposed to the electrolyte was 3.14 cm^2. Electrolyte solutions were prepared with analytical-grade NaCl and distilled water. The polarization curves were measured immediately after recording the open circuit potential (OCP) of the samples for 50 min in 3.5 wt% NaCl solution.

Table 1. Spray parameters used for PS alumina coatings.

Feeding mode	internal
Arc current (A)	500
Arc voltage (V)	50
Working gases	Ar, H$_2$
Working gases flow rate (l/min)	Ar:47.4, H$_2$:20%
Carrier gas flow rate	Ar:20%
Torch traverse speed (mm/s)	60 (16 passes)
Spray distance (mm)	100

3. Results and Discussion

3.1. Microstructure and Phase Structure

The PS alumina coating of thickness 500 - 600 m has lamella structure (**Figure 1**) formed by accumulation of molten alumina particles. The microstructure reveals the different coating defects such as pores, micro-cracks, and non-/semi-molten particles (white angular dots). The coating is well-adhered to the steel surface as there are no cracks observed at the interface.

From X-ray diffraction analysis (**Figure 2**), the starting alumina powder is mainly comprised of -alumina, while the main phase in the PS alumina coatings are due to -alumina, which is formed from -alumina phase transition at high plasma flame temperature.

3.2. Electrochemical Measurements

3.2.1. Open Circuit Potential
Figure 3 show the alumina-coated steel before and after corrosion test. The yellow color may be revealed to the corrosion products at the steel/coating interface that accumulated in the pores.

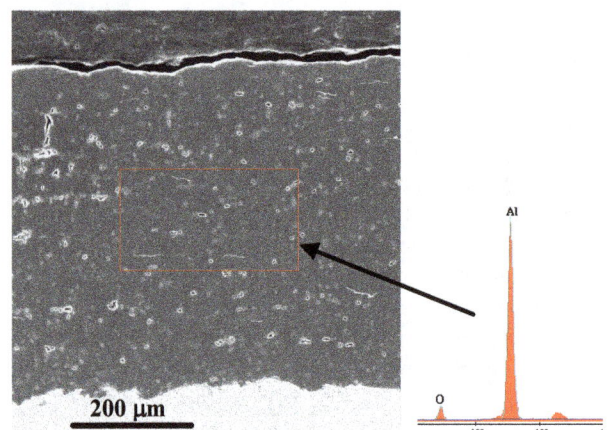

Figure 1. SEM and EDS of PS alumina coating.

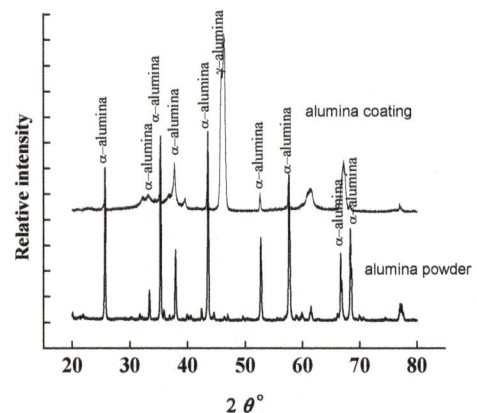

Figure 2. XRD patterns of alumina feedstock and coating.

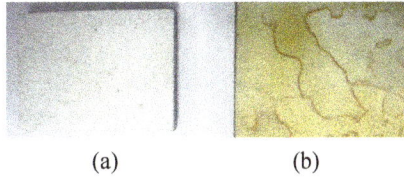

(a)   (b)

Figure 3. Photos for Al$_2$O$_3$ surface (a) before and (b) after corrosion.

Figure 4 shows the open circuit potential (OCP) curves of alumina coating and mild steel in 3.5 wt% NaCl solution at room temperature. Alumina coated steel exhibited a more positive OCP (–0.49 and –0.51 V) compared to mild steel (–0.61 V). The OCP of alumina-coated steel after corrosion testing was slightly shifted to more positive value (–0.49 V) due to the iron oxide formation inside the pores. Also the potential was in steady-state along the time of experiment. This confirms the higher corrosion resistance of alumina-coated steel compared to steel.

3.2.2. Tafel Polarization Curves

Tafel polarization curves of alumina-coated steel and mild steel are shown in **Figure 5**. The polarization curves of alumina showed a shift of corrosion potential to more positive value (–0.61 and –0.64 V) compared with mild steel (–0.92 V). The corrosion current (i_{corr}) was shifted to lower values in case of the alumina-coated steel, indicating a higher corrosion resistance. The alumina-coated steel exhibited slightly higher corrosion resistance when re-measured the same sample, mainly due to the accumulation of corrosion products particles, which may be iron oxide particles, inside the pores. The corrosion products particles blocked and sealed the pores from further penetration of corrosive solution. After corrosion test, the color of alumina coating was changed from white to rusty, due to the formation of oxides inside the pores and onto the alumina surface (**Figure 3**).

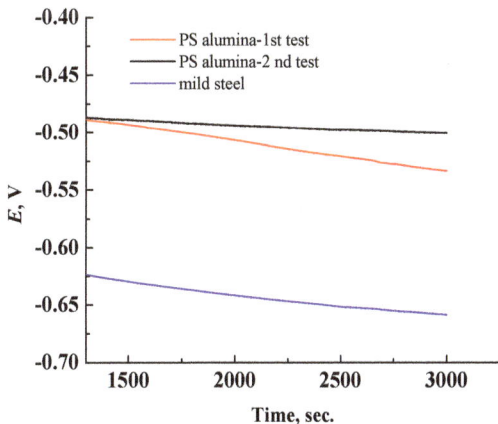

Figure 4. Open circuit curves for Al$_2$O$_3$ and mild steel.

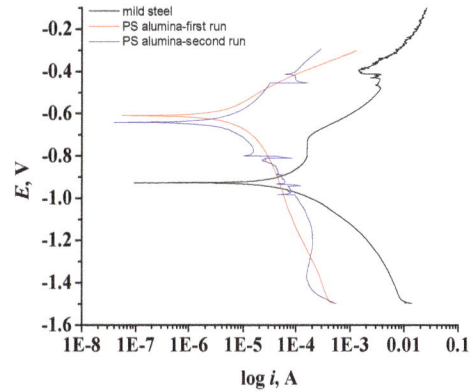

Figure 5. Tafel polarization curves for Al$_2$O$_3$ and mild steel.

Table 2 summarizes the OCP, E_{corr} and i_{corr} of coated steel and steel samples in 3.5% NaCl. The corrosion rate (CR) is calculated from the following equation [12] ASTM Standard G 102):

$$CR = K \frac{i_{corr}}{\rho A}(EW) \tag{1}$$

where: K (3272 mm/(amp-cm-year) is a constant for converting units; is density of corrode metal (steel: 7.85 g·cm^{-3}); EW (Fe: 27.93 g/equivalent) is the element equivalent weight and A is the sample surface area (3.14 cm^2).

By applying this equation on pure steel and alumina coated steel, the alumina coating was found to reduce the steel corrosion rate from 3.11 mm/year to 0.243 and 0.2 mm/year when applied alumina.

3.2.3. Electrochemical Impedance Spectroscopy

Impedance measurements of PS alumina were carried out at OCP conditions in 3.5% NaCl solution at room temperature. The impedance spectra obtained, are shown as complex impedance (Nyquist diagram) and Bode plots in **Figure 6**.

In the Nyquist diagram, the imaginary part of the impedance (Z") against real part (Z') is recognized in the form of a semicircle. The equivalent circuit that fits the impedance data consists of a resistor connected in series to a parallel connected resistor and CPE unit as shown in the equivalent circuit diagram in **Figure 7**.

Table 2 measured and calculated values of OCP, E_{corr} and i_{corr} of mild steel and coated alumina steel.

Sample ID	OCP (V)	E_{corr} (V)	i_{corr} (µA)
Mild steel	–0.64	–0.92	34
Alumina-1st	–0.49	–0.64	2.66
Alumina-2nd	–0.51	–0.61	2.2

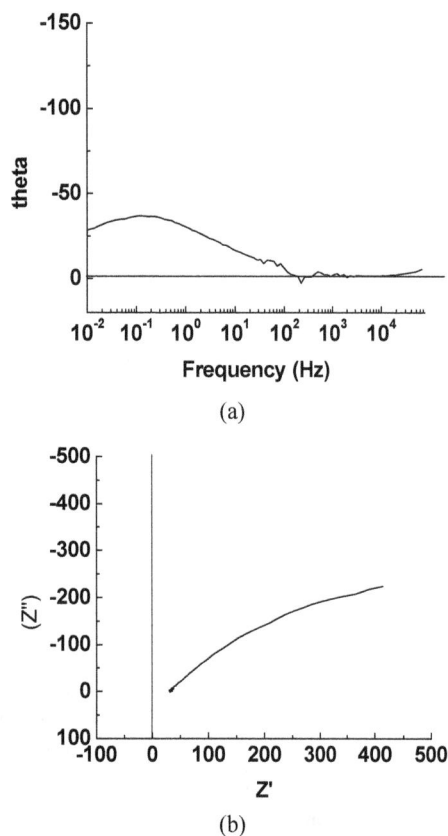

Figure 6. (a) Bode plot and (b) Nyquist plot for Al_2O_3 coating on mild steel.

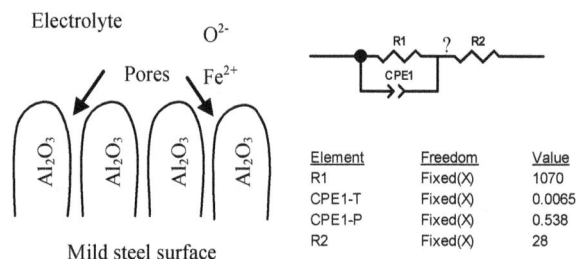

Element	Freedom	Value
R1	Fixed(X)	1070
CPE1-T	Fixed(X)	0.0065
CPE1-P	Fixed(X)	0.538
R2	Fixed(X)	28

Figure 7. Physical and equivalent circuit diagrams for Al_2O_3 coating on mild steel.

The impedance magnitude |z| (the ratio of the voltage difference amplitude to the current amplitude) increases as the frequency decreases. Over the frequency range of Max to 160 Hz the phase angle was nearly zero and the alumina layer behaves as pure resistor ($\theta = 0°$). For frequencies less than 160, the phase started to decrease from 0 towards –34.6° at 0.01 Hz.

In the equivalent circuit model, R1 represent the pores resistance with resistance equivalent to 1 KΩ. It is equivalent to the solution resistance inside the pores. R2 is the electrolyte solution resistance. CPE1 component is equivalent to double layer at the pore base which is

mainly formed from negatively (electrons) and positively (Fe^{++}) charges distribution inside the pores. The high electrical resistivity of alumina (10^{14} Ω·cm) was not represented in the model as it was considered as open circuit. The electrochemical redox reactions associated with iron corrosion are:

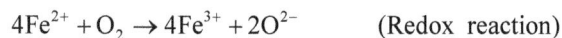

$$O_2 + 4e^- + 2H_2O \rightarrow 4OH^-$$ (cathodic reaction)

$$Fe \rightarrow Fe^{2+} + 2e^-$$ (anodic reaction)

$$4Fe^{2+} + O_2 \rightarrow 4Fe^{3+} + 2O^{2-}$$ (Redox reaction)

The corrosion current flow in and out of the coating pores and cracks which are full with electrolyte and avoid the alumina due to its high resistivity. The low impedance value (1070 Ω) of (R1) is for steel surface treated with porous alumina layer. The results reflect the low in corrosion resistance of PS alumina mainly due to the coating porosity.

4. Conclusions

1) The Tafel polarization curves demonstrated an improvement in corrosion resistance of mild steel by applying PS alumina coating.
2) Localized corrosion is likely to happen at the interface due to the porous structure of alumina layer.
3) The equivalent circuit showed the low impedance value of alumina coated mild steel due to the porous structure of alumina layer.
4) The corrosion current is only flowing in and out of the coating pores and cracks avoiding the alumina barrier due to its high resistivity.

5. Acknowledgements

The authors acknowledge the financial support from CSIRO, CMSE-Clayton.

REFERENCES

[1] M. F. Morks, Yang Gao, N. F. Fahim and F. U. Yingqing, "Microstructure and Hardness Properties of Cermet Coating Sprayed by Low Power Plasma," *Materials Letters*, Vol. 60, No. 1, 2006, pp. 1049-1053.

[2] M. F. Morks, Yang Gao, N. F. Fahim, F. U. Yingqing and M. A. Shoeib, "Influence of Binder Materials on the Properties of Low Power Plasma Sprayed Cermet Coatings," *Surface and Coating Technology*, Vol. 199, No. 1, 2005, pp. 66-71.

[3] M. F. Morks, A. Ibrahim and M. Shoeib, "Comparative study of Nanostructured and Conventional WC-Co Coatings," Proc. of The International Thermal Spray Conference, Osaka/Japan, ITSC May 2004.

[4] B. Pavitra, P. Nitin Padture and V. Alexandre, "Improved Interfacial Mechanical Properties of Al_2O_3 – 13 wt% TiO_2

Plasma-Sprayed Coatings Derived from Nanocrystalline Powders," *Acta Materialia*, Vol. 51, No. 10, 2003, pp. 2959-2970.

[5] V. P. Singh, A. Sil and R. Jayaganthan, "A Study on Sliding and Erosive Wear Behaviour of Atmospheric Plasma Sprayed Conventional and Nanostructured Alumina Coatings," *Materials and Design*, Vol. 32, No. 2, 2011, pp. 584-591.

[6] O. Tingaud, P. Bertrand and G. Bertrand, "Microstructure and Tribological Behavior of Suspension Plasma Sprayed Al_2O_3 and Al_2O_3–YSZ Composite Coatings," *Surface & Coating Technology*, Vol. 205, No. 4, 2010, pp. 1004-1008.

[7] C.-J. Li, G.-J. Yang and A. Ohmori, "Relationship between Particle Erosion and Lamellar Microstructure for Plasma-Sprayed Alumina Coatings," *Wear*, Vol. 260, No. 2, 2006, pp. 1166-1172.

[8] M. Vardelle and J. L. Besson, "Alumina Obtained by Arc Plasma Spraying: A Study of the Optimization of Spraying Conditions," *Ceramics International*, Vol. 7, No. 2, 1981, pp. 48-54.

[9] R. McPherson, "title of the article," *Journal of Material Science*, Vol. 8, 1973.

[10] R. McPherson, "On the Formation of Thermally Sprayed Alumina Coatings," *Journal of Material Science*, Vol. 15 , No. 31, 1980, pp. 41-49.

[11] R. Heung, X. Wang and P. Xiao, "Characterisation of PSZ/Al_2O_3 Composite Coatings Using Electrochemical Impedance Spectroscopy," *Electrochimica Acta*, Vol. 51, No. 8-9, 2006, pp. 1789-1796.

[12] ASTM Standard G 102

16

Effect of Conduction Pre-heating in Au-Al Thermosonic Wire Bonding

Gurbinder Singh[*], **Othman Mamat**

Department of Mechanical Engineering, Universiti Teknologi Petronas, Malaysia Bandar Seri Iskandar, Tronoh, Perak, Malaysia.

ABSTRACT

This paper presents the recent study by investigating the vital responses of wire bonding with the application of conduction pre-heating. It is observed through literature reviews that, the effect of pre-heating has not been completely explored to enable the successful application of pre-heating during wire bonding. The aim of wire bonding is to form quality and reliable solid-state bonds to interconnect metals such as gold wires to metalized pads deposited on silicon integrated circuits. Typically, there are 3 main wire bonding techniques applied in the industry; Thermo-compression, Ultrasonic and Thermosonic. This experiment utilizes the most common and widely used platform which is thermosonic bonding. This technique is explored with the application of conduction pre-heating along with heat on the bonding site, ultrasonic energy and force on an Au-Al system. Sixteen groups of bonding conditions which include eight hundred data points of shear strength at various temperature settings were compared to establish the relationship between bonding strength and the application of conduction pre-heating. The results of this study will clearly indicate the effects of applied conduction pre-heating towards bonding strength which may further produce a robust wire bonding system.

Keywords: *Conduction Pre-Heating, Intermetallic Coverage, Shear Strength, Thermosonic Wire Bonding*

1. Introduction

The revolution that began over a hundred years ago with the discovery of electricity and the rapid electrification of your society has transformed the means of communication amongst many others. The electronics industry today is driving for speedier data processing and more efficient data acquisition and transmission. With these requirements, means and methods of creating a robust and reliable bonding within an electronic chip is gaining importance.

From the three typical, wire bonding techniques mentioned, thermosonic wire bonding has been prevalent in the application of solid state inter-connect technology. This bonding method may also be known as diffusion bonding. In the process of making an interconnection, two wire bonds are formed. The first bond involves the formation of a ball with electric flame off (EFO) process.

Factors such as ultrasonic energy, temperature, and pressure may influence the quality of bonding quality [1,2]. In addition, diffusion bonding processes depend for their success on a combination of three factors. The first being the absence of contamination at the mating surfaces; secondly, the ability of at least one component to undergo sufficient plastic flow in order to develop complete contact across the interface between the two joining metals and lastly, sufficient time for diffusion to occur in the interface region in order to eliminate micro-structural instability and establish adequate bonding strength [3].

Upon application of the primary factors to form an intermetallic coverage (IMC) that makes the connection on the bond pad of a die, the wire is then lifted to form a loop and placed in contact with the desired bond area of a leadframe to form a wedge bond. In this process, the bonding temperature is one of the main bonding parameters which play an important role in the bonding [4]. Essentially, different temperature will lead to different bonding output response as different temperature conditions mean different bonding environment. In previous studies, a parabolic relationship between temperature and strength has been determined. Too low or too high temperature during bonding can lead to unsuccessful bonding or low bonding strength [5,6].

Although many studies about temperature effect in wire bonding technology has been carried out, however research findings on conduction pre-heating for the application on wire bonding is nonexistent. Therefore it is worth investigating the effects of conduction pre-heat application on thermosonic wire bonding while keeping other factors at constant. The results of this study will clearly indicate the effects of applied pre-heat towards bonding strength which may further produce a robust wire bonding system. Sixteen groups of bonding conditions which includes eight hundred data points of shear strength at various temperature settings were compared to establish the relationship between bonding strength and the application of conduction pre-heating.

2. Experimental Details

This study was carried out using a 138 KHz Thermosonic wire bonder. The bonding temperature was tuned to produce different thermosonic bonding process while all other bonding parameters such as bonding ultrasonic power, bonding force and time were as constant. The bonding parameters are listed in **Table 1**.

The diameter of gold wire (99.99% Au) used is 1 mil. The gold wire bonding was performed on a 16 mil × 16 mil die size with 1um thickness metal composition Al-99.5%, Cu-0.5%. The bond pad opening is 4 mil × 4 mils and bond pad pitch is 5 mils. The following is an outline of the gold wire bonding procedure which was carried out. Refer to **Figure 1**.

(i)   The Au wire at the capillary tip was melted to produce the Free air ball (FAB) using discharge from electric flame off (EFO)

(ii)   The capillary is lowered and the free air ball is compressed onto the Al pad for bonding.

(iii) The capillary is lifted and the Au wire is connected to the lead to produce a wedge bond.

(iv)   The Au wire is cut. (repeat step 1-4)

The bonding temperature was measured by a K-type

thermocouple sensor with a measurement range of 0°C - 500°C. A total of 16 runs were performed with bonding temperature range of 40°C to 340°C with a step of 20°C. The bonding experiments were consistently repeated for 50 times under each testing condition with heater block without conduction pre-heating and with conduction pre-heating for statistical analysis.

Upon completing the samples, the output response of the thermosonic bonding with and without conduction pre-heating were assessed using Royce 550 wire bond tester. The destructive shear strength between the gold ball and substrate is a common judgment for bondability [7,8].

To further analyze the bondability, intermetallic coverage assessment was also carried out using fuming potassium hydroxide, KOH. Cross sectioning using grinding and polishing machines were used to enable analysis on gold ball bond. Samples were observed using an optical microscope and scanning electron microscope (SEM). Thickness and diameter of the ball bond profile and Intermetallic growth was measured using micrometer scale attached to the optical microscope and the scope was calibrated using the standards provided with the scope.

3. Results and Discussion

3.1. Effect of Conduction Pre-heating on Bonding Strength in Au-Al Wire Bonding System

The result of this study showed, there was significant difference in the bonding strength response with the usage of conduction pre-heating as compared against the conventional non conduction pre-heated samples. This can be seen from the one-way analysis of ball shear strength by heater block design, as shown in **Figure 2**.

The effect of heat on wire bonding was measurable not only against ball shear strength but also by the representation

Table 1. Bonding Parameters.

US Power	Bond Force (gm)	Bond Time (ms)	Temperature (°C)
48	28	15	40°C - 340°C

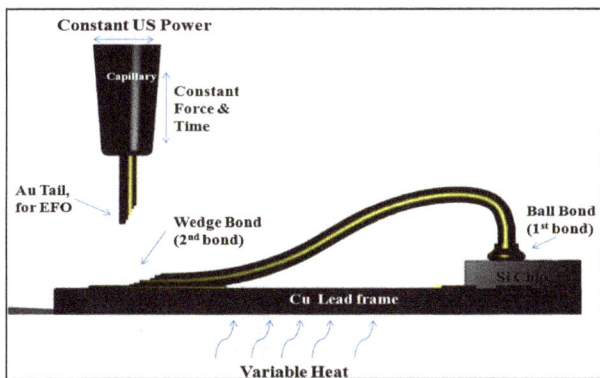

Figure 1. Thermosonic wire bonding setup.

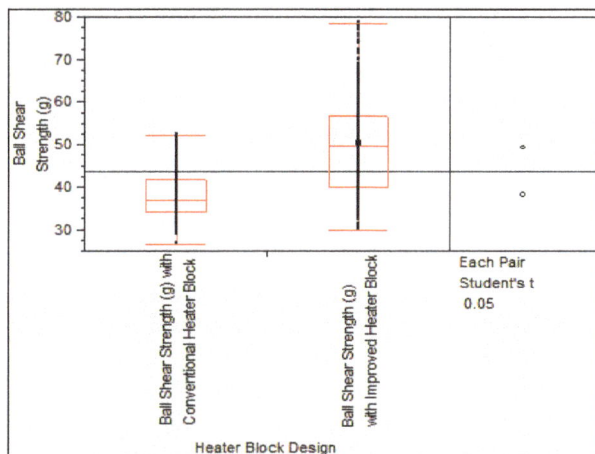

Figure 2. Ball shear strength by heater block design

of intermetallic coverage. IMC which is denoted by the darker colored area is the result of diffusion in between Au and Al. The metallic properties contained in this area consist of AuAl.

It was observed that, by using the conventional non conduction pre-heat block, too low or too high temperature may result in unfavorable intermetallic coverage. However, by using the conduction pre-heat block, it was able to produce homogeneous intermetallic coverage at all assessed levels. This observable fact can be further explained by the following related theories [9-11]

    a.   The presence of oxide and other forms of contamination may not be able to be removed with inadequate heat and may impede diffusion between the bonding metals. This may lead to unsuccessful bonding or lower bonding strength with poor IMC.

    b.   The application of appropriate and adequate temperature will soften the metal surface and accelerate the diffusion between the bonding systems.

As such, it was evident that, adequate heat supplied via conduction pre-heating is able to deliver improved intermetallic coverage. This can be seen by the results of intermetallic coverage provided in **Table 2**.

3.2. Effect of Conduction Pre-heating on IMC Voiding in Au-Al Wire Bonding System

Results of this study showed significant difference in the formation of voids with respect to different temperature levels applied with and without conduction pre-heating. From the study, we are able to understand that the characteristics of ball bond are able to be altered physically. Research carried out previously with the application of short-time Fourier transform (STFT) to input/ output power of an ultrasonic transducer is able to deduce this phenomena [12,13].

The cross sectioning image in **Figure 3** shows the difference in IMC growth with the application on non conduction pre-heat heater block and with conduction pre-heat heater block. At 180°C the IMC growth seems to be thicker (3um) with the application of conduction pre-heat. With the application of pre-heat, the IMC formation is homogeneous and there are no voids observed.

Results of study at temperature 220°C in **Figure 4** shows; the IMC thickness has grown to 3.1um and 3.4 respectively with the usage of non conduction pre-heat and pre-heat heater block. The IMC growth seen at this stage shows obvious voiding with the application of non pre-heat heater block. The IMC formation with the usage of pre-heat heater block remains to be homogeneous and without voids.

The result of this study clearly shows, the application

Table 2. Comparison of IMC

Temperature (°C)	IMC - Non Pre-heat	Approx. IMC (%)	IMC - Pre- Heat	Approx. IMC (%)
40		50		60
60		50		60
80		50		60
100		50		60
120		60		70
140		70		70
160		80		80
180		75		90
200		75		90
220		80		90
240		80		100
260		75		100
280		65		100
300		65		100
320		65		100
340		65		100

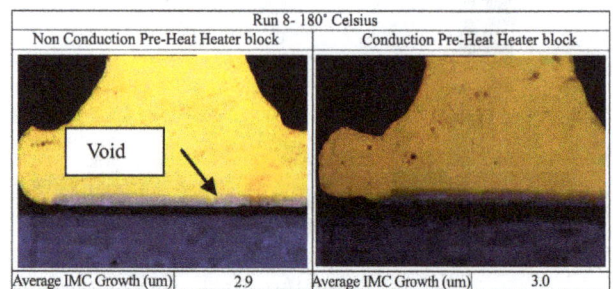

Run 8- 180° Celsius	
Non Conduction Pre-Heat Heater block	Conduction Pre-Heat Heater block
Void	
Average IMC Growth (um) 2.9	Average IMC Growth (um) 3.0

Figure 3. IMC growth at 180°C.

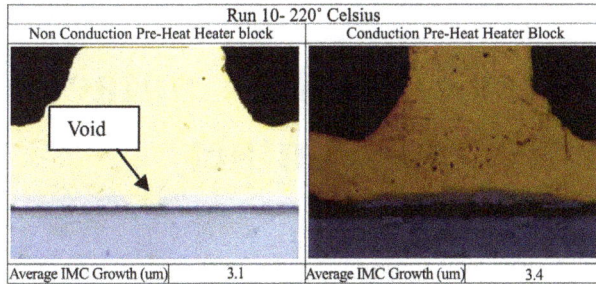

Figure 4. IMC growth at 220°C.

of conduction pre-heat does have positive impact towards improved IMC formation. Previous theories (Long *et al.*, 2008) have concluded, with the usage of adequate heating environment, the impedance of piezo-transducer is high and the consumed power of piezo-transducer is low. This effect may exert adequate physical bonding power to the bonded aluminum interface and result in IMC formation which is homogeneous and without voids.

4. Conclusions

From the study and measurements carried out, the effect of conduction pre-heating in Au-Al thermosonic wire bonding was investigated. It is clearly revealed that, adequate heat supplied via conduction pre-heating is able to deliver improved intermetallic coverage. This was evident in the significantly increased ball shear strength values and the improved intermetallic formation. In addition, high optical magnification on the cross sectioned samples also showed an improvement in the area of void formation which is critical in ensuring good bondability.

It is concluded, the application of conduction pre-heating unaccompanied by other bonding factors, has an effect in Au-Al thermosonic wire bonding. The objective of this study has been achieved. In future, further in depth studies to assess the reliability of parts with conduction pre-heating on bonding surfaces should be explored for enhanced wire bondability.

REFERENCES

[1]   J. Antony, "Improving the Wire Bonding Process Quality Using Statistically Designed Experiment," *Journal of Microelectronics*, Vol. 30, No. 2, 1999, pp. 161-168.

[2]   Z. N. Liang and F. G Kuper, "A Concept to Relate Wire Bonding Parameters to Bondability and Ball Bond Reliability," *Microelectronics Reliability*, Vol. 38, No. 6-8, 1998, pp. 1278-1291.

[3]   D. Brandon and W. Kaplan, "Joining Processes an Introduction," John Wiley and Sons, New York, 1997.

[4]   S. J. Hu., G. E. Lim and T. L. Lim, "Study of Temperature Parameter on the Thermosonic Gold Wire Bonding of High-Speed CMOS," *IEEE Journal of Transactions on Components, Hybrids and Manufacturing*, Vol. 14, No. 4, 1991, pp. 855-858.

[5]   Y. X. Wu, Z. L. Long, H. Lei and Z. Jue, "Temperature Effect in Thermosonic Wire Bonding," *Journal of Transactions on Non-ferrous Materials Society*, Vol. 16, No. 3, 2006, pp. 618-622.

[6]   Z. L. Long, H. Lei, Y. X. Wu and Z. Jue, "Study of Temperature Parameter in Au-Ag Wire Bonding," *IEEE Journal of Transactions on Electronics Packaging Manufacturing*, Vol. 31, No. 3, 2008, pp. 221-226.

[7]   H. Lei, F. L. Wang and W. H. Xu, "Bondability Window and Power Input for Wire Bonding," *Journal of Microelectronics Reliability*, Vol. 46, No. 2-4, 2008, pp. 610-615.

[8]   N. Srikanth, S. Murali and Y. M. Wong, "Critical Study of Thermosonic Copper Ball Bonding," *Thin Solid Flims*, Vol. 462-463, 2004, pp.339-345.

[9]   M. Y. Li, H. J. Ji and C. Wang, "Interdiffusion of Al-Ni System Enhance by Ultrasonic Vibration at Ambient Temperature," *Journal of Ultrasonics*, Vol. 45, No. 1-4, 2006, pp. 61-65

[10]   D. S. Liu and C. Y. Ni, "A Thermo Mechanical Study on the Electrical Resistance of Aluminum Wire Conductors," *Microelectonics Reliability*, Vol. 42, No. 3, 2002, pp. 367-374.

[11]   S. Murali, "Formation and Growth of Intermetallics in Thermionic Wire Bonds: Significance of Vacancy-Solute Binding Energy," *Alloys Compounds*, Vol. 426, No. 1-2, 2006, pp. 200-204.

[12]   L. Cohen, "Time-Frequency Analysis," Prentice-Hall, Eaglewood Cliffs, New Jersey, 1995.

[13]   D. Zhang and S. Ling, "Monitoring Wire Bonding via Time-frequency Analysis of Horn Vibration," *IEEE Transactions Electronics Packaging Manufacturing*, Vol. 26, No. 3, 2006, pp. 216-220.

17

Finite Element Analysis of Elastic-Plastic Contact Mechanic Considering the Effect of Contact Geometry and Material Properties

Abodol Rasoul Sohouli[*], Ali Maozemi Goudarzi, Reza Akbari Alashti

Department of Mechanical Engineering , Babol Noshirvani University of Technology, Babol.

ABSTRACT

Each surface of roughness has different shape of asperity which is modeled with various shapes of analytical models. In this paper, the differences among various models of shape of asperity investigate using the Finite Element Method (FEM) and various analytical models. The contact stresses in rough surfaces are calculated analytically using various asperity shape models. Finite element analysis is also carried out assuming three types of material properties namely, the linear, the elastic-perfect plastic and the elastic-nonlinear hardening. The analytical results are compared with the results obtained by the finite element method. The results illustrate for using a deterministic approach which the numerical models are suitable. In hertz model, the result of force is very big in interface of causing deformation plastic, while Model Zhao has almost same result with FEM nonlinear property model. It is observed that the results obtained from Zhao's model are generally in a better agreement with the results obtained from various finite element models especially in elastic-plastic and plastic zones, hence it may be concluded that Zhao's model can be used for analyzing the rough surfaces in contact mechanics.

Keywords: *Contact Mechanics; Micro-Contact; Finite Element Method (FEM); Zhao's Model; Chang's Model*

1. Introduction

The rough contact problem has been studied for many years as it is critically important to understand the tribological phenomenon such as friction, wear, contact fatigue, and sealing. Engineering surfaces have roughness and even highly polished surfaces possess some degrees of Roughness. When two engineering surfaces are pressed together, contact occurs at the peaks of the Surfaces where the contact pressure and subsurface stress can be extremely high, often causing plastic deformation of these spots. The contact between a deformable half-space and a rigid sphere was first solved by Hertz in 1896 [1] but this model only considers the elastic contact and disdains the effects of the roughness and the effects of the plasticity, and the real contact area is unvalued.

Greenwood and Williamson [2] pioneered the study of frictionless contact between a hemisphere and a rigid flat (the GW model) applied the Hertz contact solution to model an entire contact surface of elastic asperities. To supplement the GW model, many elastic-plastic asperity models have been devised.

The study of the deformation behavior of contact asperities and the accurate modeling of rough surfaces is important for understanding contact problems. Several theories can be applied to deal with the microcontacts of two contact surfaces [3-9].

Some new models were proposed to consider elastic-plastic contact that Chang's model [10] and Zhao's Model [11] are investigated in this paper.

In this work, some selected models are in particular reported, whose formulas, have been used to test the different models of roughness description developed in [12, 13]. The calculated contact zones and loads are compared with different numerical models.

The results illustrate for using a deterministic approach which the numerical models are suitable.

2. Models of Contact Mechanics

The contact model was first established by Heinrick Hertz in 1882, although the model considered the elastic contact only and the effects of the surface roughness and plasticity were not considered. Contact problems with rough surfaces have been modeled with stochastic techniques.

Finite Element Analysis of Elastic-Plastic Contact Mechanic Considering the Effect of Contact Geometry and Material Properties

113

The pioneering contribution to this field was done by Greenwood and Williamson, who developed a basic elastic contact model (GW model) [2]. The basic asperity GW model has been extended to other models, two of which are explained below.

2.1. Chang's Model

Chang's model or CEB model is an elastoplastic based model. This model is similar to the Greenwood and Williamson model but it considers the volume conservation [10]. Behavior of the contact is related to the interference, δ, between the two surfaces. If δ is smaller than it's critical value δc, i.e. $\delta < \delta_c$, the contact is assumed to be elastic, otherwise the contact is assumed to be plastic. The relations used for the elastic regime are based on the Hertz theory. Therefore, the contact area, Ae, and the elastic contact load, fe, are calculated as:

$$A_e = \pi a_e^2 = \pi R\delta \qquad (1)$$

$$a_e = (R\delta)^{\frac{1}{2}} \qquad (2)$$

$$f_e = \frac{4}{3}E^*R^{1/2}\delta^{3/2} \qquad (3)$$

where R and a_e are the radius of the sphere and the radius of the circular contact area respectively and E^* is the composite elasticity modulus which is expressed as follow:

$$E^* = \left(\frac{1-\nu_1^2}{E_1} + \frac{1-\nu_2^2}{E_2}\right)^{-1} \qquad (4)$$

the critical interference, δ_c, is introduced by the expression

$$\delta_c = R\left(\frac{\pi k H}{2E^*}\right) \qquad (5)$$

where the interference is, $\delta \geq \delta_c^2$, the relations are found to be.

$$A_p = \pi R\delta\left(2 - \frac{\delta_c}{\delta}\right) \qquad (6)$$

$$a_p = \left(R\delta\left(2 - \frac{\delta_c}{\delta}\right)\right)^{1/2} \qquad (7)$$

The contact load for each asperity under the plastic conditions is:

$$f_p = A_p k H \qquad (8)$$

This model is only used for fully elastic and fully plastic conditions and there are no relations developed for the elastic-plastic regime.

2.2. Zhao's Model

This model is an elastic-plastic asperity contact model for rough surfaces introduced by Zhao et al. [11]. In Zhao's model three regions of fully elastic, elastic plastic and fully plastic contacts are considered. In the elastic region, where $\delta < \delta_c$, the governing relations are similar to those of CEB model. When the interference, δ, between the two surfaces is increased i.e. $\delta \geq 54\delta_c$, the following equation is employed for calculation of the radius of fully plastic contact.

$$a_p = (2R\delta)^{1/2} \qquad (9)$$

Therefore, the surface of the contact area is:

$$A_p = \pi a_p^2 = 2\pi R\delta \qquad (10)$$

Hence, the contact load for each asperity is found to be:

$$f_p = A_p H = 2\pi R\delta H \qquad (11)$$

In the elastic-plastic region, where $\delta_c \leq \delta < 54\delta_c$, the relation between the mean pressure (P_m) and the interference (δ), in order to calculate the area of the contact and the elastoplastic load is as follow:

$$P_m = a_1 + a_2 \ln\left(\frac{\delta}{r}\right) \qquad (12)$$

where a_1 and a_2 are two constants to be determined and r is the contact radius of the asperity.

Finally, the following equations are used for the radius, R, area of the contact surface, A_{ep}, and the contact load, f_{ep}, respectively:

$$A_{ep} = \pi R\delta\left[1 - 2\left(\frac{\delta - \delta_c}{53\delta_c}\right)^3 + 3\left(\frac{\delta - \delta_c}{53\delta_c}\right)^2\right] \qquad (13)$$

$$a_{ep} = \left\{R\delta\left[1 - 2\left(\frac{\delta - \delta_c}{53\delta_c}\right)^3 + 3\left(\frac{\delta - \delta_c}{53\delta_c}\right)^2\right]\right\}^{1/2} \qquad (14)$$

$$f_{ep} = A_{ep}\left[H - H\left(1 - \frac{2}{3}k\right)\frac{\ln 54\delta_c - \ln\delta}{\ln 53}\right] \qquad (15)$$

3. Finite Element Model

The asperity is modeled in three geometrical shapes i.e. hemispherical, spherical and conical shapes against a rigid flat punch. All the three geometrical shapes have the same radius of 0.0005 m on the peak. In this analysis, axisymmetric 2-D models are used. A total of 11200 four-node bilinear axisymmetric elements were used to model each shape. The meshing of the model is refined near the region of contact in order to allow the curvature to be captured and accurately simulated during deformation. The contact force acting on each model is found from the reaction forces at the corresponding base nodes that retain the desired interference. The material property is modeled in two types, elastic-perfect plastic and nonlinear hardening. The elastic perfect plastic material is assumed to have yield stress, σ_Y, equal to 24.4 M Pa. The stress-strain relation of the elastic-nonlinear hardening material is shown in **Figure 1**.

A reference point is defined for the rigid flat punch and is allowed to move down as shown in **Figure 2**. The

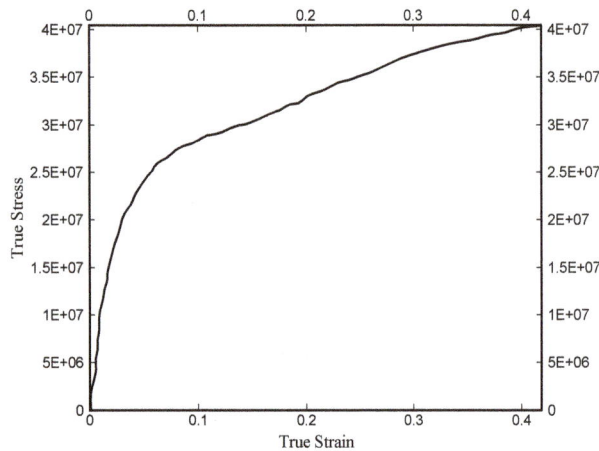

Figure 1. Nonlinear behavior property of material in ABAQUS.

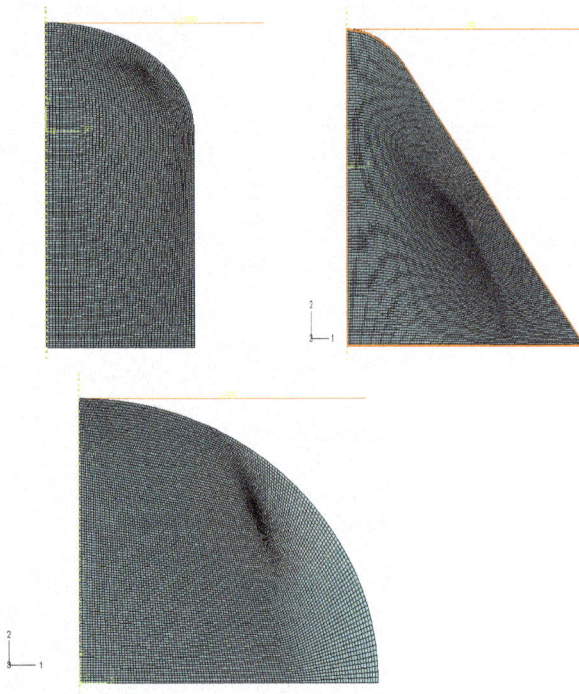

Figure 2. Finite element modeling of various shapes of asperity.

nodes at the base of all three models are fixed in all directions. The contact force acting on the model is found from the reaction forces on the reference point that retain the desired interference. Finite element meshes of the three model generated by ABAQUS are shown in **Figure 2**.

4. Results and Discussion

The results of the finite element models are presented for a variety of interferences. The contact area and the interference are normalized by the mean area of 0.785E-6 m^2 and critical interference of 0.462E-6 m respectively.

The contact forces obtained for all models are normalized by the Hertzian contact force of 0.011800474 N as obtained from the equation (3). In order to improve the computational efficiency, the radii of the asperity peaks are assumed to be constant. In **Figure 3**, the dimensionless contact force is plotted as a function of interfere.

At low interference ratios, the differences between the dimensionless contact forces in all modes are small because the contact is in the elastic state. It is interesting to note that the Chang's model predicts the least loads at big interferences. This is because of the fact that the Chang's model assumes that for all plastically deformed asperities the average pressure over the contact area is equal to kH. However, in the Zhao's model, it is assumed that for all plastically deformed asperities the average pressure over the contact area is equal to H, So the contact force for plastic zone predicted by the Zhao's model is bigger than that of the Chang's model. In the Zhao's model, transition from fully elastic deformation to fully plastic flow of the contacting asperity is modeled based on the contact mechanics theories in conjunction with the continuity and smoothness of variables across different modes of deformation. As shown in Fig. 3, the dimensionless contact force predicted by the Zhao's model is smoothly connected from the elastic state to the plastic state.

In **Figure 4**, the results of FEM analysis for the spherical asperity with both elastic-perfect plastic (EP PP) and elastic-nonlinear hardening properties (NLP) are compared with the Hertzian, Chang's model and Zhao's model. It is observed that the dimensionless contact force obtained by the FEM for both the elastic-perfect plastic and the elastic-nonlinear hardening materials follow the Zhao's model. Dimensionless contact forces for all three asperity shapes made of elastic-nonlinear hardening materials are shown and compared with three previous models in **Figure 5**. It is observed that the results obtained for both the spherical and conical asperity shapes follow the Zhao's model at all interferences while the result for the hemispherical shape follows the Chang's model. As can be seen from **Figure 5** that the results obtained for the

Finite Element Analysis of Elastic-Plastic Contact Mechanic Considering the Effect of Contact Geometry and
Material Properties

115

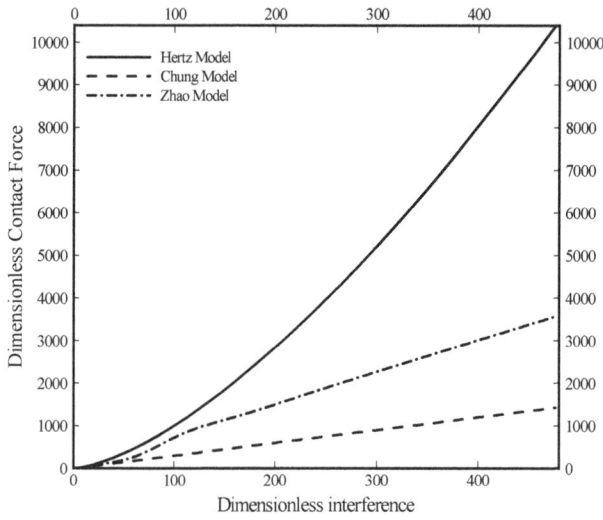

Figure 3. Dimensionless contact force versus interference for analytical models.

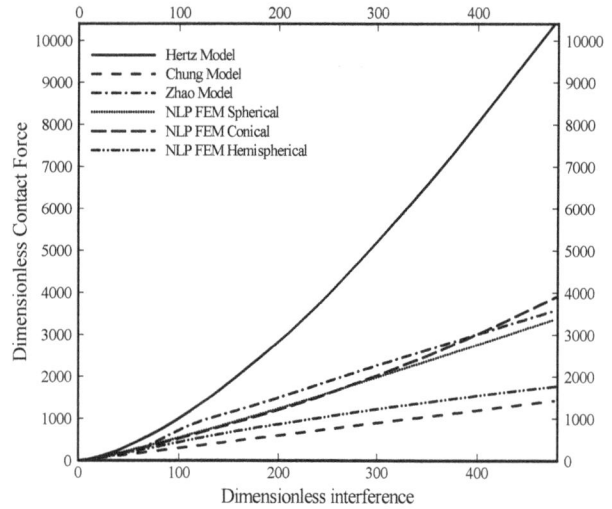

Figure 5. Dimensionless contact force versus interference for various models.

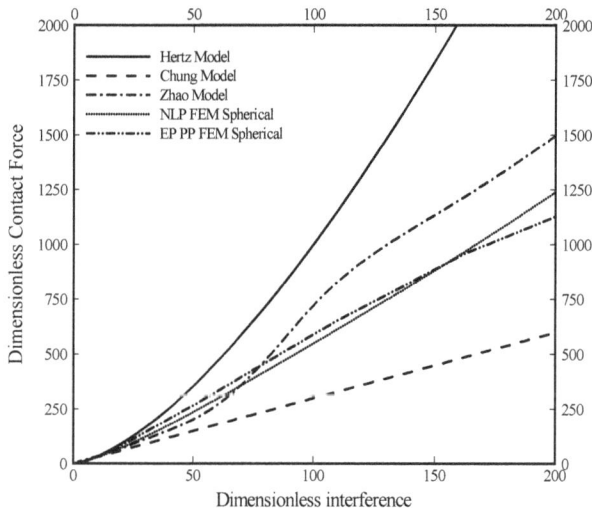

Figure 4. Dimensionless contact force versus interference for various models in medium interference.

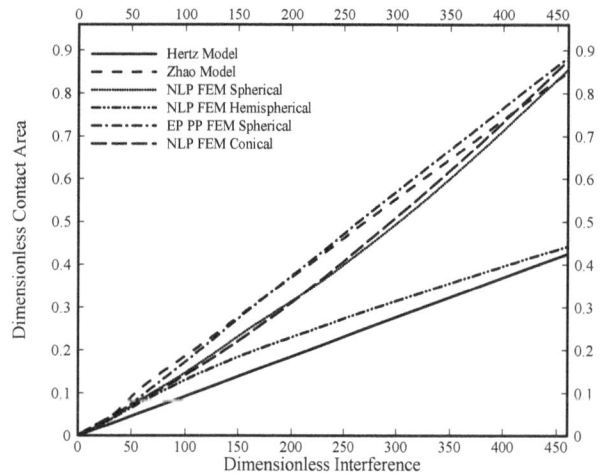

Figure 6. Dimensionless contact area versus interference for various models.

spherical asperity shape follow the Zhao's model the best. It is due to the fact that relationships drive on base assumption spherical shape. The results obtained for all models are in good agreement with the Hertzian model for low interferences.

The results of dimensionless contact areas obtained for all three asperity shapes made of elastic - nonlinear hardening material properties and spherical shape made of elastic-perfect plastic material for a variety of interferences are shown in **Figure 6**.

5. Conclusions

This study showed that dimensionless contact force obtained

for various asperity shapes are in good agreement with Hertzian model for low interferences. However the finite element results obtained for spherical and hemispherical asperity shapes are in very good agreement with the Zhao's model for nearly the whole range of interferences.

It is observed that the results obtained the results obtained for both the spherical and conical asperity shapes follow the Zhao's model at all interferences while the result for the hemispherical shape follows the Chang's model but it can be said that Zhao's model are generally in a better agreement with the results obtained from various finite element models especially in elastic-plastic and plastic zones, hence it may be concluded that Zaho's model can be used for analyzing the rough surfaces in contact mechanics.

REFERENCES

[1] K. L. Johnson, "Contact Mechanics," Cambridge University Press, Cambridge, 1985.

[2] J. A. Greenwood and J. H. Tripp, "The Contact of Two Nominally Flat Rough Surfaces," *Proceedings of the Institution of Mechanical Engineers*, Vol. 185, No. 1, 1971, pp. 624-633.

[3] J. I. McCool, "Comparison of Model for Contact of Rough Surfaces," *Wear*, Vol. 107, No. 1, 1986, pp. 37-60.

[4] L. Kogut and I. Etsion, "Elastic-Plastic Contact Analysis of a Sphere and a Rigid Flat," *Journal of Applied Mechanics*, Vol. 69, 2002, pp. 657-662.

[5] A. Majumdar and C. L. Tien, "Fractal Characterization and Simulation of Rough Surfaces," *Wear*, Vol. 136, No. 2, 1990, pp. 313-327.

[6] A. Majumdar and B. Bhushan, "Role of Fractal Geometry Inroughness Characterization and Contact Mechanics of Surfaces," *Journal of Tribology*, Vol. 112, No. 2, 1990, pp. 205-216.

[7] A. Majumdar and B. Bhushan, "Fractal Model of Elastic-Plastic Contact between Rough Surfaces," *Journal of Tribology*, Vol. 113, No. 1, 1991, pp. 1-11.

[8] B. Bhushan and A. Majumdar, "Elastic-Plastic Contact for Bifractal Surfaces," *Wear*, Vol. 153, No. 1, 1992, pp. 53-64.

[9] C. Hardy, C. N. Baronet and G. V. Tordion, "The Elasto-Plastic Indentation of a Half-Space by a Rigid Sp- here," *International Journal for Numerical Methods in Engineering*, Vol. 3, No. 4, 1971, pp. 451-462.

[10] W. R. Chang, L. Etsion and D. B. Bogy, "An Elastic-Plastic Model for the Contact of Rough Surfaces," *Journal of Tribology*, Vol. 109, 1987, pp. 257-263.

[11] Y. Zhao and L. Chang, "A Model of Asperity Interactions In Elastic-Plastic Contact of Rough Surfaces," *Journal of Tribology*, Vol. 123, No. 4, 2001, pp. 857-864

[12] E. Ciulli, L. A. Ferreira, G. Pugliese and S. M. O. Tavares, "Rough Contacts Between Actual Engineering Surfaces Part I. Simple Models for Roughness Description," *Wear*, Vol. 264, No. 11-12, 2008, pp. 1105-1115.

[13] G. Pugliese, S. M. O. Tavares, E. Ciulli and L. A. Ferreir, "Rough Contacts Between Actual Engineering Surfaces Part II. Contact Mechanics," *Wear*, Vol. 264, No. 11-12, 2008, pp. 1116-1128.

18

Oxidation Behaviour of a Newly Developed Superalloy

I. V. S. Yashwanth[1], I. Gurrappa[2*], H. Murakami[3]

[1]M. V. S. R. Engineering College, Nadargul, Hyderabad, India; [2]Defence Metallurgical Research Laboratory, Kanchanbagh PO, Hyderabad, India; [3]National Institute for Materials Science, Ibaraki, Japan.

ABSTRACT

The current paper explains the oxidation behaviour of a newly developed nickel-based superalloy in simulating aero gas turbine engine conditions. The results showed that the new superalloy is highly susceptible to high temperature oxidation. Within three of hours of oxidation, extensive oxide scales were formed. The formed oxide scales were analysed with electron dispersive spectroscopy (EDS) and morphology was studied with scanning electron microscope (SEM) for varied oxidation times. The oxidation products were determined with XRD and cross sections of all the oxidised superalloys were also studied. The elemental distribution of all the superalloys after oxidation was also studied with a view to understand and compare the characteristics of the new superalloy with other superalloys. Finally, an oxidation mechanism that is responsible for its faster degradation under elevated temperatures was established based on the results obtained with different techniques and presented in detail.

Keywords: *Gas Turbine Engine Components, Superalloys, High Temperature Oxidation*

1. Introduction

Newer materials with improved properties are essential in order to enhance the efficiency of gas turbine engines. Efforts made in this direction made it possible to develop a new superalloy for aero engines. The developed superalloy exhibits excellent high temperature strength properties [1]. For achieving enhanced efficiency, it is necessary that the developed superalloy should exhibit excellent high temperature oxidation resistance as it is detrimental at elevated temperatures. It reduces the superalloy component life by forming oxides at a faster rate, thereby reducing the load-carry capacity and potentially leading to catastrophic failure of components. Therefore, high temperature oxidation resistance of new superalloy is as important as its high temperature strength properties.

It is understood that the high temperature capability of superalloys depends on their chemistry such as nature of alloying elements and concentration of each alloying element. The major change is the addition of rhenium or of both ruthenium and iridium at the cost of chromium, which are named as the 3rd or 4th or 5th generation superalloys respectively. Therefore, the chemistry of an advanced superalloy that belongs to 3rd generation was greatly influenced by reducing chromium (Cr) content and

increasing the rhenium (Re) content. Similarly, the 4th and 5th generation superalloys contain only about 3% Cr but instead contain about 6% Re, 1.5% ruthenium (Ru) and/or iridium (Ir), which is a great contrast to the earlier generation superalloys containing about 10% Cr and no Re, Ru or Ir. These are unique elements, which can increase high temperature creep properties considerably, but make the superalloys susceptible to high temperature corrosion *i.e.* hot corrosion and high temperature oxidation. It is due to the fact that the new superalloys cannot form corrosion resistant alumina or chromia scale because of the presence of high rhenium content and small amounts of ruthenium and iridium. Their effect is similar to Mo on oxidation *i.e.* the high vapour pressure of its oxide. Therefore, the new alloying elements are harmful for high temperature corrosion resistance of advanced Ni-based superalloys.

Several failures of gas turbine engine blades were reported during service [2-7]. It was attributed primarily to high temperature corrosion of different types and established their relevant theories. Extensive amount of work was carried out in the laboratory on hot corrosion of several superalloys and established their degradation mechanisms [8-10]. It was shown that the hot corrosion of superalloys takes place through electrochemical reactions and it is an electrochemical phenomenon [10,11]. Further, high performance protective coatings were successfully identified for protection of superalloys under hot corrosion conditions [12-14]. Efforts made by other researchers in developing protecting coatings helped in understanding their behaviour [15-17]. Basically, the high temperature corrosion can be divided into oxidation and hot corrosion. The hot corrosion can be further divided into type I (800˚C - 950˚C) and type II (600˚C - 750˚C). As already mentioned above, any new superalloy material that is developed for gas turbine engine applications should exhibit excellent high temperature corrosion resistance apart from excellent mechanical properties.

In the present paper, high temperature oxidation characteristics of the newly developed superalloy is presented. XRD was used to determine the corrosion products that formed during high temperature corrosion processes. The surface morphologies were studied with scanning electron microscope (SEM) and the composition of oxidation products were analysed by EDS. Elemental distribution of the oxidised superalloys were determined with a view to establish its degradation mechanisms under oxidation conditions.

2. Experimental

The composition of newly developed advanced superalloy is presented in **Table 1**. In contrast to previous generation superalloys, the newly developed superalloy contains 6.5% rhenium and very small content of chromium, which makes it to exhibit very good high temperature strength properties [1]. Disc specimens of about 10 mm diameter and 2 mm thick were cut from the superalloy rods, polished them up to 600 SiC surface finish, cleaned with tap water followed by distilled water and then decreased with acetone. Subsequently, the cyclic oxidation studies were carried out at 1100˚C ± 10˚C for a period of 100 hours *i.e.* the specimens were placed in a furnace heated at 1100˚C for 20 hours and allowed them to cool for 3 hours after removing from the furnace. Initially, the specimens were oxidized only for three hours to determine their weight change and subsequently for every 20 hours. The weight gain measurements were made each time after cooling to room temperature.

After completion of oxidation tests, the specimens were examined for their surface morphology with Scanning Electron Microscope (SEM) and the oxidation products were analyzed by Electron Dispersive Spectroscopy (ED S).

Table 1. The chemical composition of an advanced superalloy (wt%).

Material	Ni	Cr	Co	W	Al	Ta	Re	Hf
New alloy	Bal	2.9	7.9	5.8	5.6	8.5	6.5	0.1

Different phases formed during oxidation were analyzed by X-ray diffraction technique (XRD). Cross sections of oxidized speciemens were analyzed for understanding the effect of oxidation and then elemental distribution was determined in order to evolve the degradation mechanisms of advanced superalloy under high temperature oxidation conditions.

3. Results and Discussion

Figure 1 shows as oxidized advanced superalloy. As can be seen, the superalloy was oxidized to a maximum extent by forming thicker oxide scales and subsequently spalled. It indicates that the rate of formation of oxide scales is considerably high on the newly developed superalloy leading to the formation of thick oxide scales over a period of time and subsequently spalled because of adherence problem. In essence, the modified chemistry with increased addition of rhenium and tantalum and considerably reduced chromium made the new superalloy to oxidize at a faster rate when compared to earlier generation superalloys [18]. Typical surface morphology of the oxidized superalloy is shown in **Figure 2**. As can be seen, the surface morphology observed on the alloy is the spalled area as the oxide scale was spalled because of adherence problem due to its high thickness. Almost uniform morphology with reformed oxide scale in some regions is the characteristics of new superalloy. The EDS measurements revealed that the oxide scale was consisted the oxides of base as well as alloying elements. The cross section of oxidized new alloy shows that the oxidation-affected zone is very large (**Figure 3**). Faster oxidation and subsequent spallation is the reason for observing more affected region.

The XRD patterns along with their identifications for the oxidized superalloy after definite oxidation times are demonstrated in **Figure 4** and the oxidation products were summarised in **Table 2**. The superalloy was formed

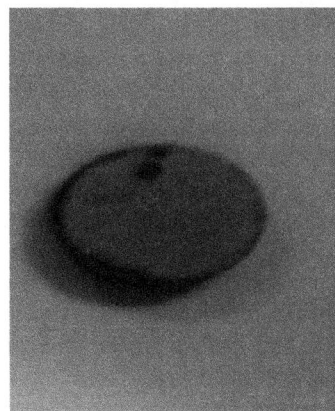

Figure 1. As oxidised advanced superalloy at 1100˚C for 100 hours.

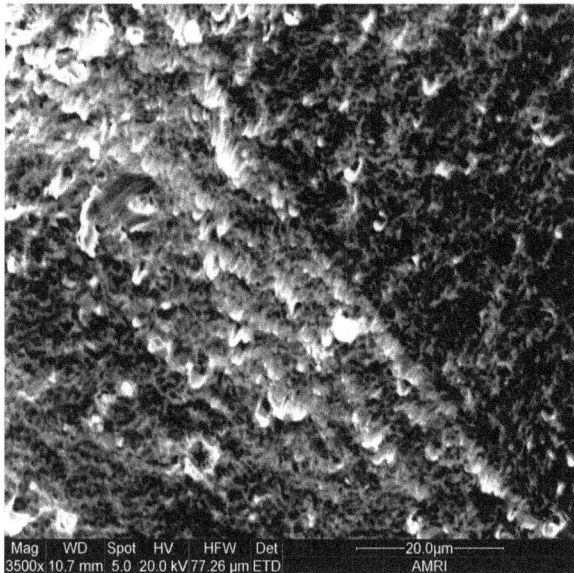

Figure 2. Surface morphology of oxidised superalloy at 1100°C for 100 hours.

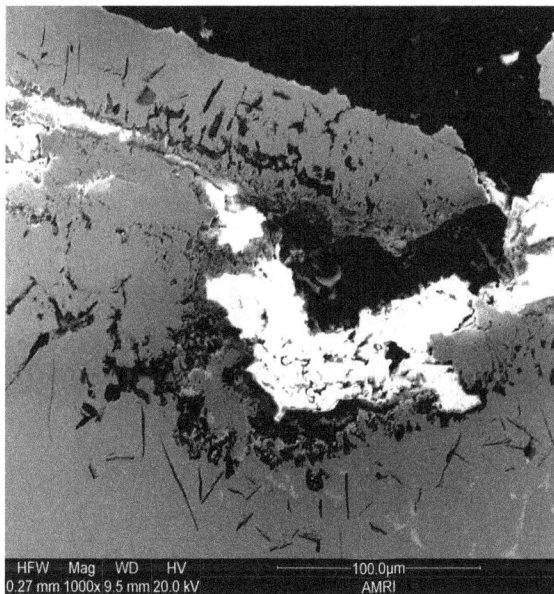

Figure 3. Cross section of oxidised superalloy at 1100°C for 100 hours.

Table 2. XRD data of oxidized superalloy.

Material	Oxidation products
New superalloy	NiO, Co_3O_4, TaO_2, WO_3, Al_2O_3

nickel oxide as well as oxides of all the alloying elements (Table 2) during high temperature oxidation. The dominant oxidation product was nickel oxide with small peaks

relating to oxides of cobalt, tantalum, tungsten and aluminium. It is due to the fact that the oxide scale was spalled because of high thickness.

The elemental distribution of an advanced superalloy after oxidation at 1100°C for 100 hours is illustrated in **Figure 5**. Discontinuous alumina in association with chromia is clearly observed on its surface. Rhenium and cobalt were present beneath the alumina and chromia scale. No nickel and cobalt oxides were present. These oxides might have spalled during cutting and polishing, as they are prominent in XRD. It demonstrates that the external scale was consisted of nickel and cobalt oxides, which were spalled after attaining maximum thickness and the internal scale was a mixture of alumina, chromia and tantalum oxide and hence observed in the cross section.

The present investigation clearly established that the newly developed superalloy is highly vulnerable to high temperature oxidation and degrades due to formation of a multilayer corrosion products consists of oxides of nickel as well as the alloying elements. During normal oxidation of superalloys, the oxides of most of the alloying elements as well as base are expected to form as given below:

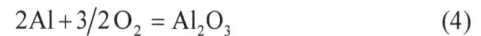

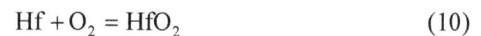

$$2Ni + O_2 = 2NiO \tag{1}$$

$$2Cr + 3/2 O_2 = Cr_2O_3 \tag{2}$$

$$3Co + 2O_2 = Co_3O_4 \tag{3}$$

$$2Al + 3/2 O_2 = Al_2O_3 \tag{4}$$

$$2W + 3O_2 = 2WO_3 \tag{5}$$

$$4Ta + 5O_2 = 2Ta_2O_5 \tag{6}$$

$$Ta_2O_5 + Al = Al_2O_3 + TaO_2 + Ta \tag{7}$$

$$Ti + O_2 = TiO_2 \tag{8}$$

$$4Re + 7O_2 = 2Re_2O_7 \tag{9}$$

$$Hf + O_2 = HfO_2 \tag{10}$$

Due to reaction with oxygen, rapid weight gain of the alloys takes place initially. After oxidation for an appropriate time, thermodynamically stable oxides such as Al_2O_3, Cr_2O_3 are formed as a dense oxide scale on the surface of superalloys. These oxide scales act as diffusion barriers for the ingress of deleterious species such as oxygen and sulphur. However, the oxidation is significantly high for the new superalloy, as it could not able to form a stable alumina or chromia scale because of change in its chemistry.

The following degradation mechanism has been proposed for the oxidation of new superalloy: diffusion processes occurring at elevated temperatures at the superalloy and environment interface act as a major degradation

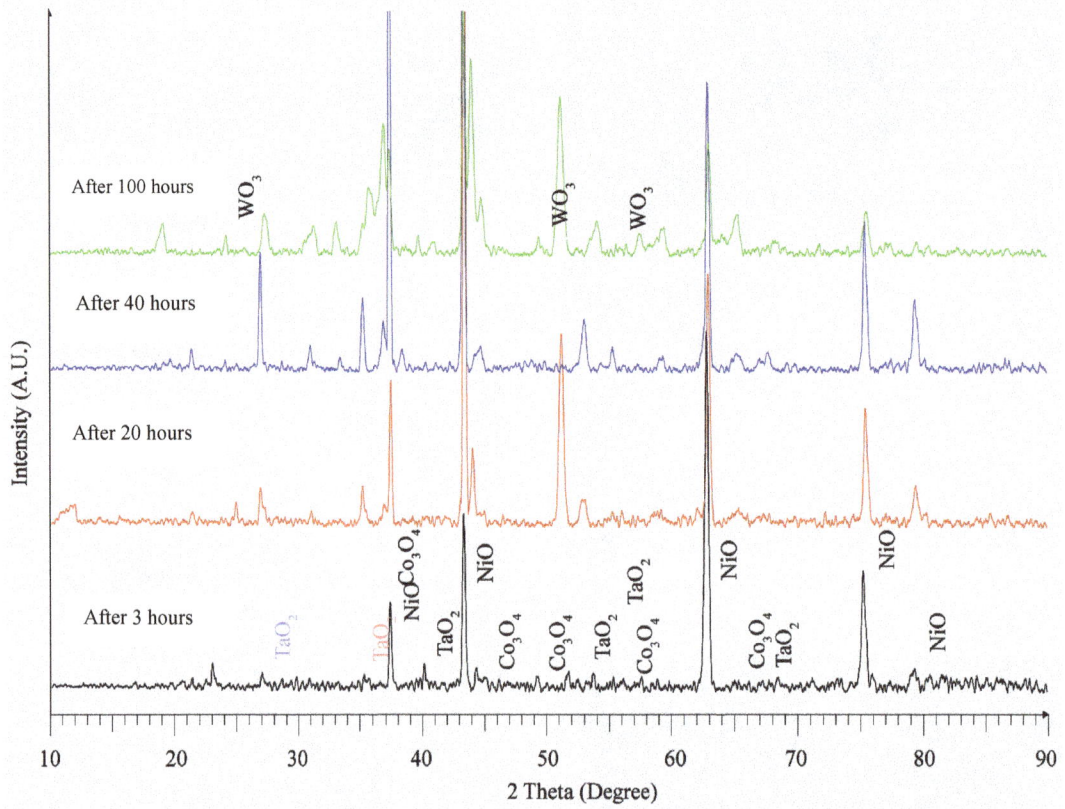

Figure 4. XRD Patterns of an advanced superalloy after oxidation at 1100°C for different times.

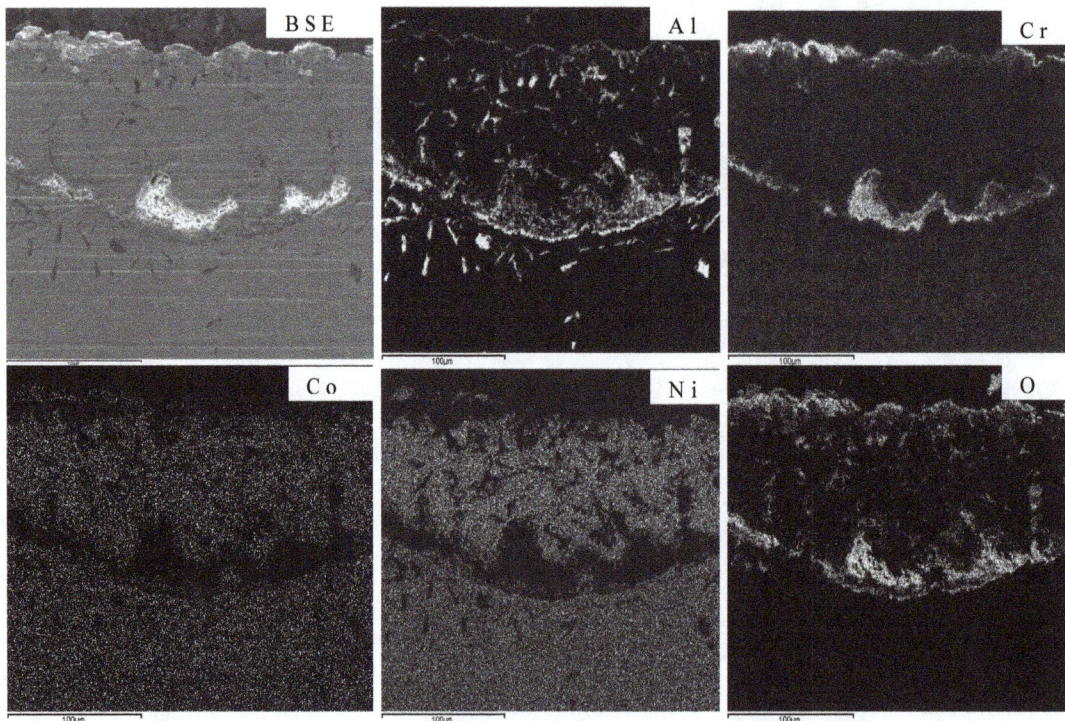

Figure 5. Elemental distribution of an advanced superalloy after oxidation at 1100°C for 100 hours.

mode of the superalloy. At the superalloy interface, diffusion occurs primarily due to thermodynamic considerations at the elevated temperatures. The new superalloy contains a number of alloying elements like Co, Ta, W, Re, Al, Cr etc. mainly to enhance high temperature mechanical characteristics. The nickel and alloying elements of the superalloy diffuse outwards leading to the formation of phases that contain non-protective elements during high temperature oxidation process. As a result, non-protective oxide scales form rather than continuous and protective pure alumina scale. It is important to mention that nickel appears to have high affinity towards oxygen at elevated temperatures and hence diffuses outwards at a faster rate, interact with oxygen ions to form nickel oxide. This process leads to extensive nickel oxide growth and forms thicker scales. Thicker scales lead to spall after reaching to certain thickness due to adherence problem. Further, excessive growth of nickel oxide scale, in association with alloying elements of superalloy like cobalt, tantalum, tungsten and rhenium, leads to the formation of non-protective oxide scales with large volumes. Consequently, spallation of oxide scale takes place because of non-adherence to the superalloys. Practical observation of extensive oxide scale growth (**Figure 5**) and subsequent spallation clearly supports the proposed degradation mechanism. A schematic representation of degradation of new superalloy during high temperature oxidation is presented in **Figure 6**.

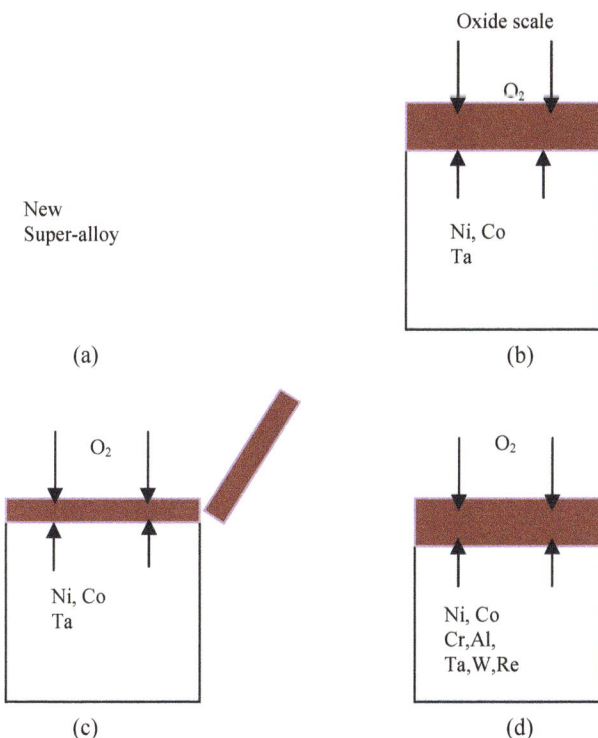

(e)   (f)

Figure 6. Sequence of degradation processes for an advanced superalloy during high temperature oxidation process. (a) Before oxidation; (b) During 3 hours of oxidation; (c) After 3 hours of oxidation; (d) During 20 hours of oxidation; (e) After 20 hours of oxidation; (f) After 40, 60, 80, 100 hours of oxidation and continuation of process leading to failure of the alloy.

4. Development of Smart Coatings

The present results clearly stresses the need to apply high performance protective coatings for its protection against high temperature oxidation as the gas turbine engines encounter oxidation problem during service. The protective coatings allow the gas turbine engines to operate at varied temperatures and enhance their efficiency by eliminating failures during service. Research in this direction has resulted in design and development of smart coatings which provide effective protection to the superalloy blades for the designed period against type I, type II hot corrosion and high temperature oxidation that are normally encountered in gas turbine engines which in turn enhances the efficiency of gas turbine engines considerably [18,19]. This is a major developmental work in the area of gas turbine engines used in aero, marine and industrial applications. Unlike the conventional/existing coatings, the smart coatings provide total protection to the superalloy components used in aero, marine and industrial applications by forming appropriate protective scales depending on the surrounding environmental conditions [18, 19].

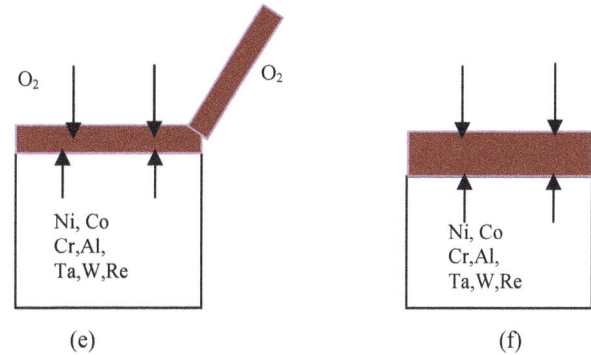

(a)   (b)

(c)   (d)

REFERENCES

[1] N. Das, US patent 5,925,198, July 1999

[2] M. R. Khajavi and M. H. Shariat, "Failure of First Stage Gas Turbine Blades," *Engineering Failure Analysis*, Vol. 11, No. 4, 2004, pp. 589-597.

[3] J. M. Gallardo, J. A. Rodrigue and E. J. Herrera, "Failure of Gas Turbine Blades," *Wear*, Vol. 252, No. 3-4, 2002, pp. 264-268.

[4] N. Eliaz, G. Shemesh and R. M. Latarision, "Hot Corrosion in Gas Turbine Components," *Engineering Failure*

Analysis, Vol. 9, No. 1, 2002, pp. 31-43.

[5]   T. J. Carter, "Common Failures in Gas Turbine Blades," *Engineering Failure Analysis*, Vol. 12, No. 2, 2005, pp. 237-247.

[6]   R. Nutzel, E. Affeldt and M. Goken, "Damage Evolution during Thermo-Mechanical Fatigue of a Coated Mono-crystalline Nickel-Bas Superalloy," *International Journal of Fatigue*, Vol. 30, No. 2, 2008, pp. 313-317.

[7]   R. S. J. Corran and S. J. Williams, "Lifing Methods and Safety Criteria in Aero Gas Turbines," *Engineering Failure Analysis*, Vol. 14, No. 3, 2007, pp. 518-526.

[8]   I. Gurrappa, "Hot Corrosion Behaviour of CM 247 LC Alloy in Na_2SO_4 and NaCl Environments," *Oxidation of Metals*, Vol. 51, No. 5-6, 1999, pp. 353-382.

[9]   I. Gurrappa, "Hot Corrosion Behaviour of Nimonic-75," *Journal of High Temperature Materials Science*, Vol. 38, 1997, pp. 1-9

[10] I. Gurrappa, "Influence of Alloying Elements on Hot Corrosion of Superalloys and Coatings-Necessity of Smart Coatings," *Materials Science and Technology*, Vol. 19, 2003, pp. 178-183

[11] I. Gurrappa, "Overlay Coating Degradation an Electro-chemical Approach," *Journal of Materials Science Letters*, Vol. 18, No. 21, 1999, pp. 1713-1717.

[12] I. Gurrappa, "Identification of Hot Corrosion Resistant MCrAlY Based Bond Coatings for Gas Turbine Engine Applications," *Surface and Coating Technology*, Vol. 139, No. 2-3, 2001, pp. 272-283.

[13] I. Gurrappa, "Hot Corrosion Behaviour of Protectuve Coatings on CM 247 LC Superalloy," *Materials and Manufacturing Processes*, Vol. 15, 2000, pp. 761-767

[14] I. Gurrappa and A. Sambasiva Rao, "Thermal Barrier Coatings for Enhanced Efficiency of Gas Turbine Engines," *Surface and Coating Technology*, Vol. 201, 2006, pp. 3016-3029

[15] R. Mevrel, "State of the Art on High Temperature Corrosion Resistant Coatings," *Materials Science and Engineering A*, Vol. 120-121, 1989, pp. 13-24

[16] R. Mobarra, A. H. Jaffari and M. Karamirezhaad, "Hot Corrosion Behaviour of MCrAlY Coatings on IN 738LC," *Surface and Coating Technology*, Vol. 201, 2006, pp. 2202-2207

[17] M. M. Warres, "Improved Aluminide McrAlX Coating Systems for Superalloys using CVD low activity Aluminizing," *Surface and Coating Technology*, Vol. 163-164, 2003, pp. 106-111

[18] I. Gurrappa, "Final Report on 'Design and Development of Smart Coatings for Aerospace Applications' Submitted to European Commission," July 2008

[19] I. Gurrappa, "Identification of a Smart Bond Coating for Gas Turbine Engine Applications," *Coating Technology Research*, Vol. 5, 2008, pp. 385-390

19

Surface and Bulk Defects in Cr-Mn Iron Alloy Cast in Metal and Sand Moulds: Characterization by Positron Annihilation Techniques

Parthasarathy Sampathkumaran[1], Subramanyam Seetharamu[1], Chikkakuntappa Ranganathaiah[2*], Jaya Madhu Raj[2], Pradeep Kumar Pujari[3], Priya Maheshwari[3], Debashish Dutta[3], Kishore[4]

[1]Materials Technology Division, Central Power Research Institute, Bangalore, India; [2]Department of Studies in Physics, University of Mysore, Mysore, India; [3]Radiochemistry Division, Bhabha Atomic Research Centre, Mumbai, India; [4]Department of Materials Engineering, Indian Institute of Science, Bangalore, India.
*

ABSTRACT

High chromium (Cr: 16% - 19%) iron alloy with 5% and 10% manganese (Mn) fabricated in metal and sand moulds by induction melting technique were investigated for defects microstructure both in the as-cast and heat treated conditions. Non-destructive techniques namely Positron Lifetime Spectroscopy and slow positron Doppler Broadening studies were employed to characterize the defects in the bulk as well as surface of the alloy and their influence of metallurgical parameters. The Positron Lifetime Spectroscopy data reveals that the defect concentration is higher for sand mould alloy samples compared to metal mould ones. The reasons for fewer defects in metal mould are attributed to faster heat transfer in the metal mould. Further, heat treatment yielded spherodization of carbides in the matrix resulting in reduced defects concentration. The S-parameter profiles from Doppler Broadening studies suggest defect concentration at the surface is less in 5% Manganese and near absence of any modification of defect structure following heat treatment in 10% Manganese sample closer to surface.

Keywords: *Cr-Mn Cast Iron, Heat Treatment, Positron Lifetime Spectroscopy, Slow Positron Beam Analysis*

1. Introduction

Metal wastage occurs in ferrous materials on account of wear and erosion in certain critical parts of thermal power generators like coal and ash handling equipments, pressure parts etc., This is usually ascribed to high percentage of alpha quartz present in the coal [1,2] and is responsible for wear damage. Therefore, it is always desired to enhance the lifespan of such parts to withstand wear & erosion [3]. Several attempts have been made in the past two or three decades to minimize wear and erosion. From these studies it is understood that the material found to be a promising wear resistant is high chromium (Cr) iron [4] since it contains hard carbides (M_7C_3) in a martensitic matrix, but fails to withstand sudden load/shock. To improve the impact resistance, manganese (Mn) is added since Mn is an austenite stabilizing agent. Earlier works [5,6] show that Mn content up to 4.4% in chromium irons has yielded higher toughness compared to Mn free irons. Unfortunately there is no literature on the use of Mn beyond 4.4% in such alloy system. Hence the

present work focuses on the use of Mn at 5% & 10% in Cr (16% - 19%) rich iron, cast in metal and sand moulds to study the defect structure in the as-cast and heat treated conditions. The slow positron beam analysis (DBAR) and conventional positron lifetime analysis (PLS) have been used for the first time to study the defect morphology in terms of defect concentration both at the surface and bulk to understand the influence of manganese addition under change of mould and heat treatment of the samples which has a direct bearing on erosion of particles from the surface and its connection to bulk material. There are only few studies using other techniques reported on this particular iron system [3-6] which is one of the most sought after materials in thermal power generators and the like.

Positron annihilation spectroscopic studies evolving experiments, namely, slow positron beam based Doppler broadening of annihilation radiation (DBAR) measurements and conventional positron lifetime measurements (PLS) are outlined in the following; Slow positron beam used for the present study consists of a Ultra high vacuum compatible sealed ^{22}Na radioisotope as the positron source. Positrons emitted from the source were thermalized by W single crystal floated at 200 V (moderator), which has negative work function for the positrons. The thermalized positrons diffuse to the surface of moderator, which are extracted by an einzel lens to the solenoid. Positrons travel in spiral motion under the magnetic field in the solenoid, which act as a velocity filter to enhance the monochromatic nature of the positron beam. Positrons in the form of beam fall on the sample in the target chamber, where a magnetic field is maintained by two Helmholtz coils. The sample holder can be floated from 0 V to 50 kV which can accelerate positrons to the required energy. Doppler broadening spectroscopy measurement as a function of implantation depth of positron beam has been evolved as a good technique for depth profile of defects at the surface of system, because the fraction of implanted positrons annihilating in two gamma photons depends upon the size and distribution of defects and electron density at the site of positron [4]. In the conventional positron lifetime technique, positrons from a ^{22}Na source are injected into the system under investigation and they thermalize very rapidly. Subsequently, they annihilate with free electrons of the medium labeled as free annihilation, or if get trapped in a defect of the system and then annihilates, is usually called trapped state annihilation [7,8]. Since positrons localize in defects and annihilate [7,8], their lifetime and intensity provide information on the nature of defects and their concentration in the bulk and as such it has been established as a novel tool for studying the microstructural behavior of metals, alloys and a wide variety of materials for more than four decades [8].

The bulk positron lifetime spectroscopy technique has been used to establish a good correlation between fatigue life ratios and PLS parameters [4] in stainless steels to predict early fatigue damage detection. The work related to correlation of erosion behavior with surface defect characteristics in Cr-Mn alloy systems does not seem to exist in literature. Hence, the usefulness of slow positron beam studies combined with PLS has been used to understand the correlation of surface defects with the microstructure in 5% and 10% Cr-Mn iron produced in sand and metal moulds in this work.

2. Experimental

The metal & sand molded test samples of size $75 \times 25 \times 6$ mm^3 were given austenitization soak at 960°C for 2 hours followed by oil quenching and then finally tempering at 200°C for 30 min with air cooling to room temperature. The as-cast and heat treated samples were subjected to defect characterization using slow positron beam analysis (DBAR) spectroscopy and conventional positron lifetime spectroscopy (PLS). The details in respect of the melting and casting procedures of Cr-Mn alloy system under investigation are covered in detail in our earlier published work [6].

2.1. Slow Positron Doppler Broadening Annihilation Radiation Measurement (DBAR)

DBAR was carried out in Cr-Fe alloy prepared as described in metal and sand moulds and in as cast and heat treated conditions using the slow positron beam facility at Radiochemistry Division, Bhabha Atomic Research Centre, India. The present beam has three components interconnected under high vacuum *viz.* (i) slow positron production (Source and moderator), (ii) focusing and transport (Einzel lens and magnetic transport) and (iii) acceleration of positrons at the target. Positrons emitted from a sealed ^{22}Na source are moderated by 1 μ thick tungsten single crystal. The thermalized positrons come to the surface as tungsten has negative work function for positrons. The positrons extracted from the moderator are focused by the Einzel lens and are guided towards the sample through a magnetically guided assembly (90 degree bent solenoid and two Helmholtz coils, respectively). The positron energy is varied by floating the sample at different voltages. The energy range of the positron beam is 200 eV - 50 keV. Doppler broadened annihilation radiation measurements were carried out using an HPGe detector having resolution of 1.7 keV at 1332 keV photo peak of ^{60}Co. A spectrum with 10^6 counts was acquired at each energy. The shape parameter, namely, S-parameter defined as the ratio of the number of counts falling in a fixed energy window (± 1 keV) centered at 511 keV to the total number of counts under the Gaussian peak, was evaluated. The S-parameter at the surface and in bulk was determined using computer program VEPFIT [5]. The variation in the S-parameter as a function of depth gives the defect depth profile in the sample. Further details of this experiment and analysis can be found from Ref [9,10].

2.2. Positron Lifetime Spectroscopy (PLS)

Positron annihilation lifetime spectra were recorded at room temperatures in the Cr-Mn iron system using a fast-fast coincidence system with BaF$_2$ scintillators coupled with photo multiplier tubes and quartz window as detectors in about 1 to 2 hours. Three Gaussian time resolution functions were used in the lifetime analysis for fast and good convergence keeping the net resolution function around 220×10^{-12} s *i.e.* (220 ps). The details of the expe-

Surface and Bulk Defects in Cr-Mn Iron Alloy Cast in Metal and Sand Moulds: Characterization by Positron
Annihilation Techniques

125

rimental procedure and analysis can be found in our earlier work [11]. All spectra were analyzed into two lifetime components with the help of the computer program PATFIT-88 [12] with proper source and background corrections. The analysis gives two lifetimes τ_1 and τ_2 with respective intensities I_1 and I_2. In the present analysis, τ_1 is fixed at 107 ps which correspond to lifetime of positrons in Fe and in the present systems Fe is the matrix. The fixed analysis will not suppress any information since the free annihilation lifetime does not provide any material information. The trapping rate (k) which is a measure of the defect concentration in the system is estimated by adopting the two state trapping model [7,8]. The parameters of the trapping model are the positron lifetimes in the free and trapped states with intensities I_1 and I_2. The rate at which transitions from the delocalized states to the localized ones happen is the trapping rate. This transition rate (positron trapping rate) k is proportional to the concentration of the defects. The positron annihilation lifetime in bulk (τ_b) and in defect (τ_d) can be determined from [7,8]

$$\lambda_b = \left(1/\tau_b\right) = \left(I_1\tau_1^{-1} + I_2\tau_2^{-1}\right) \qquad (1)$$

and $\lambda_d = 1/\tau_d = \tau_2^{-1}$, where λ_b and λ_d are the decay rate of positron from bulk and defect respectively. The positron mean lifetime, τ_m can be calculated using the formula [7,8]

$$\tau_m = \left(\tau_1 I_1 + \tau_2 I_2\right) \qquad (2)$$

The τ_d is always larger than τ_b for open-volume defects, such as vacancies, dislocations due to the decreased electron density in the defect site compared to the bulk material and hence traps positrons.

Then the trapping rate k is calculated from the equation below

$$k = \lambda_b \left(\frac{\tau_m - \tau_b}{\tau_d - \tau_m}\right) \qquad (3)$$

Also, $k = \mu C_d$, where C_d is the defect concentration and the proportionality constant μ is the specific trapping coefficient and a value of 1.1×10^{15}/sec is used in the calculation of C_d with appropriate weight fraction taken into account for the alloy system of the present study.

3. Results and Discussion

The positron annihilation lifetime and intensity are sensitive to microstructural changes due to lattice defects, such as vacancies and dislocations caused either by deformation or processing or addition of substitutional elements. They are considered to be effective non destructive evaluation parameters of the damage or microstru-

ctural changes brought about to the system under study. If lattice defects exist in materials, these defects represent regions of low electron density and become negatively charged. This in turn causes positively charged positrons to be attracted to the lattice defects or in other words, positrons get trapped in to such defects and then annihilates from these sites. When positrons annihilate from these defects, the lifetime will be larger than the bulk lifetime. Therefore, as the number of defects increases, the number of positrons trapped in such defects also increases and the intensity of trapped lifetime also increases [7,8,13].

Positron lifetime data from PLS for both as-cast (AC) and heat treated (HT) conditions are shown in **Table 1 & 2**. The evaluated S-parameter, an index of defect concentration, and the diffusion length of positrons in respective samples are given in **Table 3**. The sample designation followed in the Tables is as follows: first numeral (% Mn) followed by a letter (mould type) and lastly a number (section size). Hence, for example, 5% Mn bearing 24 mm sized metal mould sample is designated as 5M24.

Table 1. Positron Lifetime results for Cr-Mn iron as-cast (AC) samples.

AC: τ_1 was fixed at pure Fe lifetime of 107 ps								
Samples	I_1 (%)	τ_2 (ps)	I_2 (%)	τ_{mean} (ps)	τ_b (ps)	λ_b (10^9s^{-1})	k (10^9s^{-1})	$C_d = {}^*A_d$ (k/μ) (10^{16}cm^{-3})
5M24	67.5	215.3	32.5	142.20	127.91	7.818	1.528	11.71
5S24	65.9	220.8	34.1	145.80	129.81	7.703	1.643	12.59
10M24	68.6	226.2	31.4	144.43	128.21	7.799	1.546	11.81
10S24	68.3	222.5	31.7	143.61	128.07	7.808	1.538	11.74

Table 2. Positron Lifetime results for Cr-Mn iron heat treated (HT) samples.

HT: τ_1 was fixed at pure Fe lifetime of 107 ps								
Samples	I_1 (%)	τ_2 (ps)	I_2 (%)	τ_{mean} (ps)	τ_b (ps)	λ_b (10^9s^{-1})	k (10^9s^{-1})	$C_d = A_d$ (k/μ) (10^{16}cm^{-3})
5M24	69.2	197.2	30.8	134.78	124.55	8.029	1.317	10.09
5S24	67.1	212.7	32.9	141.77	127.91	7.818	1.528	11.70
10M24	69.5	224	30.5	142.68	127.27	7.857	1.489	11.37
10S24	69.3	215.7	30.7	140.37	126.58	7.900	1.446	11.04

Table 3. VEPFIT Slow positron beam DBAR data of Cr-Mn iron samples

Sample	S-parameter Bulk	S-parameter Surface	Diffusion length (nm)
5M24 (AC)	0.5585	0.5885	82.16
10M24 (AC)	0.5730	0.5960	63.51
10M24 (HT)	0.5769	0.6020	63.51

3.1. Influence of Mould type of As-Cast (AC) Samples on PLS Parameters

Considering the positron lifetime data, the mean range of positrons in the Cr-Mn iron systems for 540 keV has been found to be 32.93 μm. With this energy the positrons are expected to probe only the bulk of the system. It is known that the defect profile at the surface is not necessarily the same as that of the bulk; however, the defect profile of the bulk influences the surface properties. This issue has been reported from the surface studies carried out in various metals and semiconductors using slow positron beams [14]. In the present study, we have used both PLS and slow positron beam studies to characterize separately the defects at the bulk and the surface respectively. The lifetime in the present samples in the absence of defects should be in the range 99 - 107 ps. Further, the theoretical estimate of the expected defect lifetime for these systems assuming mono-vacancy type defects should be about 186 ps. But the second lifetime measured for the present alloy systems is in the range 197 - 220 ps which is higher and indicates that there are defects other than mono vacancy present in the sample which might have evolved in the Cr-Mn-Fe system during the solidification process in both metal and sand moulds. Although, the mean range of positrons in all these systems is the same, their defect lifetime show variation with respect to the mould variety used as explained below. For the metal and sand mould employed, the heat transfer process appears to be different in each case; a transformation in the microstructure takes place as a result of varied cooling rates prevailed in the moulds. Further, it is observed that a consistent reduction in mean lifetime for samples after heat treatment (**Table 1&2**) indicates that some of the defects are indeed annealed out. As the fraction of positrons probing the single interface is negligible due to the mean positron diffusion length of few hundred nanometers, very little trapping is expected from the interface. These facts indicate that most of the positrons are getting annihilated in the bulk state as seen from I_1 and I_2 values given in **Table 1 & 2**.

The lower positron lifetime and less number of defects exhibited by 5M24 (**Figure 1 & 3**) indicates the fact that

metal mould produces less number of defects compared to 5S24. Similarly, there is a marginal variation in lifetime and defects concentration C_d in 10M24 compared to 10S24 (**Figure 2 & 4**). In an earlier published work of the author on the same system [15,16], it was reported that the erosion volume loss is lower at all impact angles and hardness is higher for the 5% Mn and 10% Mn bearing metal mould samples compared to the sand mould ones. More clearly, for metal mould samples, because of shorter positron lifetime and less defect concentration, higher hardness and low erosion loss were observed compared to sand mould samples.

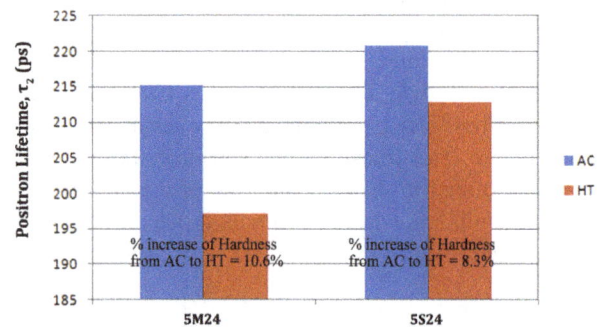

Figure 1. Plot of positron lifetime τ_2 for 5M24 and 5S24 samples.

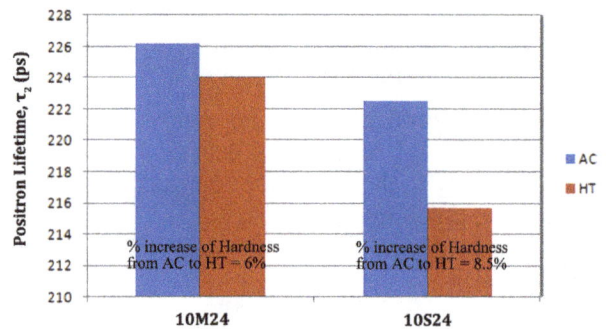

Figure 2. Plot of positron lifetime τ_2 for 10M24 and 10S24 samples.

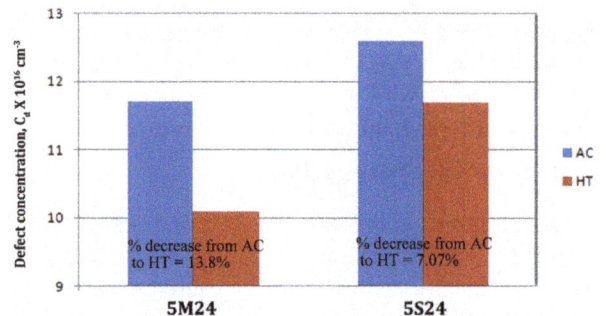

Figure 3. Plot of defect concentration C_d for 5M24 and 5S24 samples.

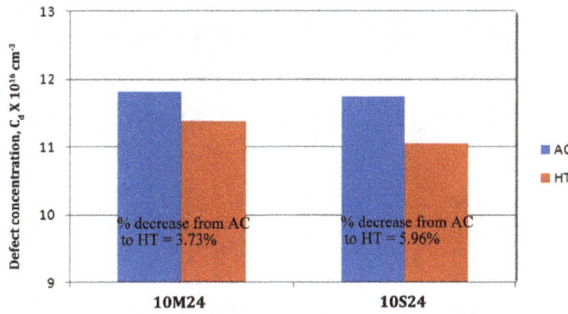

Figure 4. Plot of defect concentration C_d for 10M24 and 10S24 samples.

This shows that there is a definite correlation between defect structure from PLS, hardness and erosion results observed for these samples. These data trends have now been explained based on the heat transfer characteristics obtained during the solidification process on account of mould variety adopted. The heat transfer coefficients may be calculated using the well known equation [17] given below.

$$\frac{\partial}{\partial x}\left(\kappa \frac{\partial T}{\partial t}\right) = \rho c \frac{\partial T}{\partial t} \qquad (4)$$

where κ is the thermal conductivity, ρ is density and c is specific heat.

When the effect of solidification rate on the erosion and PLS parameters is to be looked into, equation (5) given below is used to calculate the heat transfer rate during melting and casting taking in to account the thermal conductivity of the mould type used. This equation is obtained by integrating the equation (4) with appropriate boundary conditions [17].

$$Q = \left(k.A.\Delta T\right)/L \qquad (5)$$

Here Q is Heat transfer rate, A is the cross sectional area, L is the overall thickness of the casting, and ΔT is temperature difference during solidification.

It is noted that the solidification rate prevailed in the metal mould is higher than in sand mould in view of higher thermal conductivity (κ) and lower specific heat (c) prevailed in the metal ($\kappa = 52$ W/m.k) mould compared to sand ($\kappa = 0.325$ W/m.k) mould which has lower κ and higher c. Further, the heat transfer rate is more in the metal mould compared to sand mould counterpart as it is directly proportional to 'κ' the thermal conductivity. Hence, the heat transfer coefficient will also be higher in the metal mould than in sand mould. No attempt has been made in this work to measure the temperatures at the mould wall as well on the casting periphery. These factors give credence to the present data trends thus emphasizing the fact that the metal mould samples have smaller sized carbides and less defect concentrations compared to

the sand mould counterparts.

3.2. Effect of Mould Variety on PLS Parameters in the Heat Treated (HT) Condition

From the PLS data, it is seen that the defect size and its concentration (C_d) are higher for as-cast samples irrespective of Mn content compared to the corresponding heat treated samples. Following heat treatment in 5M24 sample, the defect concentration C_d decreases by 13.8% while it is 7.07% decrease in 5S24 sample (**Figure 3**). In case of 10M24, upon heat treatment, the defect concentration C_d decreases by 3.73% while 10S24 sample showed a decrease by 5.96% (**Figure 4**). These observations are attributed to annealing out of some defects and also resulting in smaller size defects. Therefore, the changes in defect concentration correlate well with the improved erosion behavior and hardness characteristics of Cr-Mn iron systems reported earlier on the high chromium irons [18] that the erosion process is dominated by the matrix removal with carbide particles not getting damaged. Interestingly, from the earlier work of the authors [15,16], it was observed that, following heat treatment, the erosion data showed higher volume loss at all impact angles and lower hardness for 5S24 compared to 5M24 samples and the same was the trend observed in the case of 10M24 and 10S24 samples. Therefore, we can see that increased defect concentration resulted in lower hardness and higher erosion loss in the as cast samples and heat treated samples of 5S24 and 10S24 which can be attributed to longer time available for the diffusion of molten material and the evolution of bigger size carbides in the sand mould samples and vice versa is true for metal mould samples. In other words, the globular type of carbides were formed in faster cooling conditions prevailing in metal mould casting. This further verifies the correlation of the positron data with the erosion and hardness results observed by the authors.

3.3. Surface Defect Characterization Using the Slow Positron Beam Analysis (DBAR)

The basis of DBAR spectroscopy is due to the relative velocities of positron and electron pair just prior to annihilation; the energy deviation from 511 keV is dominated by the moment of the electrons. Since the electron momentum distribution at a defect site is the characteristic of that defect, the DBAR energy lines-shape is in fact a 'fingerprint' of the defect structure in material. Thus, by monitoring changes in the DBAR line-shape parameter, it is possible to track changes in the defect types and/or defect concentrations [19]. If the energy of the incident positron can be varied, the depth profile of the defects can be understood from DBAR. The variation in positron annihilation line-shape parameter S as a function of incident positron energy and the mean implantation depth of the positron beam is shown in **Figure 5**. The mean implantation depth $\langle Z \rangle$ of mono-energetic positron beam having incident energy E in a material with density ρ (gm/cc) is determined using the following equation [20]

$$\langle Z \rangle = \frac{40}{\rho} E^{1.6} \qquad (6)$$

where, $\langle Z \rangle$ is expressed in nm and E in KeV.

The positrons implanted in the sample have three paths to annihilate: (i) annihilation in the vacancy-free bulk; S_{bulk} is a characteristic parameter for the material; (ii) trapping by vacancy-type defects and annihilation there, leading to a value S_{vac} higher than S_{bulk} and (iii) diffusion to the sample surface, giving a value S_{surf} lower than S_{bulk}.

$$S(E) = f_{surf} S_{surf} + f_{vac} S_{vac} + f_{bulk} S_{bulk} \qquad (7)$$

where f_{surf}, f_{vac} and f_{bulk} are the fraction of positrons annihilating at the surface, in the vacancy-type defects, and in the bulk.

S-parameter characterizes the positrons that annihilate with low momentum electrons, mostly valence electrons and it is sensitive to open volume defects. Therefore, an increase in S-parameter can be taken as an indication of increased vacancy defects [21-23]. For low energy positrons, the changes in the S parameter as a function of positron energy provides the defects at the surface and we towards the bulk as the positron energy are increased. The S-parameter profile of 5% Mn concentration is seen to be different from other two 10% Mn [as-cast (AC) and heat treated (HT)] samples (**Figure 5**). The surface, as

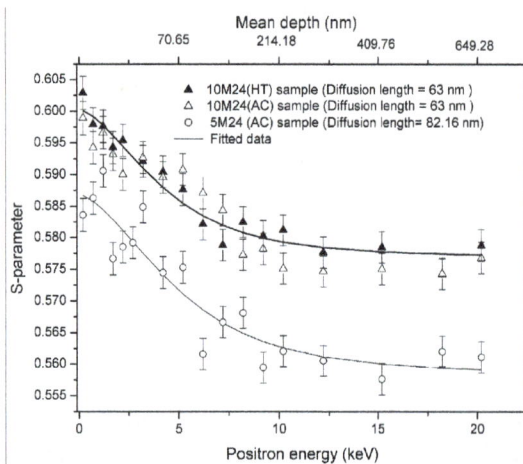

Figure 5. The variation in S-parameter as a function of incident positron energy and the mean implantation depth of the positron beam.

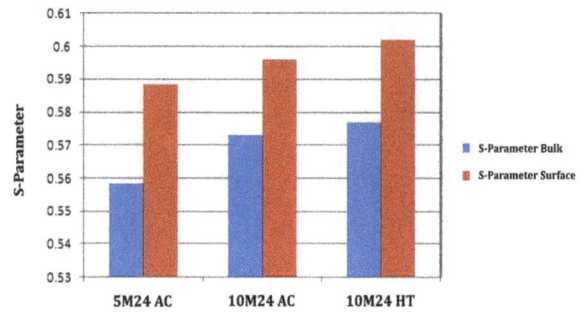

Figure 6. Plot of S-Parameter for 5M24AC, 10M24 AC and 10M24 HT samples.

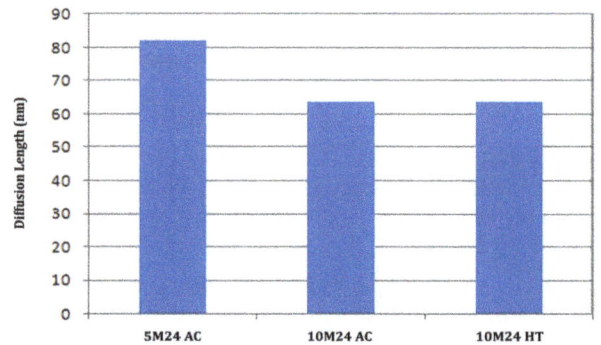

Figure 7. Plot of Diffusion length for 5M24 AC, 10M24 AC and 10M24 HT samples.

well as the bulk values of the S-parameter (S_{bulk} and S_{surf}) in 5% Mn sample is seen to be lower than the other two 10% Mn samples (**Figure 6**). It indicates that the defect concentration increases with the increase in Mn percentage both at the surface as well as in bulk. The diffusion length in 5% Mn sample is also higher compared to 10% Mn samples (**Figure 7**), suggesting a low concentration of defects in 5% Mn sample. However, the profiles of S-parameter among the two 10% Mn (AC and HT) samples are almost identical indicating near absence of any modification of defect structure near the surface following heat treatment in 10% Mn sample.

4. Conclusions

It is inferred from the above study that:

1) Defect size and their concentration are found to be less in 5M24 metal mould sample compared to sand moulded one and this further improves with heat treatment according to PLS data. 10% Mn will not change the defects concentration very much.

2) Heat treatment brings out improved microstructural transformation with some defects getting annealed out and hence defect concentration (C_d) has come down.

3) The slow positron beam analysis (DBAR) data reveals that surface, as well as the bulk values of the S-parameter in 5M24 sample is seen to be lower than the two 10M24 samples suggesting less concentration of defects at the surface. Further it is observed that the defect concentration increases slightly with the increase in Mn from 5% to 10% at the surface.

4) The PLS and DBAR techniques can be effectively used to study the defect structure in the bulk and surface of an alloy material which is very important information while designing materials with little erosion volume loss in their final industrial application.

5. Acknowledgements

The authors thank the management of Central Power Research Institute for having accorded permission to publish this paper. The authors wish to acknowledge with thanks Mr. R. K. Kumar & Mr. J. Shankar of MTD, CPRI for their support in conducting the experiments and preparation of the manuscript. One of the authors (Ki-shore) would like to thank CSIR for the award of fellowship under Emeritus Scientist Scheme.

REFERENCES

[1] P. R. Krishnamoorrthy, S. Seetharamu and P. Sampathkumaran, "Erosion Wear in Thermal Power Plants", 55th R and D Session of CBI & P, India, July 1999, p. 1.

[2] J. T. H. Pearce, "Structure and Wear Performance of Abrasion Resistant Chromium White Cast Irons," *AFS Transactions*, Vol. 126, 1984, pp. 599-622.

[3] A. Basak, J. Pening and J. Dellewyns, "Effect of Manganese on Wear Resistant and Impact Strength of 12% Chromium White Cast Iron," *AFS International Cast Metal Journal*, 1981, p. 12.

[4] Y. Kawaguchi and Y. Shirai, "Fatigue Evaluation of Type 316 Stainless Steel Using Positron Annihilation Line Shape Analysis and $\beta \pm \gamma$ Coincidence Positron Lifetime Measurement," *Journal of Nuclear Science and Technology*, Vol. 39, No. 10, 2002, pp. 1033-1040.

[5] A. Van Veen, H. Schut, M. Clement, J. M. M. de Nijs, A. Kruseman and M. R. IJpma, "VEPFIT Applied to Depth Profiling Problems," *Applied Surface Science*, Vol. 85, No. 2, 1995, pp. 216-224.

[6] P. Sampathkumaran, S. Seetharamu and Kishore, "Erosion and Abrasion Characteristics of High Manganese Chromium Irons," *Wear*, Vol. 259, No. 11-6, 2005, pp. 70-77.

[7] P. Hautojarvi, "Positrons in Solids," Springer-Verlag, Berlin, 1979.

[8] W. Brandt and A. Dupasquier, "Positron Solid State Physics," North-Holland, Amsterdam, 1983.

[9] P. K. Pujari, D. Sen, G. Amarendra, S. Abhaya, A. K. Pandey, D. Dutta and S. Mazumder, "Study of Pore Structure in Grafted Polymer Membranes Using Slow Positron Beam and Small-Angle X-ray Scattering Techniques," *Nuclear Instruments Methods Physics Research B*, Vol. 254, No. 2, 2007, pp. 278-282.

[10] M. Tashiro, Y. Honda, T. Yamaguchi, P. K. Pujari, N. Kimura, T. Kozawa, G. Isoyama and S. Tagawa, "Development of a short-Pulsed Slow Positron Beam for Application to Polymer Films," *Radiation Physics Chemistry*, Vol. 60, No. 4-5, 2001, pp. 529-533.

[11] H. B. Ravikumar, C. Ranganathaiah, G. N. Kumaraswamy and Siddaramaiah, "Influence of Free Volume on the Mechanical Properties of Epoxy/Poly (Methylmeth-Acrylate) Blends," *Journal of Material Science*, Vol. 40, No. 24, 2005, pp. 6523-6527.

[12] P. Kirkegaard, N. J. Pederson and M. Eldrup, "PATFIT-88: Riso National Laboratory Report PM-2724," Riso National Laboratory, Riso, 1989.

[13] M. J. Puska, and R. M. Nieminen, "Theory of Positrons in Solids and on Solid Surfaces," *Reviews of modern Physics*, Vol. 66, No. 3, 1994, pp.841-897.

[14] S. Shikata, S. Fujii, L. Wei and S. Tanigawa, "Effect of Annealing Method on Vacancy Type Defects in Si Implanted GaAs Studies by a Slow Positron Beam," *Journal of Applied Physics*, Vol. 31, 1992, pp. 732-736.

[15] P. Sampathkumaran, C. Ranganathaiah, S. Seetharamu and Kishore, "Effect of Increased Manganese Addition and Mould Type on the Slurry Erosion Characteristics of Cr-Mn Iron Systems," *Bulletin Materials Science*, Vol. 31, No. 7, 2008, pp. 1001-1006.

[16] S. Seetharamu, P. Sampathkumaran and R. K. Kumar, "Erosion Resistant of Permanent Moulded High Chromium Iron," *Wear*, Vol. 267, No. 1, 1995, pp. 159-167.

[17] S. N. Kulkarni and K. Radhakrishna, "Evaluation of Metal-Mould Interfacial Heat Transfer during the Solidification of Aluminium −4.5% Copper alloy Castings Cast in CO2-Sand Moulds," *Materials Science*, Vol. 23, No. 3, 2005, pp. 821-838.

[18] E. Raask, "Erosion Wear in Coal Utilization," Hemisphere Publishing Corporation, New York, 1988.

[19] R. Vaidyanathan, J. P. Schaffer and B. Thanaboonsombut, "A Doppler Positron Annihilation Spectroscopy Study of Magnetically Induced Recovery in Nickel," Atlanta, Easton, 1993.

[20] P. J. Schultz, K. G. Lynn, "Interaction of Positron beams with Surfaces, Thin Films, and Interfaces," *Reviews of Modern Physics*, Vol. 60, No. 3, 1989, pp. 701-779.

[21] M. Zhang, R. Scholz, H. Weng and C. Lin, "Defects and Voids in He ± Implanted Si Studied by Slow Positron Annihilation and Transmission Electron Microscopy," *Applied Physics A*, Vol. 66, No. 5, 1998, pp. 521-525.

[22] C. He, T. Suzuki, E. Hamada, H. Kobayashi, K. Kondo, V. P. Shantarovich and Y. Ito, "Characterization of Polymer Films Using a Slow Positron Beam," *Materials Research Innovation*, Vol. 7, No. 1, 2003, pp. 37-41.

[23] Y. C. Jean, R. Zhang, H. Cao, Jen-Pwu Yuan and Chia-Ming Huang, "Glass Transition of Polystyrene Near the Surface Studied by Slow-Positron-Annihilation-Spectro-Scopy," *Physical Reviews B*, Vol. 56, No. 14, 1997, pp. 8459-8462.

20

Improvement in Tribological Properties of Surface Layer of an Al Alloy by Friction Stir Processing

Soheyl Soleymani, Amir Abdollah-zadeh*, **Sima Ahmad Alidokht**

Department of Materials Engineering, Tarbiat Modares University, Tehran, Iran.

ABSTRACT

An innovative technique, friction stir processing (FSP) was employed to modify the surface layer of Al5083 alloy. The FSP passes of 1 to 4 were applied on alloy samples. The processed samples were subjected to microstructural analysis and dry sliding wear test. FSP resulted in microstructural refinement and improvement in wear resistance of Al5083. Moreover, the results indicated that the more number of FSP passes were found to be more effective in improvement of wear resistance, due to more microstructural refinement. It was also found that the load bearing capacity of samples significantly improved with increasing the number of FSP passes.

Keywords: *Friction Stir Processing, Wear, Grain Refinement, Friction Coefficient*

1. Introduction

Surface properties such as wear resistance can determine the service life of components in many industrial applications. Components which are produced by aluminum and its alloys, exhibit poor tribological properties leading even to seizure under detrimental conditions. Hence, There is a strong drive to develop new Al-based materials with greater resistance to wear and better tribological properties [1,2]. Recently, much attention has been paid to FSP that is known as a surface modification technique [3,4]. FSP which was developed based on the principle of friction stir welding (FSW), is remarkably simple. A rotating tool with a pin and shoulder is inserted into a single piece of material and results in significant microstructural changes in the processed zone, due to intense plastic deformation. FSP has been proved to be an effective way to refine the microstructure of aluminum alloys, and thereby improve the mechanical properties [5,6]. Previous investigations [7,8] have indicated that a fine grain structure affects the tribological properties of surface layer. Prasada *et al*. [9,10] reported that the grain refinement leads to the improvement of wear resistance and load bearing capacity of Al-7Si alloy. Chandrashekharaiah and Kori [11] also reported similar results for different Al-based alloys. One of the major problems associated with Al5083 like other aluminum alloys, is their relatively poor wear resistance which limits their tribological performance [12].

The aim of this study is to investigate the effects of multiple FSP passes on microstrutural and tribological properties of Al5083 alloy. Besides, the effects of the applied normal force during sliding have been studied in order to find out the wear mechanisms in details.

2. Experimental Procedure

The base metal was commercially Al5083 rolled plates of 3 mm thickness with a nominal composition of 4.3Mg –0.68Mn–0.15Si–bal. Al (in wt pct). FSP was applied on the surface of the alloy by a tool made of steel H-13 with a shoulder of 20 mm diameter and a pin of 6 mm diagonal length and 2.8 mm height.

Samples were subjected to 1 to 4 FSP passes along the same direction with rotation rate of 1250 rpm and travel speed of 50 mm/min in room temperature.

Microstructural observations were performed on specimens by transmission electron microscopy (TEM). The hardness of the surface composite layers was measured using 31.25 KgF. Wear behavior of the specimens was evaluated by using a pin-on-disk tester at room temperature. Pin specimens with 5 mm diameter were cut from the processed zone of each sample and ground on emery paper up to grade 320. Counterparts were Discs made of AISI D3 steel with hardness of 58 HRc. The tests were carried out at normal loads of 1 to 5 KN and the rotation speed of 60 rpm. The friction force was recorded automatically against sliding distance by the tester software. The wear weight loss was measured with an accuracy of

± 0.01 mg. The worn surfaces of samples were studied using scanning electron microscopy (SEM).

3. Results and Discussions

Figure 1 shows typical TEM microstructure of processed zone for the 1-pass and 4-pass samples. **Figure 2** also shows the variations of grain size of base metal and samples processed at 1 to 4 FSP passes. The results indicated that FSP has led to the grain refinement. Moreover, increasing the number of FSP passes has resulted in decreasing grain size of processed zone. Previous investigations [13,14] have indicated that in FSP/W, a continuous dynamic recrystallization phenomenon occurs due to the tool pin disruptive mechanical action and the frictional heat produced. This phenomenon can lead to intensive microstructural refinement. Increasing the number of FSP passes probably has led to occurrence of more dynamic recrystallization.

Figure 3 shows the mean hardness of base metal and processed zone of 1 to 4-pass samples. As can be seen in this figure, FSP has led to the increase in the hardness of

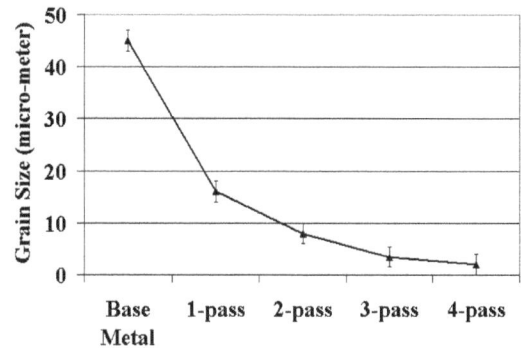

Figure 2. Variations of grain size of base metal and samples processed under 1 to 4 FSP passes.

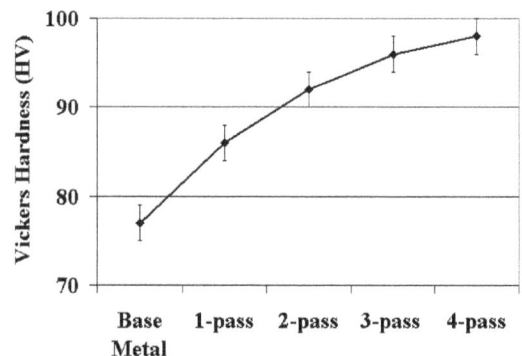

Figure 3. Variations of mean Vickers hardness (HV) of processed zone of base metal and samples produced by different number of FSP passes.

material in comparison to as-received sample. Furthermore, increasing the number of FSP passes has resulted in more increase in hardness of samples. These results are in good agreement with microstructural observations and may be attributed to the increase in yield strength of material owing to microstructural refinement in accordance with Hall-Petch relation [5,6].

Figure 4 shows wear weight loss of base metal and 1 to 4-pass FSP samples. The results indicate that FSP has led to the decrease in wear weight loss compared with as-received metal and hence improving wear resistance. This can be attributed to the increase in hardness due to microstructural refinement. Besides, as can be seen in this figure, wear weight loss of samples decreases with the increase in the number of FSP passes. This observation is in good agreement with the trend observed for the hardness of the samples (**Figure 3**). Increasing the hardness of metal can lead to occurrence of less plastic deformation during sliding [9,10]. The results indicate that FSP can lead to the improvement of wear resistance as well as hardness especially for higher number of FSP passes due to the improvement of microstructural characteristics.

Figure 5 shows the variations of friction coefficient

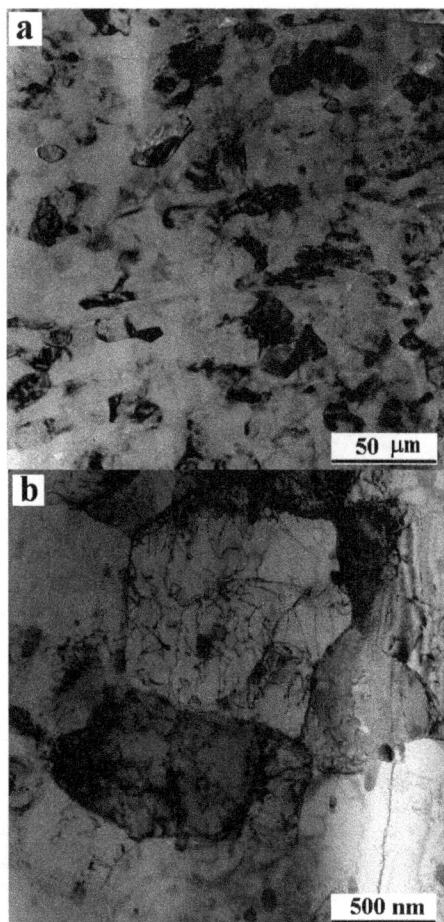

Figure 1. TEM micrographs of the surface of (a) 1-pass and (b) 4-pass samples.

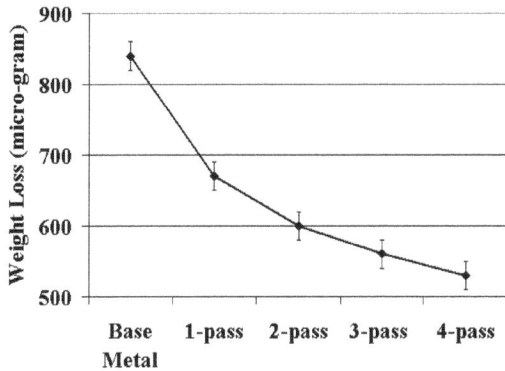

Figure 4. Variations of weight loss of base metal and samples produced by different number of FSP passes.

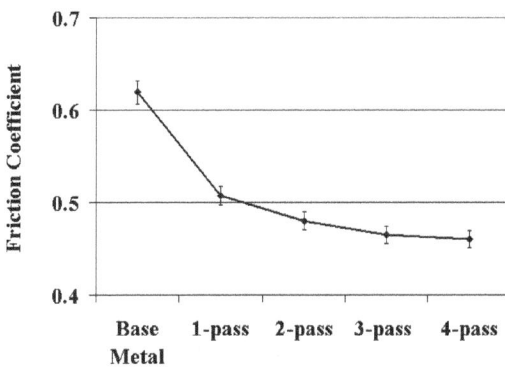

Figure 5. Variations of mean friction coefficient of processed zone of base metal and samples produced by different number of FSP passes.

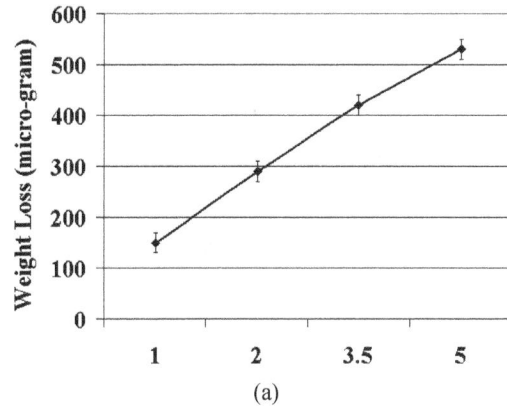

(a)

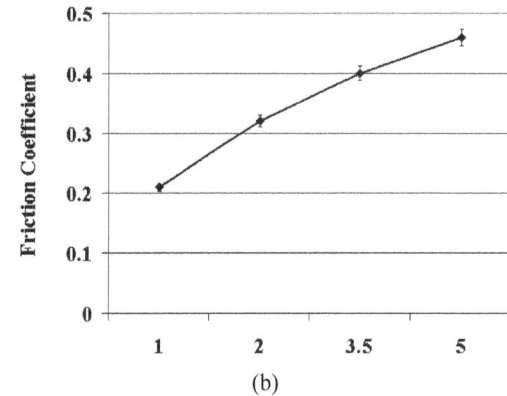

(b)

Figure 6. Variations of (a) weight loss and (b) friction coefficients of 4-pass samples worn under different applied forces (KgF).

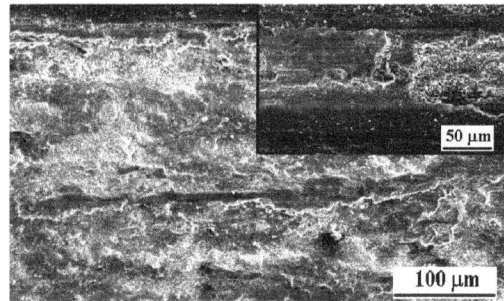

Figure 7. SEM micrograph of surface of base metal sample worn under 5 KN applied force.

for different samples. As can be seen in this figure, FSP has led to the decrease in friction coefficient of samples in comparison to base metal. Moreover, increasing the number of FSP passes has led to the decrease in the amount of mean friction coefficient. These results can be attributed to the differences in the extent of localized plastic deformation at real contact areas of samples and counterface. Harder surfaces of FSP samples have led to less plastic deformation and lower friction coefficient. It seems that higher number of passes results in decreasing plastic deformation on the surface which may be attributed to the microstructural refinement due to FSP [9].

Figure 6(a) shows the variations of weight loss of 4-pass samples worn under different applied forces. As can be seen in this figure, wear weight loss decreases with decreasing the applied force. **Figure 6(b)** also shows that with the decrease in the applied force, friction coefficient decreases. These observations are in good agreement with the wear laws and may be related to the decrease in contact area between counter surfaces which results in decreasing in the weight loss and friction coefficient [15,16].

Figure 7 illustrates the surface of base metal. The signs

of adhesion mechanism can be seen in different parts of the surface of as-received metal sample. The lower hardness of as-received metal has led to the adhesion mechanism. The high friction coefficient of as-received sample may be due to the increase in the friction force needed to overcome the highly adhesive contact between tested surface and counterpart. The high wear weight loss of as-received sample can be the result of the adhesion mechanism operating on the surface during dry sliding. The results indicated that as-received sample has higher level of surface damages among test samples. These may be attributed to lower hardness and higher mean grain size of this sample.

Figure 8 shows the worn surfaces of FSP samples. As can be seen in this figure, delamination and abrasion are active wear mechanisms for the samples produced by 1 to 4 FSP passes and worn under 5 KN applied force. When the number of FSP passes increases, the hardness of samples increases and the amount of grooves and cracks which are associated with delamination mechanism, also decrease and hence 4-pass sample shows the lowest level of surface damages (**Figure 8(d)**). The decrease in the signs of delamination mechanism with increasing FSP passes may be attributed to the improvement of metal resistance to cracks nucleation and growth. This may be due to the increase in toughness of material

resulted by grain refinement [7,8].

As mentioned earlier, friction coefficient of samples decreases with increasing the number of passes. This can lead to the decrease in the level of shear stresses applied to subsurface layers of material during sliding. This phenomenon results in decreasing the probability of nucleation and growth of subsurface cracks during sliding and leads to diminishing the signs of delamination mechanism. Therefore, the wear weight loss decreases with increasing FSP passes due to the higher resistance of material to cracks propagation.

The surfaces of 4-pass samples worn under different applied forces have been illustrated in **Figure 9**. Decreasing the applied force during sliding, has led to the decrease in the existence of subsurface cracks and width

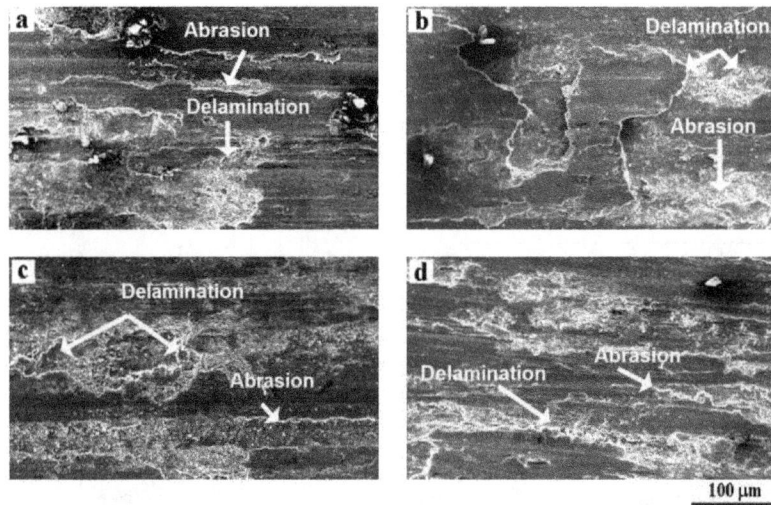

Figure 8. SEM micrographs of surfaces worn under 5 KN applied force for the specimens made from: (a) 1-pass; (b) 2-pass; (c) 3-pass and (d) 4-pass samples.

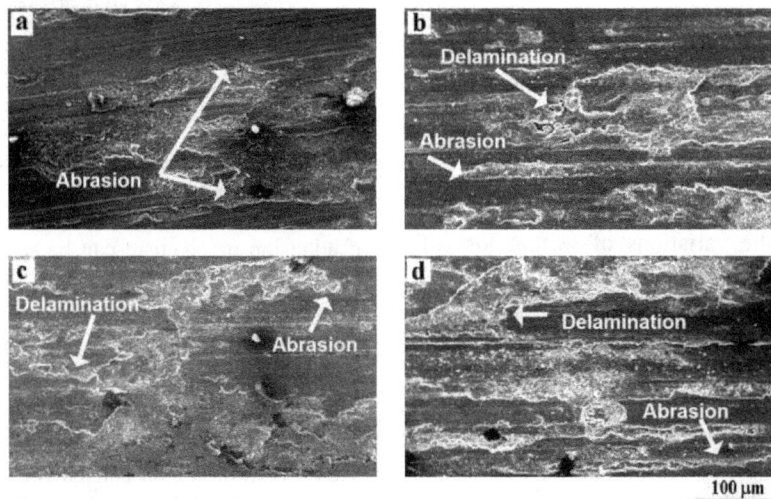

Figure 9. SEM micrographs of the surface of 4-pass samples worn under different applied forces: (a) 1 KN; (b) 2 KN; (c) 3.5 KN and (d) 5 KN samples.

and depth of worn tracks. Besides, for the samples worn under 1 KN applied force, the signs of delamination mechanism disappeared and abrasion was the only active wear mechanism.

As the shear stresses applied to subsurface layers of material during sliding depend directly on the normal load and friction coefficient [16], decreasing the normal load results in decreasing shear stresses and leads to the decrease in the probability of cracks propagation. Similar trends were also observed for 1 to 3-pass samples and can explain diminishing the signs of delamination mechanism with decreasing the applied force.

On the other hand, increasing the number of FSP passes leads to the decrease in friction coefficient which results in decreasing the subsurface shear stresses and signs of delamination mechanism. Therefore, observations indicate that the effects of decreasing the applied force on the worn surface of samples are similar with those related to higher FSP passes. It seems that for the samples with higher number of FSP passes, surface layer can tolerate higher applied forces during sliding. Hence, increasing the number of passes can promote wear resistance and improve the load bearing capacity of samples.

4. Conclusions

In the present investigation, an attempt has been made to study the effects of FSP and the number of passes on hardness and wear resistance of Al5083 alloy. The obtained results can be summarized as follows:

1) FSP resulted in grain refinement of the alloy. Moreover, increasing the number of FSP passes led to the more decrease in the size of grains. These observations could be due to the recrystallization phenomenon.

2) FSP has led to the increase in the hardness of the alloy. Increasing the number of FSP passes also led to the improvement of hardness. These can be attributed to the microstructural refinement.

3) FSP resulted in the improvement of wear resistance of the alloy. Moreover, increasing the number of FSP passes led to the improvement of wear resistance. This can be due to the increase in hardness and microstructural refinement.

4) Wear weight loss and friction coefficient decreased with decreasing the force applied during sliding. Moreover, decreasing the applied force led to diminishing surface grooves and scratches and the signs of delamination mechanism.

5) Increasing the number of FSP passes led to the decrease in weight loss and friction coefficient and also surface damages. This trend has been observed for the samples worn under lower applied forces and indicates that more FSP passes can result in the improvement of load bearing capacity of the alloy during sliding.

REFERENCES

[1]   A. P. Sannino and H. J. Rack, "Dry Sliding Wear of Discontinuously Reinforced Aluminium Composites: Review and Discussion," *Wear*, Vol. 189, No. 1, 1995, pp. 1-19.

[2]   B. C. Pai, T. P. D. Rajan and R. M. Pillai, "Aluminium Matrix Composite Castings for Automotive Applications," *Indian Foundry Journal*, Vol. 50, No. 9, 2004, pp. 30-39.

[3]   S. H. Aldajah, O. O. Ajayi, G. R. Fenske and S. David, "Effect of Friction Stir Processing on the Tribological Performance of High Carbon Steel," *Wear*, Vol. 267, No. 1-4, 2009, pp. 350-355.

[4]   M. Barmouz, P. Asadi, M. K. Besharati Givi and M. Taherishargh, "Investigation of Mechanical Properties of Cu/SiC Composite Fabricated by FSP: Effect of SiC Particles' Size and Volume Fraction," *Materials Science and Engineering*: A, Vol. 528, No. 3, 2011, pp. 1740-1749.

[5]   R. S. Mishra and Z. Y. Ma, "Friction Stir Welding and Processing," *Materials Science and Engineering*: R: Reports, Vol. 50, No. 1-2, 2005, pp. 1-78.

[6]   M. L. Santella, T. Engstrom, D. Storjohann and T. Y. Pan, "Effects of Friction Stir Processing on Mechanical Properties of the cast Aluminum Alloys A319 and A356," *Scripta Materialia*, Vol. 53, No. 2, 2005, pp. 201-206.

[7]   K. G. Basavakumar, P. G. Mukunda and M. Chakraborty, "Influence of Grain Refinement and Modification on Dry Sliding Wear Behaviour of Al-7Si and Al-7Si-2.5Cu Cast Alloys," *Journal of Materials Processing Technology*, Vol. 186, No. 1-3, 2007, pp. 236-245.

[8]   K. G. Basavakumar, P. G. Mukunda and M. Chakraborty, "Influence of Grain Refinement and Modification on Microstructure and Mechanical Properties of Al-7Si and Al-7Si-2.5Cu Cast Alloys," *Materials Characterization*, Vol. 59, No. 3, 2008, pp. 283-289.

[9]   A. K. Prasada, K. Das, B. S. Murty and M. Chakraborty, "Effect of Grain Refinement on Wear Properties of Al and Al-7Si Alloy," *Wear*, Vol. 257, No. 1-2, 2004, pp. 148-153.

[10] A. K. Prasada, K. Das, B. S. Murty and M. Chakraborty, "Microstructure and the Wear Mechanism of Grain-Refined Aluminum during Dry Sliding Against Steel Disc," *Wear*, Vol. 264, No. 7-8, 2008, pp. 638-647.

[11] T. M. Chandrashekharaiah and S.A. Kori, "Effect of Grain Refinement and Modification on the Dry Sliding Wear Behaviour of Eutectic Al-Si Alloys," *Tribology International*, Vol. 42, No. 1, 2009, pp. 59-65.

[12] B. Venkataraman and G. Sundararajan, "Correlation between the Characteristics of the Mechanically Mixed Layer and Wear Behaviour of Aluminium, Al-7075 Alloy

and Al-MMCs," *Wear*, Vol. 245, No.1-2, 2000, pp. 22-38.

[13] K. V. Jata and S. L. Semiatin, "Continuous Dynamic Recrystallization during Friction Stir Welding of High Strength Aluminum Alloys," *Scripta Materialia*, Vol. 43, No. 8, 2000, pp. 743-749.

[14] J. Q. Su, T. W. Nelson, R. Mishra and M. Mahoney, "Microstructural Investigation of Friction Stir Welded 7050-T654 Aluminium," *Acta Materialia*, Vol. 51, No. 3, 2003, pp. 713-729.

[15] G. W. Stachowiak, "Wear—Materials, Mechanisms and Practice," John Wiley & Sons Limited, England, 2005.

[16] ASM Handbook, "Friction, Lubrication, and Wear Technology," Vol. 18, ASM International, Materials Park, Ohio, 1994.

21

The Behaviour of Superalloys in Marine Gas Turbine Engine Conditions

I. Gurrappa[1*], I. V. S. Yashwanth[2], A. K. Gogia[1]

[1]Defence Metallurgical Research Laboratory, Kanchanbagh PO, Hydereabad; [2]M.V.S.R. Engineering College, Nadargul, Hyderabad, India.

ABSTRACT

This paper presents hot corrosion results carried out systematically on the selected nickel based superalloys such as IN 738 LC, GTM-SU-718 and GTM-SU-263 for marine gas turbine engines both at high and low temperatures that represent type I and type II hot corrosion respectively. The results were compared with advanced superalloy under similar conditions in order to understand the characteristics of the selected superalloys. It is observed that the selected superalloys are relatively more resistant to type I and type II hot corrosion when compared to advanced superalloy. In fact, the advanced superalloy is extremely vulnerable to both types of hot corrosion. Subsequently, the relevant reaction mechanisms that are responsible for slow and faster degradation of various superalloys under varied hot corrosion conditions were discussed. Based on the results obtained with different techniques, a degradation mechanism for all the selected superalloys as well as advanced superalloy under both types of hot corrosion conditions was explained. Finally, the necessity as well as developmental efforts with regard to smart corrosion resistant coatings for their effective protection under high temperature conditions was stressed for their enhanced efficiency.

Keywords: *Marine Gas Turbines, Superalloys, Hot Corrosion, Degradation Mechanism, Smart Coatings*

1. Introduction

Improved efficiency is the requirement for all types of modern gas turbines. In particular, achieving enhanced efficiency for marine gas turbines is a major challenge as the surrounding environment is highly aggressive. This aspect depends not only on the design but also on the selection of appropriate materials for their construction. Between the two, selection of materials plays a vital role as the materials have to perform well for the designed period under severe marine environmental conditions. Hot corrosion in a marine environment causes the materials to degrade at a significantly faster rate and causes catastrophic failures. It is important to mention that hot corrosion becomes a limiting factor for the life of components in marine gas turbines. Hence, the focus is on selection of appropriate materials and coatings.

Therefore, advanced materials with considerably improved properties are essential in order to enhance the efficiency of modern gas turbine engines. Efforts made in this direction made it possible to develop an advanced superalloy which exhibits excellent high temperature strength properties [1]. Application of high performance protective coatings over the conventional superalloys is an alternative approach to enhance the efficiency. As mentioned above, the issue is complicated for marine applications by the aggressivity of the environment [2-4]. Thus, the hot corrosion resistance of superalloys is as important as their high temperature strength in gas turbine engine applications [5-11]. An exhaustive review with recent developments as well as fundamentals of hot corrosion in gas turbine engines is available elsewhere [12].

Systematic studies were carried out on the selected superalloys in marine environments and at different elevated temperatures, which simulates the marine gas turbine engines. Comparative studies with an advanced superalloys were also carried out under similar environmental conditions in order to determine the nature of degradation and to establish the possible reaction mechanisms that cause the selected superalloys to corrode under marine environmental conditions before suggesting suitable high performance protective coatings for their effective protection.

2. Experimental

The selected superalloys for the present investigation are presented in **Table 1**. It is to be noted that the selected

Table 1. The chemical composition of selected superalloys (wt%).

Superalloy	Ni	Cr	Co	W	Al	Ta	Ti	Mo	Re	Hf	Fe	Mn	Si	Cu
GTM-SU-263	Bal	20	20	-	0.6	1.3	2.4	6.0	-	-	0.7	0.6	0.4	0.2
GTM-SU-718	52.5	18.5	9.0	6.0	0.5	6.5	0.9	3.0	-	-	19.0	0.2	0.2	5.1 Nb
IN 738 LC	Bal	16	8.5	2.6	3.4	8.5	3.4	1.75	-	-	-	0.2	0.3	0.9 Nb
Advanced alloy	Bal	2.9	7.9	5.8	5.6	8.5	-	-	6.5	0.1	-	-	-	-

superalloys contain no rhenium but sufficient amount of chromium, while newly developed alloys contains about 6.5% rhenium and a very small amount of chromium. The modified chemistry with high contents of rhenium and tantalum makes the advanced superalloys to exhibit very good high temperature strength properties [1]. Small discs of about 3 mm thick were cut from all the superalloys, grounded up to 600 grit surface finish and cleaned with distilled water followed by acetone. Subsequently, the hot corrosion studies were carried out by a salt coating test, in which the specimens were coated with chloride and vanadium containing salts prior to hot corrosion studies at 900°C and 700°C, which represents type I and II hot corrosion respectively. Hot corrosion tests were also carried out at 800°C. The weight change data was recorded initially after 3 hours and later for every 20 hours. Each time, the specimens were washed with hot distilled water, dried and then recorded the weight. After recording the weight, the specimens were re-coated each time with the salt mixture and hot corrosion studies were carried out for a total period of 100 hours.

After completion of hot corrosion tests, the specimens were examined for surface morphology with Scanning Electron Microscope (SEM) and the corrosion products were analyzed by Electron Dispersive Spectroscopy (EDS) and X-ray diffraction techniques. Cross sections of all the corroded specimens were analyzed for understanding the effect of hot corrosion and then elemental distribution was determined in order to evolve their degradation mechanisms.

3. Results and Discussion

Figures 1, **2** and **3** show as hot corroded superalloys in chloride as well as vanadium containing environments under type I and type II hot corrosion conditions. As can be seen, all the selected superalloys were severely corroded at both the temperatures and environments. The corrosion is more severe under type I when compared to type II conditions in the both the environments. It indicates that all the superalloys are highly susceptible to hot corrosion. It is important to notice that the advanced superalloy is more vulnerable to hot corrosion (**Figure 4**) even under type II hot corrosion conditions *i.e.* at 700°C. Of course, the corrosion is much severe under type I

Figure 1. As hot corroded superalloy GTM-SU-263 in chloride and vanadium containing environments under type I and type II conditions.

conditions *i.e.* 900°C. The advanced alloy degrades at a very faster rate making it difficult to recognize over a period of time as evidenced from the experiments. It is clearly indicating that the modified chemistry of the advanced superalloy could not improve its hot corrosion resistance. However, it exhibits very good high temperature strength characteristics. Whereas other selected superalloys exhibits relatively better hot corrosion resistance both under type I and type II conditions and their high temperature strength is less when compared to the advanced superalloy.

Figure 5 demonstrates the weight change data for all the selected superalloys at 800°C for a period of 100 hours. The data revealed that all the superalloys were

Figure 2. As hot corroded superalloy GTM-SU-718 in chloride and vanadium containing environments under type I and type II conditions.

Figure 3. As hot corroded superalloy IN 738 LC in chloride and vanadium containing environments under type I and type II conditions.

Figure 4. As hot corroded advanced superalloy in chloride containing environments under type II conditions.

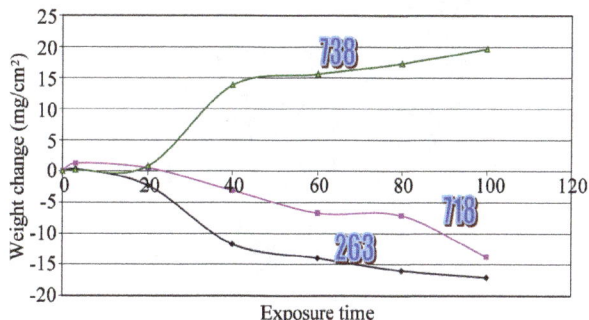

Figure 5. Weight change data for different superalloys at 800°C in vandium environment.

followed the same trend up to 20 hours and thereafter IN 738 LC superalloy forms the scale at a steady rate while other two superalloys form thicker scales and spalled subsequently. This behavior is same for every cycle and entire exposure period. The results obtained at other studied temperatures like 700°C and 900°C in both the environments exhibited similar behavior. It is clearly indicating that all the superalloys are not resistance to hot corrosion at all the temperatures in both the environments.

The surface morphologies of all the selected superalloys as well as advanced superalloy were studied and typical morphology of advanced superalloy at 900°C is presented in **Figure 6**. The surface morphology is different for different superalloys under the selected environmental conditions. EDS measurements revealed that the corrosion products contain sulphides and oxides of nickel and alloying elements of superalloys like Co, Cr, W, Ti, Ta, Re etc. The cross sections of hot corroded superalloys revealed that the corrosion-affected zone is large for all the superalloys. Among them, the affected zone is more for the advanced superalloy indicating that severe corrosion took place during the hot corrosion process for the advanced superalloy when compared to other selected superalloys (**Figure 6**).

The elemental distributions of all hot corroded superalloys including the advanced superalloy were studied in detail and typical elemental distribution of advanced superalloy at 900 and 700°C are illustrated in **Figures 7** and **8** respectively. In case of selected superalloys for marine gas turbines, which contains good amount of chromium could form continuous chromia scale on their surface. Particularly, for IN 738 LC, the continuous and protective chromia was clearly visible. They also promoted alumina as well as titania scales. However, diffusion of sulphur and oxygen into the superalloys was clearly observed. While SU 263 and 718, that contain less chromium could not form continuous chromia scale. Thin alumina scale was observed on the surface of superalloys. Small amounts of sodium and chlorine were also present

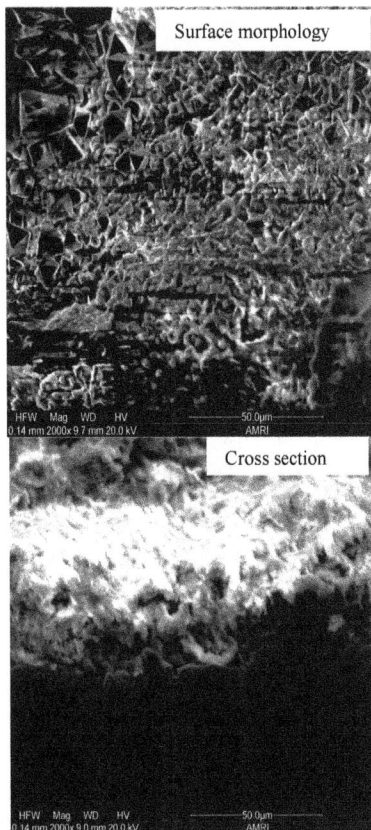

Figure 6. Surface morphology and cross section of advanced superalloy after type I hot corroison in chloride environment.

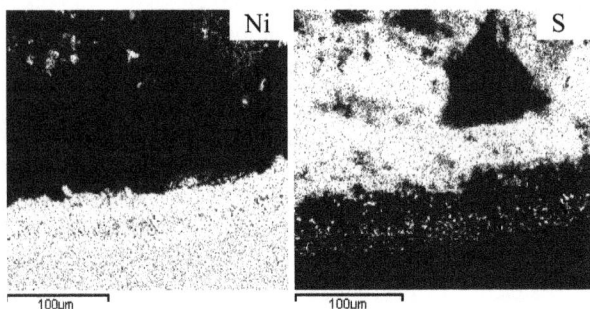

Figure 7. Elemental distribution of an advanced superalloy after hot corrosion at 900°C for 80 hours.

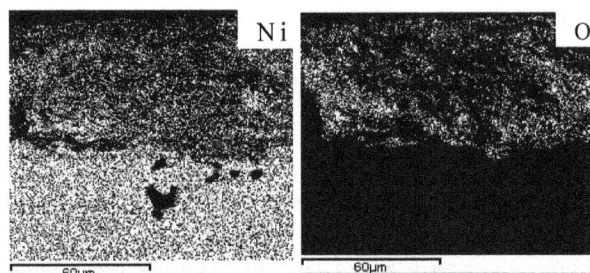

Figure 8. Elemental distribution of an advanced superalloy after hot corrosion at 700°C for 80 hours.

in the corrosion products but not diffused into the superalloys. However, diffusion of sulphur and oxygen into the superalloys was noticed.

The elemental distribution of hot corroded advanced superalloy at 900°C and 700°C showed extensive presence of oxygen, sulphur and sodium in the corrosion products. Considerable diffusion of sulphur into the superalloy was clearly observed at 900°C (**Figure 7**) while oxygen at 700°C (**Figure 8**). Rhenium and tungsten were present in the corrosion products at 900°C and they were present in the corrosion affected zone of advanced superalloy that was hot corroded at 700°C. Ta and Hf were seen in the corrosion affected region. It is important to mention here that neither alumina nor chromia formation was observed on the superalloy. It is due to the fact that chromium content in the adavanced superalloy is considerably low. At the same time, other alloying elements could not form any protective oxide scales. In essence, it is concluded that the advanced superalloy is highly susceptible to hot corrosion, though it exhibits excellent high temperature strength properties. It is important to mention here that the selected superalloys are also vulnerable to both types of hot corrosion but the intensity of attack is less. Among the selected superalloys, the IN 783 LC is more resistant while SU 263 is moderate and SU 718 is more susceptible. It clearly stresses the need to apply high performance protective coatings for their protection against hot corrosion both at low and high temperatures *i.e.* type II and type I as the marine gas turbine engines encounter both the problems during service. The protective coatings allow the marine gas turbine engines to operate at varied temperatures and enhance their efficiency by eliminating failures during service.

In essence, the present results clearly revealed that the selected superalloys as well as advanced superalloy are highly vulnerable to hot corrosion. The results further revealed that the advanced superalloy corrodes much faster when compared to selected superalloys. It is attributed to the fact that the tungsten which is the alloying element added along with other alloying elements in order to obtain high temperature strength characteristics of the superalloys, forms acidic tungsten oxide (WO_3) due to which fluxing of protective oxide scales such as alumina and chromia takes place very easily. This type of acidic fluxing is self-sustaining because WO_3 forms continuously that cause faster degradation of superalloys under marine environmental conditions at elevated temperatures. The degradation mechanism is explained in two steps as follows:

a) The tungsten present in the superalloys reacts with the oxide ions present in the environment and forms tungsten ion

$$WO_3 + O^{2-} = WO_4^{2-}$$

b) As a result, the oxide ion activity of the environment decreases to a level where acidic fluxing reaction with the protective alumina and chromia can occur

$$Al_2O_3 = Al^{3+} + O^{2-}$$

$$Cr_2O_3 = Cr^{3+} + O^{2-}$$

A similar reaction mechanism occurs if the superalloys contain other refractory elements like vanadium and molybdenum.

The following section describes an electrochemical phenomenon that explains the selected superalloys as well as advanced superalloy degradation process in detail under hot corrosion conditions:

Hot corrosion of all superalloys take place by oxidation of base as well as alloying elements like nickel, cobalt, chromium, aluminium, tantalum, rhenium etc. at the anodic site and forms Ni^{2+}, Co^{3+}, Cr^{3+}, Al^{3+}, Re^{4+}, Ta^{5+} ions etc. while at the cathodic site, SO_4^{2-} reduces to SO_3^{2-} or S or S^{2-} and oxygen to O^{2-}. Since the metal ions i.e. Ni^{2+}, Co^{3+}, Cr^{3+}, Al^{3+}, Re^{4+}, Ta^{5+} ions etc. are unstable at the elevated temperature and therefore reacts with the sulphur ions to form metal sulphides. The metal sulphides can easily undergo oxidation at elevated temperatures and form metal oxides by releasing free sulphur $(MS + 1/2O_2 = MO + S)$. As a result, sulphur concentration increases at the surface of superalloys and enhances sulphur diffusion into them and form sulphides inside the superalloys. The practical observation of sulphides in hot corroded superalloys specimens clearly indicates that the electrochemical reactions took place during their hot corrosion process. Simultaneously, the metal ions react with oxide ions that are evolved at the cathodic site leading to the formation of metal oxides. The metal oxides dissociate at elevated temperatures to form metal ions and oxide ions. As a result, oxygen concentration increases at the surface and thereby diffuses into the superalloys. Practical observation of oxides in hot corroded superalloys specimens is a clear indication that the electrochemical reactions took place during the hot corrosion process.

Therefore, the hot corrosion of all the alloys i.e. selected and advanced superalloy, is electrochemical in nature and the relevant electrochemical reactions are shown below:

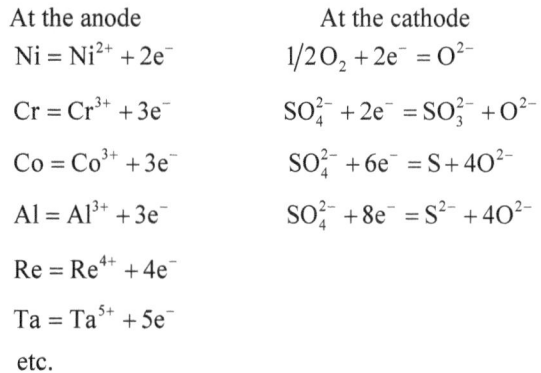

At the anode	At the cathode
$Ni = Ni^{2+} + 2e^-$	$1/2O_2 + 2e^- = O^{2-}$
$Cr = Cr^{3+} + 3e^-$	$SO_4^{2-} + 2e^- = SO_3^{2-} + O^{2-}$
$Co = Co^{3+} + 3e^-$	$SO_4^{2-} + 6e^- = S + 4O^{2-}$
$Al = Al^{3+} + 3e^-$	$SO_4^{2-} + 8e^- = S^{2-} + 4O^{2-}$
$Re = Re^{4+} + 4e^-$	
$Ta = Ta^{5+} + 5e^-$	

etc.

Figure 9 illustrates an electrochemical model showing that all the studied superalloys degradation is electrochemical in nature. Similar mechanism is applicable to other superalloys and their families.

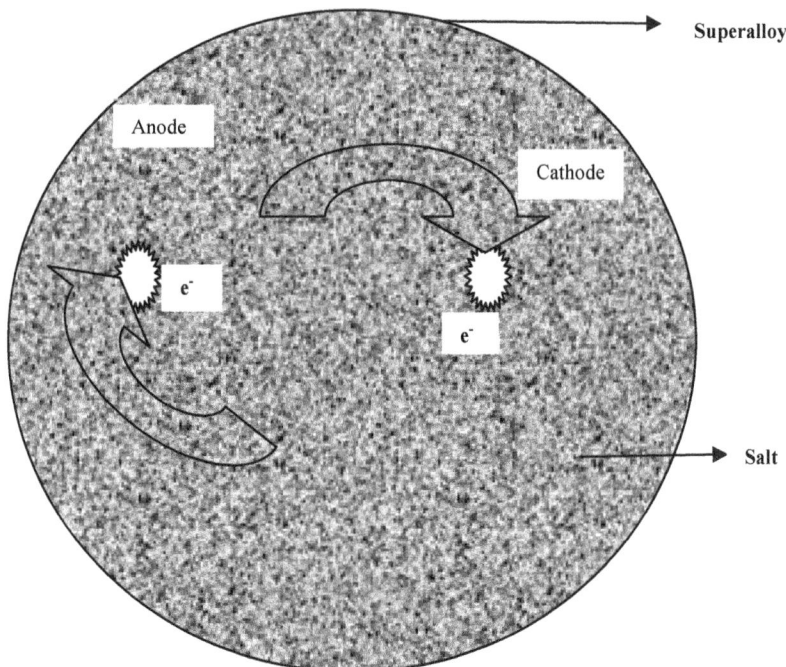

Figure 9. An electrochemical model showing the hot corrosion of selected and advanced superalloys is an electrochemical process.

The template is used to format your paper and style the text. All margins, column widths, line spaces, and text fonts are prescribed; please do not alter them. You may note peculiarities. For example, the head margin in this template measures proportionately more than is customary. This measurement and others are deliberate, using specifications that anticipate your paper as one part of the entire proceedings, and not as an independent document. Please do not revise any of the current designations.

4. Design and Development of Smart Coatings

Recent extensive research has resulted in design and development of smart coatings which provide effective protection to the superalloy blades for the designed period against type I and type II hot corrosion that are normally encountered in marine and industrial gas turbine engines which in turn enhances their efficiency considerably [13,14]. The same coating can be applied to aero gas turbine engine components due the fact that hot corrosion is a concern when they move at low altitudes across the sea and provide total protection. This is a major developmental work in the area of gas turbine engines used in aero, marine and industrial applications. Unlike the conventional/existing coatings, the smart coatings provide total protection to the superalloy components used in aero, marine and industrial applications by forming appropriate protective scales depending on the surrounding environmental conditions [12-14].

REFERENCES

[1] N. Das, US patent 5,925,198, July 1999

[2] M. R. Khajavi and M. H. Shariat, "Failure of First Stage Gas Turbine Blades," *Engineering Failure Analysis*, Vol. 11, No. 4, 2004, pp. 589-597.

[3] J. M. Gallardo, J. A. Rodrigue and E. J. Herrera, "Failure of Gas Turbine Blades," *Wear*, Vol. 252, No. 3-4, 2002, pp. 264-268.

[4] N. Eliaz, G, Shemesh and R. M. Latarision, "Hot Corrosion in Gas Turbine Components," *Engineering Failure Analysis*, Vol. 9, No. 1, 2002, pp. 31-43.

[5] M. Konter and M. Thumann, "Materials and Manufacturing of Advanced Industrial Gas Turbine Components," *Journal of Materials Processing Technology*, Vol. 117, No. 3, 2001, pp. 386-390.

[6] J. Stringer, "High Temperature Corrosion of Superalloys," *Materials Science and Technology*, Vol. 3, 1987, pp. 482-493

[7] A. S. Radcliff, "Factors Influencing Gas Turbine Use and Performance," *Materials Science and Technology*, Vol. 3, 1987, pp. 554-561

[8] R. F. Singer, "New Materials for Industrial Gas Tubines," *Materials Science and Technology*, Vol. 3, 1987, pp. 726-732

[9] I. Gurrappa, "Hot Corrosion Behaviour of CM 247 LC Alloy in Na_2SO_4 and NaCl Environments," *Oxidation of Metals*, Vol. 51, No. 5-6, 1999, pp. 353-382.

[10] C. J. Wang and J. H. Lin, "The Oxidation of MAR M247 Superalloy with Na_2SO_4 Coating," *Materials Chemistry and Physics*, Vol. 76, No. 2, 2002, pp. 123-129.

[11] I. Gurrappa and A. S. Rao, "Thermal Barrier Coatings for Enhanced Efficiency of Gas Turbine Engines," *Surface and Coating Technology*, Vol. 201, No. 6, 2006, pp. 3016-3029.

[12] I. Gurrappa, I. V. S. Yashwanth, A. K. Gogia, H. Murakami and S. Kuroda, *Interlational Materials Revews.* (In Press)

[13] I. Gurrappa, "Identification of a Smart Bond Coating for Gas Turbine Engine Applied Lications," *Journal of Coating Technology Research*, Vol. 5, No. 3, 2008, pp. 385-390.

[14] I. Gurrappa, "Final Report on Design and Development of Smart Coatings for Aerospace Applied Lications," European Commission, July 2008

22

Interfacial Actions and Adherence of an Interpenetrating Polymer Network Thin Film on Aluminum Substrate

Weiwei Cui[1,2], Dongyan Tang[1,*], Jie Liu[1], Fan Yang[1]

[1]Department of Chemistry, Harbin Institute of Technology, Harbin, China; [2]College of Materials Science and Engineering, Harbin University of Science and Technology, Harbin, China.

ABSTRACT

The interpenetrating polymer networks (IPN) thin film with the –C=O group in one network and the terminal –N=C=O group in another network on an aluminum substrate to reinforce the adherence between IPN and aluminum through interfacial reactions, were obtained by dip-pulling the pretreated aluminum substrate into the viscous-controlled IPN precursors and by the following thinning treatment to the IPN film to a suitable thickness. The interfacial actions and the adhesion strengths of the IPN on the pretreated aluminum substrate were investigated by the X-ray photoelectron spectroscopy (XPS), Fourier transform infrared spectroscopy (FTIR) and strain-stress(σ-ε) measurements. The XPS and FTIR detection results indicated that the elements' contents of N, O, and Al varied from the depths of IPN. The interfacial reaction occurred between the –N=C=O group of IPN and the AlO(OH) of pretreated aluminum. The increased force constant for –C=O double bond and the lower frequency shift of –C=O stretching vibration absorption peak both verified the formation of hydrogen bond between the –OH group in AlO(OH) and the –C=O group in IPN. The adherence detections indicated that the larger amount of –N=C=O group in the IPN, the higher shear strengths between the IPN thin film and the aluminum substrate.

Keywords: *Interfacial Action, Adherence, Interpenetrating Polymer Network (IPN), Aluminum, Thin Film*

1. Introduction

Adsorption of polymer onto the surface of metal is of great importance in such applications as the adhesive bonding, corrosion protection, colloid stabilization and many other areas [1,2]. But the weak linkage between the two different materials affects the composites' stability, and therefore restricts their long-term applications [3,4]. Reinforcing the interfacial adhesion between metal and polymer is of worldwide scientific and technical significance [5,6].

In recent years, the role of polymers containing nitrogen- or oxygen-functionalities in polymer-metal adhesion has been an important subject [7,8]. It has been reported that the adhesion strength between polymer and metal could be enhanced significantly by the incorporation of nitrogen- or oxygen-containing functionalities into polymer [9-11]. This illustrates that the reactions between polymer and metal can improve the adhesion effectively.

However, there usually exist the complex structures for polymer, the determination of the polymer reaction sites with metals facilitates the choice of metal/polymer systems, especially when specific properties such as interfacial stability and adhesion are required [12,13].

2. Experimental

2.1. Pretreatment of Aluminum Substrate

Aluminum substrate (1.5 cm × 2 cm) were obtained by cutting, wiping the aluminum sheet with acetone, drying and then immersing in boiling deionized water for 3 min. After removal from the water, the sheet was allowed to dry under ambient conditions for 10 min before further treatment [14]. Then it was dipped into the dichromate salt/concentrated sulfuric acid lotion for 0.5 min - 2 min, rinsed with deionized water and sodium carbonate solution for 10min. After that, it was rinsed again with large amount of deionized water, scrubbed with acetone and put into the boiled water for 3 min, then dried in air.

2.2. Preparation of an IPN Thin Film on an Aluminum Substrate

The pretreated aluminum substrate was dipped into the IPN precursors (with fixed mass ratio (1:1) of cross linker (trimethylolpropane) to chain expander (1,4-butylene glycol), both were dehydrated at 110°C for 4 hours before use), fixed mass rate (100:2:1) of vinyl ester resin to initiator (benzoyl peroxide) and catalyzer (cobalt naphthenate), and variable mol ratio of prepolymer (-NCO) to curing agent (-OH)). Through controlling the viscous of precursors and the lifting speed of the aluminum substrate into the precursors, the IPN thin film on the aluminum substrate(represent as IPN-Al) can be obtained. Then the IPN-Al was place into a heat oven at 80°C for 1 hour. After that, it was immersed into ethyl acetate for 30min and then scrubbed with acetone to the required thickness.

2.3. The Viscous Determination and the Morphology Observation

The viscous of IPN precursors were determined by a NDJ-1 rotational viscometer (Shanghai Optical Apparatus Co., CN). The thickness of IPN thin film was measured by a HG-1060 thickness gauge (Peijing Measuring & Cutting Tool Co. CN). The surface micro morphologies of the aluminum substrate (before and after treatment) were detected by a XSZ-H metallographic microscope (Nippon Electric Co., JP).

2.4. The Interfacial Actions Detection and the Adhesion Strengths Measurements

The surface elemental compositions of the IPN and the aluminum substrate were determined by a PHI 5700 ESCA XPS systems (Physical Electronics, USA) employing an Al Kα X-ray source (energy is 1486.6°eV) and a precise hemispherical electronic energy analyzer. The transfer energy mode was fixed at 12.5 kV and 250 W with an operating chamber pressure of approximately 10^{-6} Torr. The survey spectrum was recorded at a constant energy of 187.8 eV. The instrumental error in terms of the binding energy was within ± 0.1eV. Data were recorded at different incident angles after neutralization of the charges. The existence of groups and their interaction types or sites of the IPN-Al interface were detected by an AVATAR 360 infrared spectrometer (Nicolet Co. USA) with the solution values of 4 cm^{-1} and an incident angle of 22°. The adhesion strength tests were preceded by applying the IPN precursors onto the edge of an aluminum substrate and then overlapping by another aluminum substrate onto the precursors. The lap area would be 20 mm in width and 12.5 ± 0.5 mm in length. This assembly

was tightly clamped under 1MPa. The adhesion strength was measured by a Z050 electrical universal tester (Zwick Co. DN) according to the GB7124-1986 standard method at a pull rate of 5 mm/min. The lap shear strength was calculated by dividing the strength by the lap area. Three sets of samples for each variable component ratios of the IPN were tested, and the average was reported as the lap shear strength.

3. Results and Discussion

3.1. The Micro Morphologies of the Pretreated Aluminum Substrate Surface

The observation of the aluminum substrate (before and after treatment) with the 100 times magnification coefficient by the metallographic microscope was given in **Figure 1**.

As shown in **Figure 1**, the flat surface of the aluminum with metal textures and metal shines(a) was changed into a coarse and porous surface(b) after treatment.

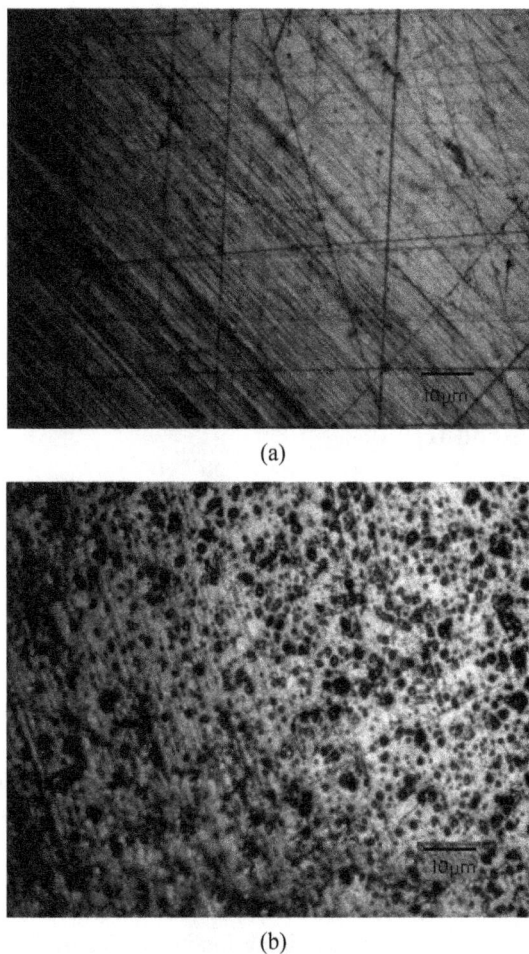

(a)

(b)

Figure 1. The metallographic microscope photos; (100 times magnification) of aluminum; (before treatment (a) and after treatment(b)).

Since the actual surface area(a') after treatment became larger than the original surface area(a) for the aluminum, so the roughness factor(φ, which equals to a'/a) was larger or equal to 1. According to the literature [15], the roughness factor also equals to the product of $\cos\theta$ (θ refers to the contact angle of the IPN precursors onto the original aluminum) with the surface tension of liquid (γ_{lv}, which equals to the minus of the surface tension of solid to the interface tension of liquid-solid). Since $\cos\theta' \geq \cos\theta$, so the wetting ability of coarse surface ($\Delta Fi'$) equals to $\varphi(\gamma_{sv} - \gamma_{sl})$ or $\gamma_{lv}\varphi\cos\theta$. The adhesion work of the coarse surface (W_A') equals to $\gamma_{lv} + \varphi(\gamma_{sv} - \gamma_{sl})$ or $\gamma_{lv}(1 + \varphi\cos\theta)$. While under a very well wetting condition, $\theta < 90°$ and $\cos\theta > 0$, so $\Delta Fi' > \Delta Fi$ (ΔFi, the original wetting ability) and $W_A' > W_A$ (W_A, the original work). Thus the porous aluminum in diameters of 0.5 μm - 2 μm increased the surface areas greatly and thus produced intermolecular forces and electrostatic attractions when contacted with the IPN. Further, the possibility of the formation of chemical bonds within the interface between the IPN and the aluminum increased apparently [16].

3.2. The Relationship of the Thickness of the Thin Film with the Viscous of the IPN Precursors

The relationship of the thickness of the IPN thin film with the viscous of the IPN precursors was shown in **Figure 2**. The lower viscous of the IPN precursors led to the thinner IPN film onto the aluminum substrate (which was suitable for the following interfacial reaction detections). But lower viscous meant less crosslink degree within the each network of IPN, and also meant a lower strength values between the IPN and the aluminum substrate. So in present studies, the thin film obtained by the dip-pulling method was then thinned by a following solvent extraction and acetone scrubbing.

3.3. The Elemental States and the Elemental Contents of the IPN-Al on the Different Depths of the Interface

The XPS spectra of C1s at different incident angles with different depths were shown in **Figure 3**(the incident angles were at 45°, 75° and 90°, respectively) by setting Al_2O_3 with the binding energy of 73.82eV as the standard to neutralize the charges. The content of carbon decreased with the increase of depths from the surface of IPN to the interface of IPN-Al. And the binding energy of carbon shifted toward the higher frequencies gradually with the increase of the incident angles. These illustrated that the chemical environment of carbon at the IPN-Al interface has been changed to form the chemical bond at the higher frequency ranges.

Changes of the contents of O、C、Al、N at the different depths of IPN-Al interface were shown in **Figure 4** by the XPS detection results at different incident angles.

If the incident angle(θ) is 90°, and the depth is represented as d, then the depth at θ is $d\sin\theta$. When θ is 90°, d is approximately 10nm. The content of carbon decreased, whereas the content of oxygen, aluminum, and nitrogen increased with the increase of the depths. These results suggested that, the higher contents of N and O appeared at places closer to the Al substrate. Therefore, the higher reactive activities could be achieved at the interface between the nitrogen- and oxygen containing functionalities in IPN and the aluminum.

3.4. The Interfacial Actions and the Adhesion Strengths of IPN-Al

The comparisons of the characteristic peaks in FTIR spectra of IPN and IPN-Al were listed in **Table 1**. The absorption peak at 1411.71 cm^{-1}, which assigned to the

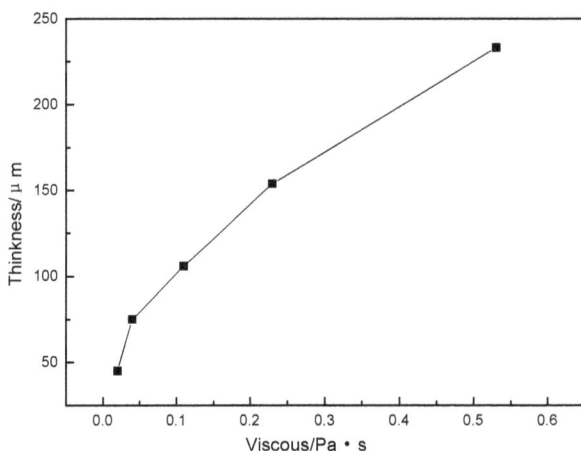

Figure 2. The relationship of the thickness of the IPN thin film with the viscous of the IPN precursors.

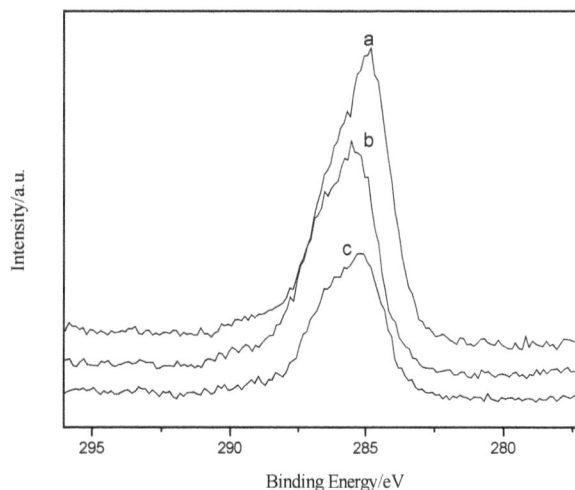

Figure 3. XPS spectra of C1s of the IPN-Al interface at different incident angles(a-45°; b-75°; c-90°).

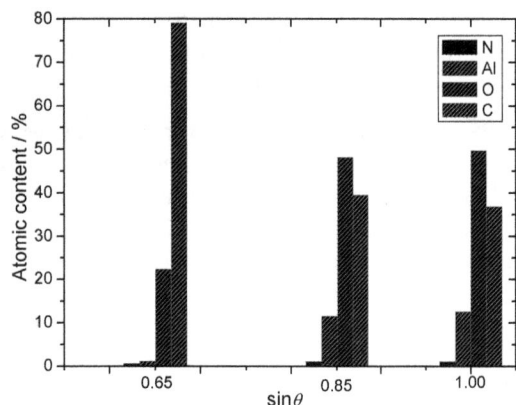

Figure 4. Relationship between the elemental contents and the incident angles of IPN-Al.

Table 1. Attribution of the characteristic absorption peaks in FTIR spectra of IPN and IPN-Al.

Wavenumber shiftness /cm^{-1}	Intensity	Attribution of the characteristic absorption peaks
1731.85→1726.65 (IPN) (IPN-Al)	Strong	Ester C=O, amide II bond, C=O stretching vibration
1411.71→none (IPN) (IPN-Al)	Weak	N=C=O symmetrical stretching Vibration
1237.99→1227.25 (IPN) (IPN-Al)	Strong	Ester C-O stretching vibration

weak symmetrical stretching vibration of –N=C=O, disappeared in the spectrum of IPN-Al, verified the reaction between the little amount of –N=C=O group remained in IPN with the AlO(OH) in aluminum substrate. Since the formation of hydrogen bond could decrease the force constant for –C=O double bond, and further produce the lower frequency shift for the –C=O group. So the lower frequencies shift of the –C=O stretching vibration peak from 1731.85 cm^{-1} for IPN to 1726.65 cm^{-1} for IPN-Al, the –C-O stretching vibration peak from 1237.99 cm^{-1} for IPN to 1227.25 cm^{-1} for IPN-Al both inferred the formation of hydrogen bond between the –C=O in IPN and the –OH in AlO(OH).

The comparisons of the amide II bond stretching vibration peaks in FTIR spectra were shown in **Figure 5** to verify the formation of hydrogen bond by the –C=O with the –OH group in AlO(OH) further.

The –C–N stretching vibration peak at 1508.14cm^{-1} in IPN was divided into two peaks in IPN-Al. The formation of hydrogen bond decrease the electronic cloud density of –C=O to stabilize the carbon atom, so the electronic cloud density of –C–N increased. So the increased force constant for C–N bond further led to the higher frequency shift for –C–N absorption peak. The –C–N stretching vibration peak at 1517.78 cm^{-1} was induced by the formation of hydrogen bond. The reaction could be represented as follows:

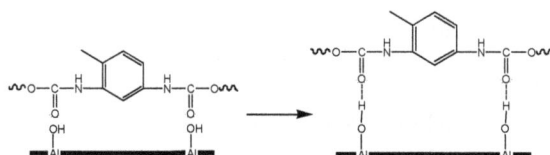

But the obvious higher reactive ability of the –N=C=O with the AlO(OH) than that –C=O in urethane with the AlO(OH) prohibited the reaction between the AlO(OH) and the –C=O mostly, so there only existed a weak hydrogen bond at IPN-Al interface.

The adhesion strengths between aluminum and IPN with different component ratios were shown in **Figure 6** and the calculated results were listed in **Table 2**.

The adhesion strengths of IPN-Al with more –N=C=O group were larger than that with less one. This indicated that the more chemical reactions between more –N=C=O group and pretreated aluminum resulted in a significant increase in adhesion strength. The reduced curing time led by the increased curing agent amount, reduced the elasticity and reinforced the brittleness of IPN. So the shear failure types of IPN changed from the interface failure (in **Figure 6(b)**) to the cohesion failure (in **Figure 6(a)**).

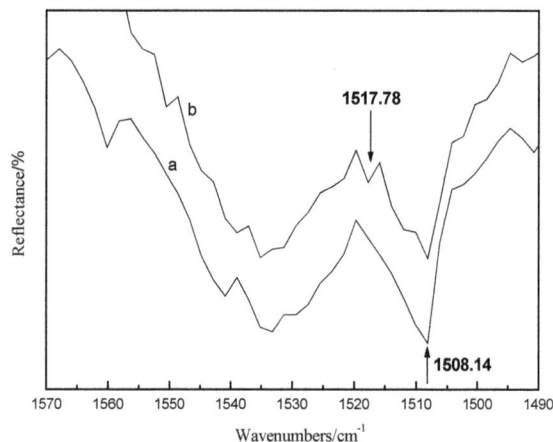

Figure 5. Comparisons of amide II absorption peaks in FTIR spectra of IPN (a) and IPN-Al (b).

Table 2. The shear strengths with different component ratios of prepolymer to curing agent in IPN.

Curing agent/ Prepolymer	Shear strength /kPa	Average Shear strength/kPa	Average deviation	Standard deviation
a) 1.628(mol)	343.66 314.08 269.52	309.09	26.38	30.473
b) 1.385(mol)	452.72 450.10 417.61	440.14	15.02	15.969

(a)

(b)

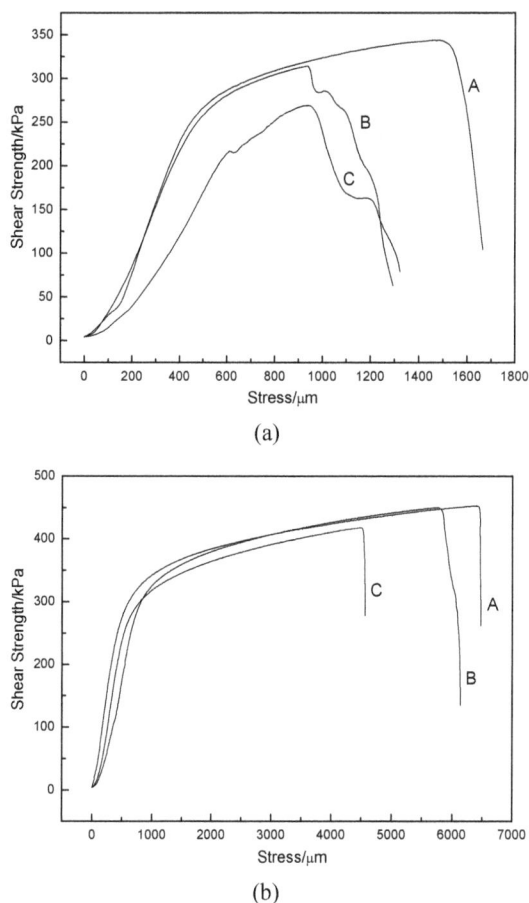

Figure 6. The shear strength curves of IPN-Al with curing agent to prepolymer(mol) in IPN of (a)-1.628 and (b)-1.385.

4. Conclusions

The pretreated porous aluminum increased the surface areas greatly and thus increased the possibilities to form the chemical bonds at interface of the IPN thin film and the aluminum substrate apparently by the intermolecular force and the electrostatic attractions. The higher reactive activities can be achieved at the interface between the nitrogen- and oxygen containing functionalities in IPN and the aluminum. The chemical bond between –N=C=O group and AlO(OH), and the weak hydrogen bond between the –N=C=O group and the –OH group (in AlO (OH)) can be formed. These interactions reinforced the shear strength apparently, especially when the IPN contained a larger amount of –N=C=O group.

5. Acknowledgements

This work was supported by Program for New Century Excellent Talents In Universities (NCET-08-0165).

REFERENCES

[1] D. H. Kim, K. H. Kim, W. H. Jo and J. Kim, "Studies on Polymer-Metal Interfaces, 3a An Analysis of Interfacial Characteristics between Aminefunctionalizedpolystyrene/ Copper and between Hydroxyl-Functionalized Polystyrene/Copper," *Macromolecular Chemistry and Physics*, Vol. 201, No. 18, 2000, pp. 2699-2704.

[2] E. Sabatini, J. C. Boulakia, M. Bruening and I. Rubinstein, "Thioaromatic Monolayers on Gold: A New Family of Self-Assembling Monolayers," *Langmuir*, Vol. 9, No. 11, 1993, pp. 2974-2981.

[3] J. F. Watts, A. Rattana and M.-L. Abel, "Interfacial Chemistry of Adhesives on Hydrated Aluminum and Hy-Drated Aluminum Treated with an Organosilane," *Surface and Interface Analysis*, Vol. 36, No. 11, 2004, pp. 1449- 1468.

[4] M. Öhman and D. Persson, "An Integrated in Situ ATR-FTIR and EIS Set-Up to Study Buried Metal-Polymer Interfaces Exposed to an Electrolyte Solution," *Electrochimica Acta*, Vol. 52, No. 16, 2007, pp. 5159-5171.

[5] M. R. Alexander, S. Payan and T. M. Due, "Interfacial Interactions of Plasma-Polymerized Acrylic Acid and an Oxidized Aluminium Surface Investigated Using XPS, FTIR and Poly(Acrylic Acid) as a Model Compound," *Surface and Interface Analysis*, Vol. 26, No. 13, 1998, pp. 961-973.

[6] G. Y. Seoung, I. Y. Kim, I. K. Sun and S. J. Kim, "Swelling and Electroresponsive Characteristics of Interpene-Trating Polymer Network Hydrogels," *Polymer International*, Vol. 54, No. 8, 2005, pp. 1169-1174.

[7] L. J. Atanasoska, S. G. Anderson, H. M.Meyer, Z. Lin and J. H. Weaver, "Aluminum/Polyimide Interface Formation: An X-ray Photoelectron Spectroscopy Study of Selective Chemical Bonding," *Journal of Vacuum Science & Technology A*, Vol. 5, No. 6, 1987, pp. 3325-3333.

[8] A. Selmani, "Theoretical Investigation of Chemical Bonding at Aluminum/Polyimide Interface," *Journal of Vacuum Science & Technology A*, Vol. 8, No. 1, 1990, pp. 123-126.

[9] A. Calderone, R. Lazzaroni and J. L. Bre´das, "A Theoretical Study of the Interfaces between Aluminum and Poly(Ethylene Terephthalate), Polycaprolactone, and Polystyrene: Illustration of the Reactivity of Aluminum Towards Ester Groups and Phenyl Rings," *Macromolecular Theory and Simulations*, Vol. 7, No. 5, 1998, pp. 509-520.

[10] J. van den Brand, W. G. Sloof and H. Terryn, "Correlation between Hydroxyl Fraction and O/Al Atomic Ratio as Determined from XPS Spectra of Aluminum Oxide Layers," *Surface and Interface Analysis*, Vol. 36, No. 1, 2004, pp. 81-88.

[11] M. R. Alexander, G. Beamson and C. J. Blomfield, "In-

teraction of Carboxylic Acids with the Oxyhydroxide Surface of Aluminum: Poly(Acrylic Acid), Acetic Acid and Propionic Acid on Pseudoboehmite," *Journal of Electron Spectroscopy and Related Phenomena*, Vol. 121, No. 1-3, 2001, pp. 19-32.

[12] H. Hu, J. Saniger, J. Garcia-Alejandre and V. M. Castaño, "Fourier Transform Infrared Spectroscopy Studies of the Reaction between Polyacrylic Acid and Metal Oxides," *Materials Letters*, Vol. 12, No. 4, 1991, pp. 281-285.

[13] J. Yu, M. Ree, Y.H. Park, T.J. Shin, W. Cai, D. Zhou and K.-W. Lee, "Adhesion of Poly(4,4-Oxydiphenylene Pyromellitimide) to Copper Metal Using a Polymeric Primer: Effects of Miscibility and Polyimide Precursor Origin," *Macromolecular Chemistry and Physics*, Vol. 201, No. 5,

2000, pp. 491-499.

[14] P. G. Roth and F. J. Boerio, "Surface-Enhanced Raman Scattering from Poly(4-Vinyl Pyridine)," *Journal of Polymer Science Part B: Polymer Physics*, Vol. 25, No. 9, 1987, pp. 1923-1933.

[15] W. Possart, C. Bockenheimer and B. Valeske, "The State of Metal Surfaces after Blasting Treatment Part I: Technical Aluminum," *Surface and Interface Analysis*, Vol. 33, No. 8, 2002, pp. 687-696.

[16] M. Öhman, D. Persson and C. Leygraf, "In Situ ATR-FTIR Studies of the Aluminium/Polymer Interface upon Exposure to Water and Electrolyte," *Progress in Organic Coatings*, Vol. 57, No. 1, 2006, pp. 78-88.

23

Importance of Surface Preparation for Corrosion Protection of Automobiles

Narayan Chandra Debnath

Department of Physics, Institute of Chemical Technology (ICT), Mumbai, India.

ABSTRACT

An overview of science and technology of pretreatment process suitable for automotive finishing with cathodic electro-deposition primer is presented in details in this paper. Both the theoretical principles and practical aspects of tricationic phosphating process that are used in automotive industry are discussed in details. The characteristic features of phosphate coatings of both conventional high zinc phosphating formulations and modern tricationic phosphating formulations on steel surface are compared in details by SEM, EDX and XRD techniques. The corrosion protections of the phosphated and painted steel panels were evaluated by both salt spray test and electrochemical impedance spectroscopy (EIS). The analysis of impedance data in terms of pore resistance (Rpo), coating capacitance (Cc) and breakpoint frequency (f_b) as a function of salt spray exposure time provides a clear insight into the mechanism of superior corrosion resistance provided by the modern tricationic phosphating formulations compared with conventional high zinc phosphating formulations.

Keywords: Pretreatment Process; Tricationic Phosphating Formulations; SEM; XRD; EDX; Electrochemical Impedance Spectroscopy; Corrosion Protection

1. Introduction

The importance of surface preparation for corrosion protection of automobiles need not be over emphasized because the durability of the phosphated and painted metal surface depends quite critically on the quality of cleaning, stabilization of cleaned surface and physico-chemical characteristics of the phosphate coating that is deposited on clean surface by chemical conversion process prior to painting of the car body. Industrial surface preparation process generally consists of five processing zones viz: degreasing, derusting, surface activation, phosphating and passivation and these pretreatment chemicals may be used either in spray mode, dip mode or in spray cum dip mode. Modern car manufacturing plants mostly use spray cum full dip mode for its obvious advantage for ensuring satisfactory cleaning and deposition of uniform phosphate coating in the areas of car body which are not normally accessible by spray mode of application. If the car body consists of mixed metal combination for different parts of auto body like mild steel and coated steel, then the in-line derusting stage is eliminated from the pretreatment line. Since the phosphate coating is deposited on a metals surface as a result of interfacial reaction between the metal surface and the phosphating solution, the surface composition of the steel and the method of cleaning will have considerable effect on the structure, composition and morphology of phosphate coating which in turn will affect the final corrosion resistance of the phosphated and painted systems. The structure and composition of the phosphate coating and also its rate of growth depends broadly on the three following factors:

- Structure of the clean metal surface *i.e.* microstructure and chemical composition of the surface.
- Design of the phosphating and other chemicals used in different pretreatment stages.
- Parameters of the processing baths viz: temperature, concentration, pressure and time of reaction etc.
- The quality of water used for bath preparation and in the rinsing stages after different stages of processing.

The automotive finishing technology has undergone significant changes in the past decades because of demand for car with higher corrosion resistance and better quality of surface finishes [1-16]. The key factors that contributed significantly to the improvement of higher corrosion resistance are the development of new substrates with better inherent corrosion resistance and higher strength, introduction of cathodic electrophoretic paints for priming the car body and development of low temperature tri-cationic phosphating formulation (45°C to 50°C) which are suitable for depositing excellent phosphate coating on multi metal autobody system containing steel, coated steel and aluminium alloys. The main characteristic features of modern tri-cationic phosphating formulations containing ions of Zn, Mn, Ni is the superior alkali resistance of the resulting phosphate coating which make these formulations highly suitable for operation in cathodic ED bath. This superior alkali resistance of phosphate coating results from the development of additional crystal phases like Phosphophyllite ($Zn_2Fe(PO_4)_2 \cdot 4H_2O$), Phosphomagnellite ($Zn_2Mn(PO_4)_2 \cdot 4H_2O$) and Phosphonicolite ($Zn_2Ni(PO_4)_2 \cdot 4H_2O$) in the coating besides the Hopeite ($Zn_3(PO_4)_2 \cdot 4H_2O$) phase. The higher the value of P/P + H ratio, better is the alkali resistance of the phosphate coating in CED bath leading to superior corrosion resistance of the phosphated and electropainted auto body system [1]. Here, "P" stands for the total Phosphophyllite phases and "H" stands for Hopeite phase present in the deposited coating. The basic chemical reactions on steel surface in a phosphating bath are described below:

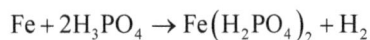

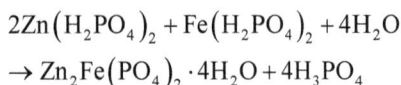

$$Fe + 2H_3PO_4 \rightarrow Fe(H_2PO_4)_2 + H_2$$

$$2Zn(H_2PO_4)_2 + Fe(H_2PO_4)_2 + 4H_2O$$
$$\rightarrow Zn_2Fe(PO_4)_2 \cdot 4H_2O + 4H_3PO_4$$
$$\text{(Coating-Phosphophyllite)}$$

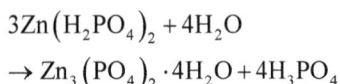

$$3Zn(H_2PO_4)_2 + 4H_2O$$
$$\rightarrow Zn_3(PO_4)_2 \cdot 4H_2O + 4H_3PO_4$$
$$\text{(Coating-Hopeite)}$$

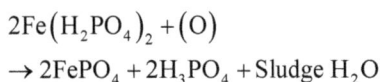

$$2Fe(H_2PO_4)_2 + (O)$$
$$\rightarrow 2FePO_4 + 2H_3PO_4 + Sludge \ H_2O$$

The coating deposited on steel surface consists of two phases viz: Phosphophyllite and Hopeite as described above. And the sludge, which is a byproduct of the phosphating reaction, settles down on the bottom of the phosphating bath [10-13].

In this work, we discuss in the details the science and technological aspects of a modern tri-cationic phosphateing process which is suitable for deposition of excellent phosphate coating on multi-metal auto body assembly consisting of steel and coated steel and also compatible with cathodic electrodeposition (CED) primers. A large number of experimental techniques like SEM/EDX, XPS, XRD and AAS, have been used in this work to characterize the chemical composition of steel surface used by automotive manufacturers and also to characterize the morphology, chemical composition, phase composition and coating weight of the phosphate coating deposited on steel surface. Electro chemical impedance spectroscopy (EIS) and Salts spray tests (ASTM B117) have been used for evaluation of overall corrosion resistance of the

phosphated and painted steel surface as a function of time and for understanding the underlying mechanism of protection and degradation of the coating system on steel surface over extended period of exposure to corrosive atmosphere.

2. Pretreatment Process Sequence Used in a Modern Automotive Finishing Plant

The outline of a 14 stages pretreatment line used in autotive finishing plant is shown in **Figure 1**. It may be noted that the derusting stage along with post derusting rinse stages have been eliminated from this line because the car body processed in this PT line has a mixed metal combination of steel and electrogalvanised steel in diffrent parts. It may be noted that a combination of spray and dip rinse stages makes effective cleaning between different stages and minimizes the carry over of chemicals to the next stage. The processing parameters of different stages of pretreatment plant are summarized in **Table 1**. However, under laboratory conditions, the pretreatment process can be implemented in five litre baths and generally the steel panels of 6" × 4" are used for depositing phosphate coating which can be used for different physico-chemical characterization viz: morphology, phase analysis, chemical analysis and coating weight determination and evaluation of corrosion resistance (ASTM B117) after depositing paint coating of appropriate thickness.

The design of the chemicals used at prephosphating stages viz. degreasing, derusting and surface activation and the corresponding bath parameters all will have considerable effect on the uniformity, morphology, coating weight and quality of phosphate coating deposited during the phosphating stage [17-19]. The physical structure and chemical composition of the phosphate coating, in turn, will affect the corrosion resistance of the phosphated and painted system. Thus it is very important to maintain the bath parameters at the recommended values at every stage of processing to get the right quality of phosphate coating.

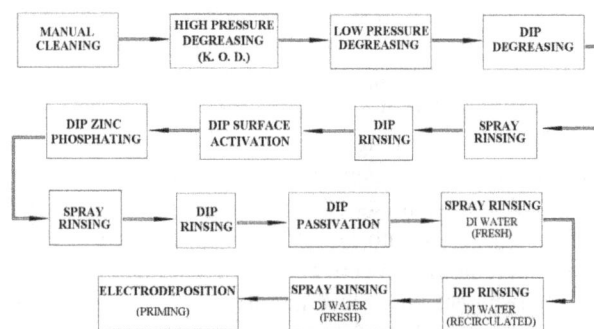

Figure 1. Pretreatment process sequence used in a modern automotive finishing plant.

Table 1. A 14-stage pretreatment process for automotive finishing.

Sr. No.	Process Sequence	Mode of Operation	Temperature (°C)	Time of Processing (min.)	Chemical Bath Parameters
1	Manual cleaning with solvent		RT	5	
2	Knock-off Degrease (or High Pressure Degrease)	Spray pressure (4 - 6 bars)	RT	1	
3	Low Pressure Degrease	Spray pressure (0.7 bars)	47°C	1	
4	Dip Degrease	Dip	55°C	4	
5	Rinsing with mains water	Spray	RT	1	
6	Rinsing with mains water	Dip	RT	4	
7	Surface Activation	Dip	RT	4.5	pH = 7.86; Ti = 22 ppm
8	Phosphating	Dip	50°C	4.5	FA = 2.8 - 3.2; TA = 26 - 30 Toner = 1.8 - 2.5 ml
9	Rinsing with Mains water	Spray	RT	1	
10	Rinsing with Mains water	Dip	RT	4	
11	Passivation	Dip	RT	4	
12	Rinsing with Fresh D. I. water	Spray	RT	1	
13	Rinsing with Recirculated D. I. water	Dip	RT	4	
14	Rinsing with fresh D. I. water	Spray	RT	1	

↓

Wet entry into Cathodic Electrocoat Bath

Apart from chemicals, the water quality used in bath make-up as well as at rinse stages plays a very critical role in maintaining the stability of the different baths as well as the quality of the phosphate coating. The post passivation rinse stage is very critical in pretreatment process because soluble salts of chloride, sulfate and ammonia if not removed thoroughly from car body, will promote blistering under a paint film. In order to minimize this possibility, water supply should be free from harmful salts as far as practical. Normally, deionised water (DI) is used for bath make up and the replenishments in surface activation stage, passivation stage and post passivation rinse stages. For other treatment stages like decreasing, derusting, phosphating and other rinse stages mains water may be used provided it conforms to the specification given in **Table 2** [7-12]. To ensure minimum carryover of harmful ions to the subsequent stage of CED bath the phosphated surface after passivetion stage is given two or three DI water rinses. Fresh water is used in the last stage, whereas recirculated water is used in first two stages. The conductivity of recirculated water should not exceed 25 µS/cm [1]. The rinse bath should be discarded once the conductivity exceeds the limit of 25 µS/cm.

In order to ensure consistently good quality of cleaning and phosphating of car body the following factors in different stages are very important.

2.1. Degreasing Stage

The degreasing zone normally consists of at least two stages. The first stage is usually a spray stage known as knock-off-degrease (K.O.D.) and that is followed by a dip stage. The advantage of having two stages is that major portion of the oil, dirt etc. will be removed by high pressure spray impact in the first stage leaving relatively lower load for the dip stage to clean. In **Figure 1**, the degreasing zone consists of three stages viz. two spray stages and one dip stage for efficient cleaning of car body. In order to ensure maximum efficacy of degreasing stage, misting spray should be provided between K.O.D. and dip degreasing stage and also between dip degreasing and next rinsing stage to prevent the drying of the car bodies during transition from one stage to the next stage. Continuous oil separating systems should also be installed for high volume production of pretreatment line.

2.2. Surface Activation Stage

The purpose of this stage is to refine the crystal size of zinc phosphate coating and to control the coating weight during phosphating stage. Modern surface activation che-

Table 2. Water specification for bath make-up and rinse stages.

Water Type used in PT line	Specifications	Usage
I. De-ionised (DI) water	(i) Conductivity 5 µs/cm (ii) pH 6.5-7.2	Surface activation stage passivation stage post passivation rinse stage (spray stage)
II. Recirculated Deionised water	Conductivity ≤ 25 µs/cm	Post passivation rinse stage (dip stage)
III. Mains water	1) Total chloride and sulphate 70 ppm maximum (calculated as $Cl^- + SO_4^{-2}$) 2) Total alkalinity-200 ppm max. (calculated as $CaCO_3$) 3) Both together should not exceed 225 ppm	Degreasing, derusting stage phosphating stage all other rinse stages

micals are weakly alkaline colloidal dispersion of titanium complex. This treatment leads to the formation of large number of finer crystallites of titanium compound on the metal surface which act as crystal nuclei for the growth of fine zinc phosphate crystals during the phosphating stage. Greater is the number of nucleating centres on the surface, more will be the inter-crystalline collisions and consequently finer will be the crystals and more compact will be the phosphate coating and better will be the paint adhesion and corrosion resistance of the phosphated and painted system [14,19]. The efficacy of surface activation bath is critically dependent upon:

- pH of the bath.
- The concentration of titanium.
- For better colloidal stability of the surface activation bath, it must be made with deionised (DI) water and the activation bath should have a circulation rate of 3 to 4 tank turn-over/hour.
- According to the results published by Yoshihara [1]. The ideal pH of the activation bath should be in the range of 8.5 to 9.5 and Ti concentration should be minimum 10 ppm. We have observed that within the range of 10 to 30 ppm, the coating weight remains essentially constant, **Table 1** shows the Ti concentration in the bath as 22 ppm.
- The grain refining action depends on amount of Ti adsorbed on the metal surface and for steel surface the adsorption is inversely proportional to the amount of segregated carbon on the surface.

2.3. Phosphating Stage

For most effective functioning of modern tricationic zinc phosphating formulation the following points are very important:

- The coating formation reaction is largely dependent upon the free acid (FA), total acid (TA), concentration of oxidizing agents or toners, temperature, deposition time etc. In **Table 1** the phosphating bath parameters are given. The phosphate coating weight on mild steel substrates generally lies in the range of 2.8 to 3.2 g/m².
- The circulation of phosphating bath solution is an essential requirement for ensuring the deposition of uniform phosphate coating and normally a circulation rate of 3 to 4 tank turn-over/hour is recommended. The direction of solution flow will be opposite to that of the moving car body to be phosphated. In order to ensure consistent good quality of phosphate coating on car body the phosphating bath should be provided with a continuous sludge removal system like filter press to minimize the accumulation of sludge in the phosphating bath. Usually, the sludge containing phosphating bath solution is pumped to Tilted Plate Separator (TPS) where a major portion of the sludge will be separated and collected at the bottom of the separator. The comparatively clear supernatant liquid from the TPS is then pumped through the filter press (containing a series of filters) where the phosphating bath solution will be completely free from sludge and then pumped back to the main phosphating tank. The total sludge separating unit will be under continuous operation as long as the phosphating plant is running and at least 60 to 70 percentages of the filters should always be in working condition for effective sludge removal. Yoshihara *et al.* [1] recommends a sludge concentration of about 300 ppm maximum at any stage in the phosphating bath.
- The total surface area processed per hour in a given volume of bath solution is very important for maintaining the chemical equilibrium of the phosphating bath. For light and medium coating weight zinc phosphating bath, the optimum recommended rate of processing is about 2 sq. ft/hour/4.5 liters of bath solution.
- Phosphating solution should be heated indirectly by external plate heat exchanger by using low pressure hot water. The temperature differential between the phosphating solution and hot water should not exceed 10°C. This will prevent the formation of scale on the heat exchanger plate.
- For high production volume automotive plant, the

auto dosing system of the chemicals and accelerators (toners) to the phosphating bath is usually recommended to ensure consistently satisfactory phosphate coating on car bodies.

2.4. Passivating Rinse

In order to improve the corrosion resistance of phosphate coating on steel surface, it is useful to give a final rinse with chromium containing solutions. This treatment which is known as passivation process provides additional stability to the phosphated surface by partial sealing of the pores in zinc phosphate coating. The trend, so far has been to use passivating solutions containing a mixture of hexavalent and trivalent chromium ions for best results. The latest trend, however, is to use Chrome free formulations for passivation purpose and a number of Zirconia based products are now available for use in automotive finishing industry.

3. Characterization of Phosphate Coating

In order to establish the structure-property-performance correlation of the phosphate coating with its protective value and also to carry out failure analysis of the Metal/Phosphate/ED primers interfaces, it is very important to characterize the phosphate coating at microscopic level [6-27]. The key physico-chemical characteristics parameters of phosphate coating on a metal surface are:

- Coatings morphology (crystal size, shape, orientation and coating compactness).
- Crystal phases of the coating.
- Coating weight.
- Coating composition.
- Chemical stability.

Usually coatings morphology and crystal size are determined by Scanning Electron Microscopy (SEM) and the crystal phases are determined by X-ray Diffraction technique (XRD)and elemental analysis on the surface coating may be done by Energy Dispersive X-ray analysis (EDX). Coating weight can be determined by chemical methods by dissolving the coating in dilute chromic acid solution and the phase composition ("P" ratio) may be determined by chemical methods like Atomic Absorption Spectroscopy (AAS) or by XRD techniques [12-21]. The chemical stability of the phosphate coatings may be determined by exposing the coatings in dilute solution of sodium hydroxide and checking the extent of solubility [19,21]. The final corrosion performance of phosphated and painted surface maybe evaluated by accelerated tests like salts spray (ASTM-B117) and also by Electrochemical impedance Spectroscopy (EIS) [22-27]. In automotive industry, salts spray tests are widely used to evaluate the performance of phosphated (PT) and Electro-deposited primer (ED) coating. For mild steel surface,

the normal specification for anodic electrocoat (AED) process is that the coating system (PT + AED) should pass 600 hrs of salts spray tests (with 20 μm primer thickness) while for cathodic electrocoat system (PT + CED) with similar thickness of primer film, the coating system should pass a minimum of 1000 hrs of salts spray test.

The structure, composition and coating weight of phosphate coating deposited on a metal substrate is a function of several factors:

1) Structure and chemical composition of the metal surface.

2) Design of the phosphating chemical.

3) Mode of application *i.e.* dip or spray.

4) Bath parameters like free acid, total acid, concentration of oxidizing accelerators or toners, temperature, time of coating deposition and loading rate *i.e.* total surface area processed per hour in a given volume of phosphateing solution.

The effect of some of these parameters on the structure, morphology and performance of phosphate coating on steel surface have been reported in details in some of our earlier works [14,16-19]. In the following section we highlight some key results which are important for the both the development of new phosphating formulations as well as for solving the quality problems encountered during their application to industrial metal finishing process.

3.1. Morphology and Chemical Composition of Phosphate Coating

The application of SEM, EDX and XRD techniques provides a comprehensive idea about the physical structure, Chemical composition and nature of the coating deposited on metal surface. The uniformity of crystal size and compactness of the coating is very important for adhesion of the phosphate coating to the metal surface as well as the adhesion of the paint coating or organic coating to the phosphated surface. In **Figures 2(a)-(c)**, the morphology of three different types of phosphate coating deposited on steel surface from three phosphating formulations 1, 2, 3 are shown. The formulations 1 and 2 are conventional high temperature immersion type phosphating formulations (70˚C) with relatively high zinc content 1) and calcium modified zinc phosphating formulations; 2) leading to uniform compact coating with nodular shaped and spherical shaped crystals respectively. In contrast, formulation; 3) which is a tricationic low temperature phosphating formulation (45˚C - 50˚C) deposits a highly uniform compact coating with cubic shaped crystals. The most important aspect of all these coating is that all three formulations provide highly uniform, thin and compact coating on steel surface but average crystal size varies in the range of from 4 - 10 mi-

(a)

(b)

(c)

Figure 2. (a) SEM micrograph of phosphate coating deposited on steel surface (formulation I); (b) SEM micrograph of calcium modified phosphate coating deposited on steel surface (formulation II); (c) SEM micrograph of phosphate coating deposited on steel surface from tricationic phosphating formulation III.

crons for different formulations. Further, the coating composition of all these formulations varies because of difference in the formulations. The corrosion performance of these three phosphating formulations are dis-

cussed in the last section. **Figure 3** provides an example of poor phosphate coating on steel surface which was not properly cleaned at degreasing stage from formulation III and hence undesirable in production line.

The morphology of phosphate coating on zinc coated steel and aluminium substrates from formulation III are shown in **Figures 4** and **5** respectively. It is quite evident that the coating morphology is very compact on the former but not so satisfactory on aluminium substrate. The XRD diffractograms of phosphate coating on steel and zinc coated steel surface are shown in **Figures 6(a)** and **(b)** respectively. It is quite evident that the coating on steel surface consists of both Phosphophyllite and Hopeite phases whereas on zinc coated steel surface the it consists of only Hopeite phase as expected. Similary EDX spectra of phosphate coatings on steel surface from all three formulations are shown in **Figure 7**. The different elements present in the coating are quite evident from the spectra.

Figure 3. SEM micrograph of poor phosphate coating from formulation III on steel surface which is not properly cleaned.

Figure 4. SEM micrograph of zinc phosphate coating deposited on zinc coated steel surface from tricationic phosphating formulation III.

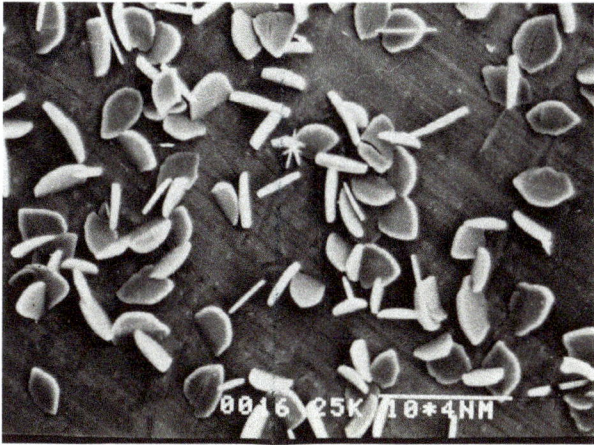

Figure 5. SEM micrograph of phosphate coating on aluminium substrate from tricationic phosphating formulation III.

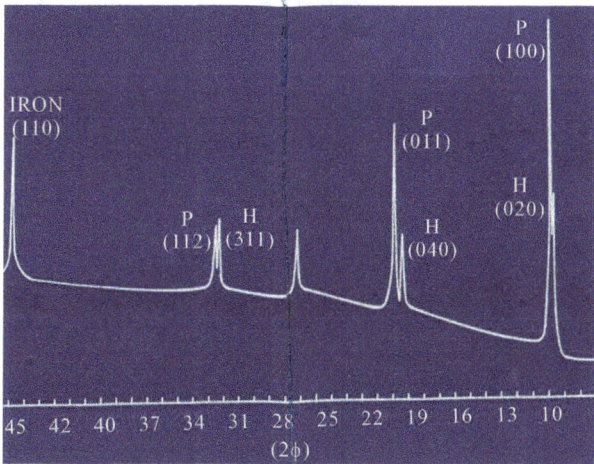

(a)

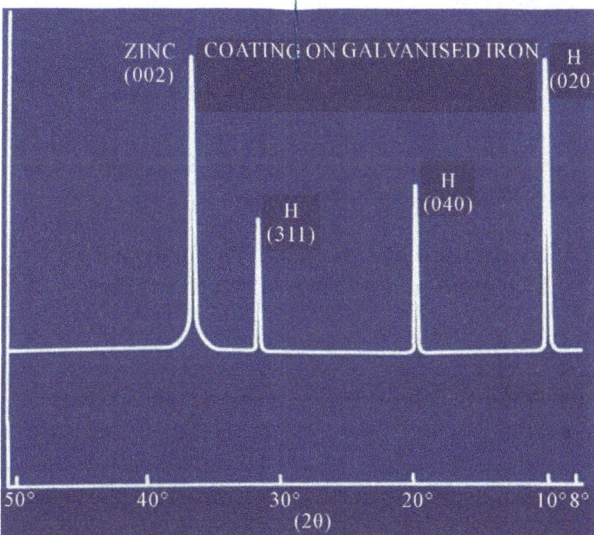

(a)

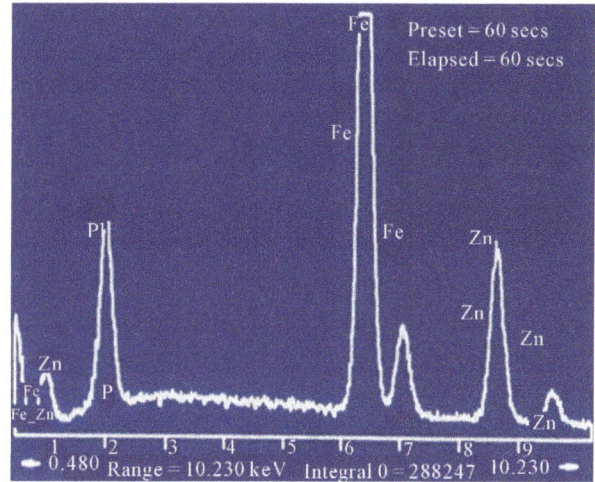

(b)

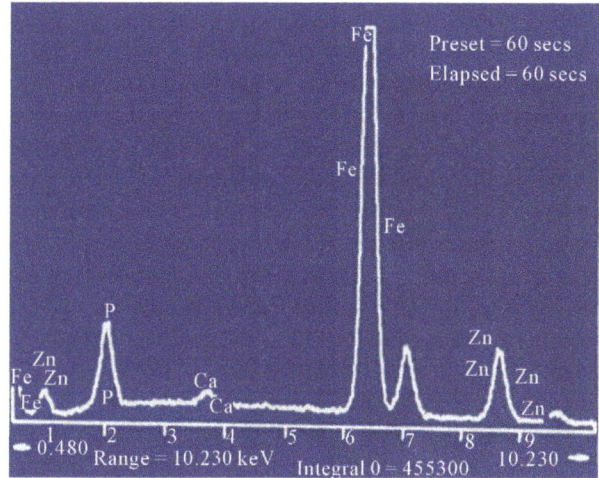

(b)

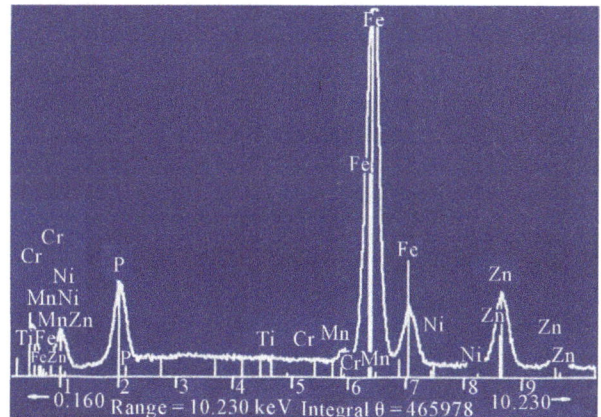

(c)

Figure 6. (a) X-ray diffractogram of phosphate coating on steel surface and (b) X-ray diffractogram of phosphate coating on zinc coated steel surface.

Figure 7. (a) EDX spectra of phosphate coating from formulation I (b) from formulation II and (c) formulation III.

It may be noted that the phosphate coating layer is the most critical link in the chain of multiple coating layers that are deposited on car body during autobody finishing and thus the integrity of metal/phosphate interface is extremely important for better corrosion resistance of car body. An example of excellent steel/phosphate interface is provided by the SEM micrograph shown in **Figure 8**. The dark part in the micrograph is steel substrate and the bright part is zinc phosphate coating.

3.2. Effect of Surface Composition on Quality of Phosphate Coating on Steel Surface

In order to address the problem of variation of coating quality viz. coating morphology and coating weight on steel panels supplied by different steel manufacturers, a systematic work was done on a set of 12 panels procured from different suppliers and phosphated under laboratory conditions and their coating quality was evaluated by SEM technique. The results were classified into four grades A, B, C, D depending on the quality of phosphate coating. Both bulk and surface composition of these samples were determined by Vacuum Emission Spectroscopy and XPS technique respectively and the data revealed that even though the bulk chemical composition of all the panels is essentially quite similar as shown in **Table 3**, there is substantial difference in the surface composition of the four panels classified under different grades. **Figure 9(a)** and **(b)** shows the XPS results (Fe $2p_{3/2}$ and $C1_s$ spectra) of steel surface for four samples S2, S4, S6 and S8. The surface Fe/C ratio decreases systematiccally form 0.41 to 0.15 as the coating morphology degrades systematically from the best (A) to the worst (D) as shown in **Figures 10(a)-(c)** and summarized in **Table 4**. The SEM picture corresponding to the D-grade steel is not shown here as it is completely amorphous coating without any structure. It was thus established that surface Fe/C ratio is a very important index and can be used as a reliable criterion for grading the steel panels and to distinguish the good steel from bad steel as far as the phosphatibility is concerned. More details of this were published in an earlier publication [16].

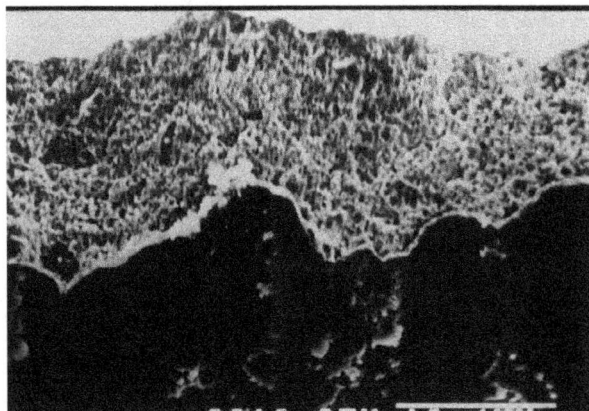

Figure 8. SEM micrograph of steel/phosphate coating interface. Dark part is metal and bright portion s phosphate coating.

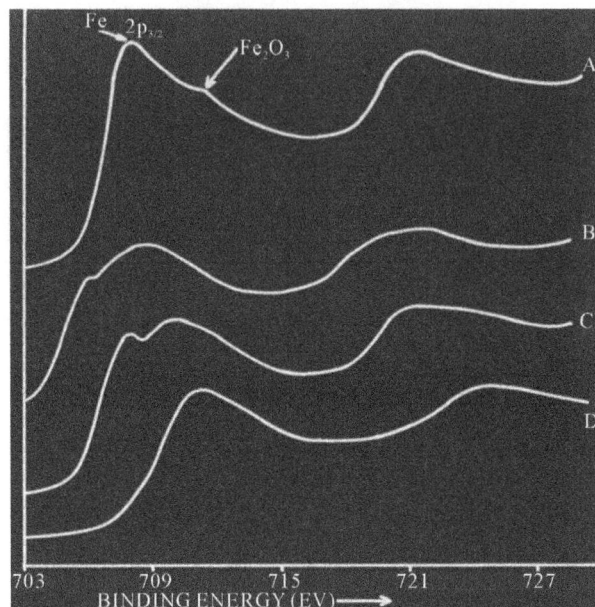

(a)

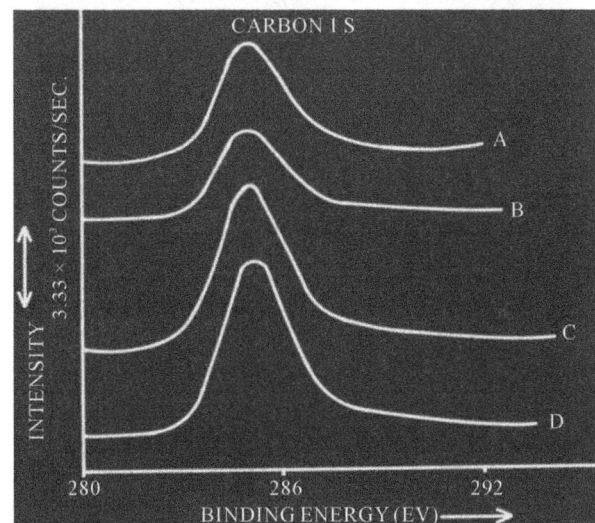

(b)

Figure 9. XPS results of (a) Fe2p$_{3/2}$ (b) C1s spectra of steel surface of the four steel samples.

3.3. Evaluation of Corrosion Performance of Phosphated and Painted Steel Surface

In order to evaluate the comparative corrosion performance of tricationic phosphating formulation(III) with two conventional immersion type zinc phosphating formulations (I and II), three sets of mild steel panels (6" × 4") were cleaned, phosphated by immersion process at the recommended parameters of the each phosphating formulations in the laboratory, keeping the degreasing, derusting and passivation stages identical. The details of phosphating process parameters for formulations I and II are already reported in earlier work [17,18]. The phosphate panels were subsequently coated with an alkyd

Table 3. Bulk composition of steel samples used in this study.

Sample No.	Elements → Fe	C	Mn	S	Si	Ni	Cr	Al
S2	99.18	0.07	<0.1	0.024	0.01	0.04	0.11	<0.012
S4	98.89	0.08	0.14	0.024	<0.01	0.01	0.12	>0.111
S6	98.77	0.059	<0.10	0.023	<0.01	0.01	0.15	>0.111
S8	99.37	0.153	0.26	0.022	0.05	0.01	0.16	>0.111

Table 4. Surface analysis of steel samples.

Sample No.	Rating of phosphate coating quality	Surface Fe/C ratio
S6	A	0.41
S4	B	0.37
S2	C	0.22
S8	D	0.15

(a)

(b)

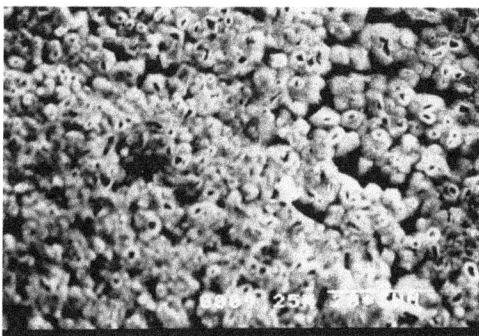

(c)

Figure 10. SEM micrographs of phosphate coating on steel sample (a) S6; (b) S4; (c) S2.

based stoving clear to 20 micron thickness by spray process and then baked at 150°C for 30 min. The mild steel panels used in this study were procured from an automobile manufacturer and were cut out from a single sheet to minimize the variation on the substrate quality.

R_Ω = Solution Resistance.
Rpo = Pore Resistance.
Rct = Charge Transfer Resistance.
Cc = Coating Capacitance.
Cdl = Double Layer Capacitance.
$Z\omega$ = Warburg Impedance.

The corrosion performance of the phosphated and alkyd coated panels with20 micron thickness in salts spray test (ASTM-B117) were monitored periodically both visually as well as by Electro-chemical Impedance Spectroscopy (EIS),over a period of 600 hrs. The impedance measurements were carried out on these panels at different interval of exposure time over a frequency range of 10^{-2} Hz to 10^5 Hz. The amplitude of the signal was 5 mV. The impedance measurements were carried out at open circuit potential using a "Schlumberger 1255 Frequency Response Analyzer" (FRA) operated under computer control. The FRA was connected to the electro chemical cell through "EG&G potentiostat/Galvanostat 273" More details of the experimental set up for impedance measurement have been reported in an earlier publication [26].

In order to interpret the impedance data we have used an equivalent circuit model of painted metal/solution interface as shown in **Figure 11**. The impedance data were analyzed in terms of three coating parameters viz. Pore resistance (Rpo), coating capacitance (Cc) and breakpoint frequency (fb) [26]. The results of variation of Rpo, Cc and f_b as a function of salt spray exposure time are shown in **Figures 12-14** respectively. The numerical values of Rpo, Cc and (fb) at 0 hr, 100 hrs and 300 hrs of salts spray exposure are tabulated in **Table 5**. A comparison of Rpo values after 300 hrs of salts spray expo-

Table 5. Variation of Rpo, Cc and f_b values of different phosphate coatings on steel surface coated with 20 μm thick alkyd coating with salts spray exposure time.

Phosphating		Salt Spray Exposure Time	
System	0 hrs	100 hrs	300 hrs
		Pore Resistance, Rpo (Ohm·cm^2)	
I	2.39×10^7	7.42×10^5	8.9×10^4
II	2.08×10^7	4.07×10^5	3.6×10^4
III	2.08×10^8	4.16×10^6	1.62×10^6
		Coating Capacitance, Cc (F·cm^{-2})	
I	6.42×10^{-10}	3.16×10^{-8}	3.23×10^{-7}
II	1.48×10^{-9}	4.2×10^{-8}	8.12×10^{-7}
III	6.9×10^{-11}	9.3×10^{-10}	2.95×10^{-8}
		Break Point frequency, f_b (in Hz)	
I	19.3	2706	8447
II	94.3	4965	27,994
III	7.5	120	1345

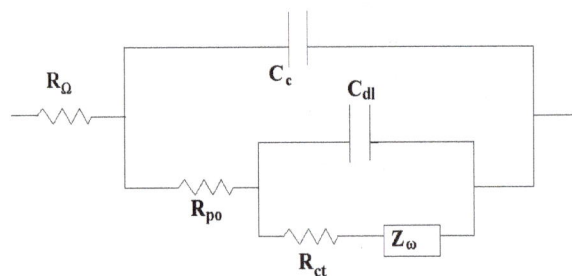

Figure 11. Equivalence circuit model for painted metal/solution interface.

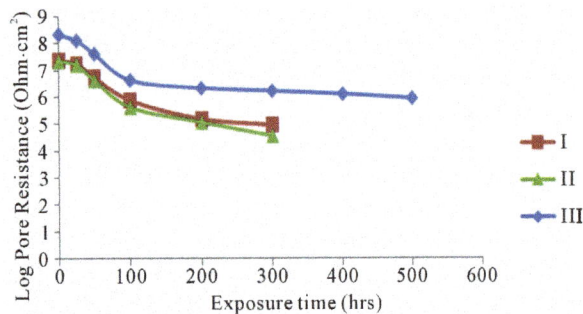

Figure 13. Coating capacitance (Cc) as a function of salt spray exposure time.

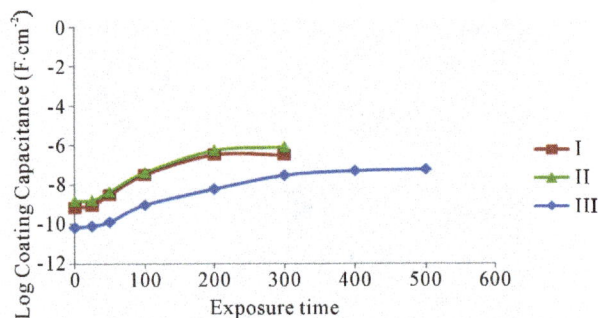

Figure 12. Pore resistance (Rpo) as a function of salt spray exposure time.

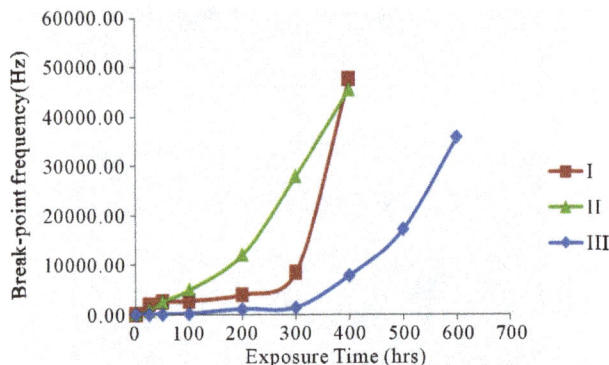

Figure 14. Break point frequency (fb) as a function of salt spray exposure time.

sure clearly shows the superior performance of tricationic phosphating formulation III (Rpo—1.62×10^6 Ohm·cm^2) compared with formulation I (Rpo—8.9×10^4 Ohm·cm^2) and formulation II (Rpo—3.6×10^4 Ohm·cm^2). As shown in **Figure 12**, the Rpo values for formulation III remain quite steady at this high value even after 500 hrs of salts spray exposure, while for other two formulations, Rpo values fall sharply indicating the rapid degradation of the protective value of those phosphate coatings.

Similarly, **Figure 13**, where log coating capacitance is plotted against exposure time, indicates that for tricationic formulation (III) the increase in Cc is relatively much less compared with formulations I and II. The increase in Cc with exposure time can be attributed to the formation of blisters due to water ingress underneath the film. After a certain exposure time, there was no further increase in the coating capacitance values, either it remained constant or started decreasing. This may be attributed to simultaneous occurrence of two opposing phenomena:

At long exposure times, the ingress of water and accumulation of corrosion products underneath the paint film exert pressure from inside the blister and the blister breaks. This process decreases the capacitance.

2) The nucleation and growth of some blisters continue even at long exposure times. This process increases the capacitance.

The break-point frequency versus exposure time plots (**Figure 14**) clearly show three distinct stages in coating failure process: water ingress, coating disbonding and blister growth.

Since the break-point frequency is proportional to the area of delamination, the performance of various coatings system could be assessed by comparing the "f_b" values at a particular exposure time to salts spray environment. For example as shown in **Table 5**, f_b value for formulation III after 300 hrs. of exposure is 1345 Hz which is much lower compared with corresponding values 8447 Hz and 27,994 Hz for formulation I and II respectively, indicating clearly that formulation III offers much superior corrosion resistance (minimum area of delamination) followed by phosphating formulation I and II, which was also corroborated by visual observation of the panels from salts spray test [26].

The other point to note is the induction times for steep increase in break-point frequency values for this particular coating system (**Figure 14**) which are approximately 300, 150 and 400 hours for phosphating formulations I, II and III respectively which is again a clear indication of the superior adhesion and corrosion performance of phosphating formulation III.

Thus, superior performance of formulation III may be attributed primarily to the difference between chemical composition, compactness and superior alkali resistance of the phosphate coating compared with formulation I and II. The superior alkaline resistance of tricationic phosphating formulation is attributed to the presence of higher level of additional crystal phases like phosphophyllite, phosphomangallite and phosphonicollite besides Hopeite phase in phosphate coating on steel surface.

4. Acknowledgements

I would like to thank my collegues Mr. G. N. Bhar and Mr. P. K. Roy of ICI India, R and D Center for Paints, Kolkata, India for their contribution to this work. I would also like to thank Mr. Nikhilesh Chaudhary of Material Science Department, Indian Association for the Cultivation of Science, Kolkatta, India for all the SEM micrographs and Dr. S. Badrinarayan of National Chemical Laboratory, Pune, India for surface analysis of steel panels and finally to Ms. Gayatri Devi and Prof. V. S. Raja of IIT Mumbai, India for evaluation of phosphate and painted coatings by EIS spectroscopy. Finally, I would also like to thank Miss Vaishali Shinde and my research students, Dr. Shilpa Vaidya, Dr. Priyanka Bhat and Dr. Rohan Jadhav of ICT for putting this paper together in the present form.

REFERENCES

[1]   T. Yoshihara and H. Okita, *Transactions of the Iron and Steel Institute of Japan*, Vol. 23, 1983, p. 984.

[2]   D. B. Freeman, *Product Finishing*, Vol. 6, 1987.

[3]   D. C. Gordan, *Corrosion Prevention & Control*, Vol. 7, 1984.

[4]   N. Satoh, "Effects of Heavy Metal Additions and Crystal Modification on the Zinc Phosphating of Electrogalvanized Steel Sheet," *Surface and Coatings Technology*, Vol. 34, No. 2, 1987, pp. 171-181.

[5]   "Corrosion in the Automotive Industry," Metals Hand-Book, Vol.13, 9th Edition, p. 1011.

[6]   E. L. Ghali and R. J. A. Potvin, "The Mechanism of Phosphating of Steel," *Corrosion Science*, Vol. 12, No. 7, 1972, pp. 583-594.

[7]   Corrosion, Ed. L. L. Shreir, Newnes ButterWorth, London, Vol. 2, 1976, pp. 16:19-16:26

[8]   ASM Handbook, "Surface Engineering," ASM International Ohio, USA, Vol. 5, 1996, Phosphate Coatings, pp. 378-404.

[9]   ASM Handbook, Corrosion, ASM International, Ohio, USA, Vol. 13 (1996), T. W. Cape, Phosphate Conversion Coating, pp. 383-388.

[10] D. B. Freeman, "Phosphating and Metal Pretreatment," Woodland Faulkner, Cambrige, 1986.

[11] W. Machu, "Handbook of Electropainting Technology," Electrochemical Publications Ltd., UK, 1978.

[12] W. Rausch, "The Phosphating of Metals," AFM International, USA, 1990.

[13] T. S. N. S. Narayanan, "Surface Pretreatment by Phosphate Conversion Coatings—A Review," *Reviews on Advanced Materials Science*, Vol. 9, No. 2, 2005, pp. 130-177.

[14] N. C. Debnath, Paint India Annual, 1985/86, pp. 19-25.

[15] S. Maeda, *Journal of Coatings Technology*, Vol. 55, 1983, p. 43.

[16] N. C. Debnath and P. K. Roy, *Transactions of the Institute of Metal Finishing*, Vol. 74, No. 1, 1996, p. 17.

[17] N. C. Debnath, G. N. Bhar and S. Roy, *Jocca*, Vol. 72, 1989, p. 492.

[18] P. K. Roy and N. C. Debnath, *Surface Coatings International*, Vol. 76, No. 5, 1993, p. 214.

[19] G. N. Bhar, N. C Debnoth and S. Roy, "Effects of Clacium Ions on the Morphology and Corrosion Resistance of Zinc-Phosphated Steel," *Surface Coatings International*, Vol. 35, No. 1-2, 1988, pp. 171-179.

[20] J. P. Servais, B. Schmitz and V. Leroy, *Material Performance*, Vol. 27, 1988, 56.

[21] W. J. Van Ooij and A. Sabota, *Journal of Coatings Technology*, Vol. 61, 1989, 778, 51.

[22] B. A. Cooke, "Organic Coatings. Science & Technology," In: G. D. Parfitt and A. V. Patsis, Eds., Marcel Dekker lnc., New York, Vol. 7, 1984, pp. 197-222.

[23] S. Maeda, T. Asa and M. Yamamoto, *Organic Coatings*: *Science and Technology*, Vol. 7, 1984, pp. 223-247.

[24] M. Mansfield, M. W. Kendig and S. Tsai, "Evaluation of Corrosion Behavior of Coated Metals with AC Impedance Measurements," *Corrosion*, Vol. 38, No. 9, 1982, pp. 478-485.

[25] N. Tang, W. J. van Ooij, G. Gorecki, "Comparative EIS Study of Pretreatment Performance in Coated Metals," *Progress in Organic Coatings*, Vol. 30, No. 4, 1997, pp. 255-263.

[26] V. S. Raja, R. Gayatri Devi, A. Venugopul, N. C. Debnath and J. Giridhar, "Evaluation of Blistering Performance of Pigmented and Unpigmented Alkyd Coatings Using Electrochemical Impedance Spectroscopy," *Surface and Coatings Technology*, Vol. 107, No. 1, 1998, pp. 1-11.

[27] N. C. Debnath and G. N. Bhar, *European Coatings Journal*, No. 4, 2002, p. 1.

24

Epitaxial Ge Growth on Si(111) Covered with Ultrathin SiO$_2$ Films

Alexander A. Shklyaev[1,2], Konstantin N. Romanyuk[1,2], Alexander V. Latyshev[1,2]

[1]A. V. Rzhanov Institute of Semiconductor Physics of SB RAS, Novosibirsk, Russia; [2]Novosibirsk State University, Novosibirsk, Russia.

ABSTRACT

The epitaxial growth of Ge on Si(111) covered with the 0.3 nm thick SiO$_2$ film is studied by scanning tunneling microscopy. Nanoareas of bare Si in the SiO$_2$ film are prepared by Ge deposition at a temperature in the range of 570°C - 650°C due to the formation of volatile SiO and GeO molecules. The surface morphology of Ge layers grown further at 360°C - 500°C is composed of facets and large flat areas with the Ge(111)-c(2 × 8) reconstruction which is typical of unstrained Ge. Orientations of the facets, which depend on the growth temperature, are identified. The growth at 250°C - 300°C produces continuous epitaxial Ge layers on Si(111). A comparison of the surface morphology of Ge layers grown on bare and SiO$_2$-film covered Si(111) surfaces shows a significantly lower Ge-Si intermixing in the latter case due to a reduction in the lattice strain. The found approach to reduce the strain suggests the opportunity of the thin continuous epitaxial Ge layer formation on Si(111).

Keywords: Ge/Si Heterostructures; Epitaxial Growth; Surface Morphology; Scanning Tunneling Microscopy

1. Introduction

Further development of optoelectronics and photonics can be associated with the fabrication of integrated devices based on III-V semiconductors grown on Si substrates [1,2]. The preparation of continuous thin GaAs layers by their growth on the bare Si surfaces is impeded because of the large lattice mismatch (~4%) between GaAs and Si. The growth occurs through the Stranski-Krastanov growth mode, leading to the formation of three-dimensional islands. Since the lattice constants of Ge and GaAs are almost equal, Ge or SiGe layers can be used as an intermediate layer between Si and GaAs [3-5]. Moreover, Ge layers on Si improve the performance of integrated Si circuits due to a greater hole mobility in Ge. For these purposes, the preparation of continuous thin Ge layers with good crystalline quality on Si is required.

The Ge growth on Si(111) proceeds through the formation of a Ge wetting layer with the thickness of 2 - 3 bilayers (BL) [6-8]. Further Ge deposition leads to the three-dimensional island formation, the shape and size of which strongly depend on the growth temperature [7,9,10]. The growth is accompanied by the introduction of

threading dislocations into the islands and a dislocation network into the Ge layer at the interface between Ge and Si [11-13]. When the growth temperatures are above 500°C, the lattice strain causes the significant intermixing of Ge and Si atoms [14-16]. The strain, being most strong at the island edges, induces the formation of deep trenches around large flat islands [17-19]. All these factors prevent the formation of continuous thin Ge layers on Si(111).

Nanocontact epitaxy has been recently proposed to grow continuous layers of semiconductor materials on Si despite the large difference in their lattice constants [20,21]. The method is based on the use of Si surfaces covered with the ultrathin SiO$_2$ film. Ge deposition on such surfaces at rather high temperatures results in the formation of bare Si nanoareas with the size depending on the amount of deposited Ge and temperature [22-25]. The distance between the areas is 7 - 10 nm. The areas of bare Si in the SiO$_2$ film appear due to the reaction of deposited Ge with SiO$_2$ producing the volatile SiO and GeO molecules [22,23]. The bare Si nanoareas serve for the epitaxial growth of semiconductor materials which can form a continuous layer due to the island nucleation and growth over the residuals of the SiO$_2$ film. The nanocontact epitaxy is studied here with respect to revealing the possibility of thin continuous epitaxial Ge layer formation on Si(111) substrates covered with ultrathin SiO_2 films. The influence of the technological parameters, such as growth temperature and Ge coverage on the structure and the surface morphology of the grown Ge layers, is examined using scanning tunneling microscopy.

2. Experimental Details

The experiments were carried out in an ultrahigh-vacuum chamber with the base pressure of about 1×10^{-10} Torr. The chamber was equipped with a scanning tunneling microscope (STM) manufactured by Omicron. A Knudsen cell with a BN crucible was used for Ge deposition at the rate from 0.5 to 1.1 BL/min [1 bilayer (BL) $\approx 1.44 \times 10^{15}$ atoms/cm^2] which was calibrated with the STM for the Ge wetting layer growth on the Si(111) surface. After electrochemical etching, the sharp W STM tips were modified by tip apex cut from several sides using the 30 kV Ga ion beam of a separate Zeiss 1540 XB cross beam scanning electron microscope. The opening angle of sharpened STM tips was less than 25°. The STM images of the surfaces covered with Ge were usually obtained with the sample bias voltage of 2.4 V and the constant current between 3 and 30 pA.

A $10 \times 2 \times 0.3$ mm^3 sample was cut from an n-type Si(111) wafer with a miscut angle of < 10' and the resistivity of 5 - 10 $\Omega \cdot$cm. Clean Si surfaces were prepared by flash direct-current heating at 1200°C. The sample temperature was measured using IMPAC IGA 12 pyrometer. To grow the ultrathin SiO_2 film, the sample temperature was set to 400°C and raised to 550°C for 10 min after oxygen had been introduced into the chamber at the pressure of 2×10^{-6} Torr. The SiO_2 film, prepared at the similar conditions, was previously investigated to be 0.3 - 0.5 nm thick and mainly composed of silicon dioxide (SiO_2) and it also contained Si atoms at different oxidation stages [26]. However, it is named here as the SiO_2 film. After the SiO_2 film growth, the chamber was pumped to the pressure of ~1×10^{-9} Torr for 10 min and then the Ge crucible temperature was set to the range from 1110 to 1150°C. The sample temperature was being maintained for 5 min after finishing the deposition. Image processing and correction software were employed to reduce distortion of the STM images caused by effects of thermal drift of the STM tip against the sample and to obtain statistical characteristics of the surface morphology, such as stereographic projections of surface areas and portions of the areas as a function of their inclination angle with respect to the sample surface.

3. Surface Morphology at a Relatively Small Ge Coverage

The 4 BL Ge deposition on Si(111) covered with the

ultrathin SiO_2 film results in the appearance of Ge islands, as shown in **Figure 1**, without the formation of Ge wetting layers. The interaction of Ge with SiO_2 occurs through the reaction [22,23]

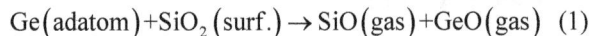

$$Ge(adatom) + SiO_2(surf.) \rightarrow SiO(gas) + GeO(gas) \quad (1)$$

producing volatile SiO and GeO molecules at temperatures above 430°C. Reaction (1) has a strong temperature dependence that is characterized by the activation energy of 2 - 3 eV [27]. This leads to the appearance of bare Si surface areas in the SiO_2 film. Then other reactions

$$Ge(adatom) + Si(surf.) \rightarrow Ge-on-Si(surf.) \quad \text{and}$$

$$Ge(adatom) + Ge-on-Si(surf.) \rightarrow Ge(island) \quad (2)$$

start competing with reaction (1). These are the attachment of deposited Ge atoms to the bare Si areas, giving nucleation and growth of the epitaxial Ge islands. Reactions (2) include the surface diffusion and are characterized by the activation energy of about 1 eV [28].

The density of Ge island arrays and the lateral island size are independent of the growth temperature up to about 570°C, whereas the density decreases and the lateral size increases with the temperature in the higher temperature range between 570°C and 650°C. **Figure 2** shows that at 590°C the island size increases with the increasing Ge coverage so that, at a coverage of 3 BL, the separation between some islands disappears and the islands start to merge. This causes the appearance of large variations in the island size under the further Ge deposition [**Figure 3(a)**]. At higher temperatures the larger islands continue to grow, while smaller islands can decrease in size and even disappear [**Figure 3(b)**]. This feature looks like Oswald ripening that occurs under the lattice strain, which increases with Ge coverage.

The temperature dependence of the Ge island formation can be explained by a large difference in the activation energies of reactions (1) and (2). The STM data suggest that the rate of reaction (2) is substantially higher

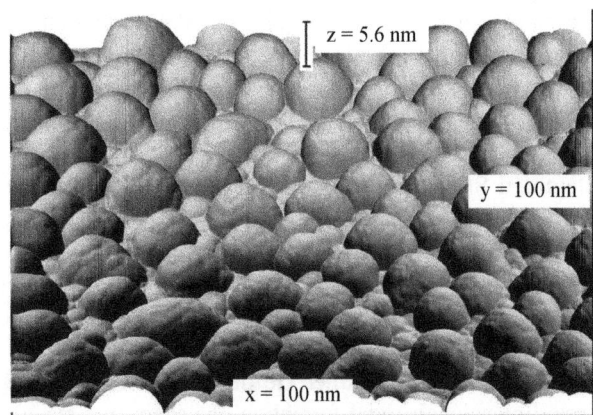

Figure 1. STM image of the oxidized Si(111) surface covered with 4 BL of Ge at 450°C.

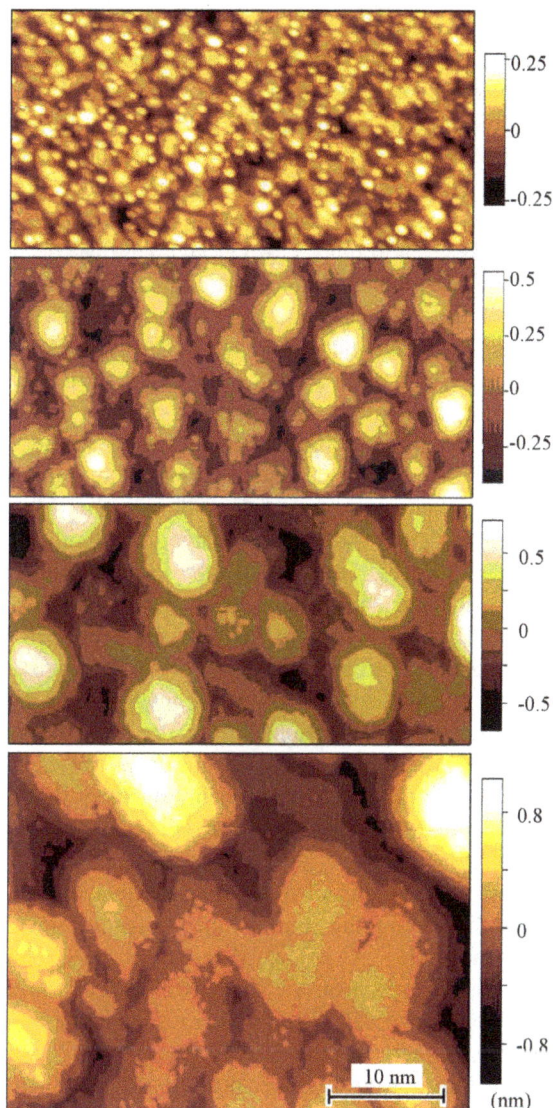

Figure 2. STM images of the oxidized Si(111) surface (a) before and after Ge deposition at the amount of (b) 1, (c) 2 and (d) 3 BL at 590°C. The images have the same lateral scale.

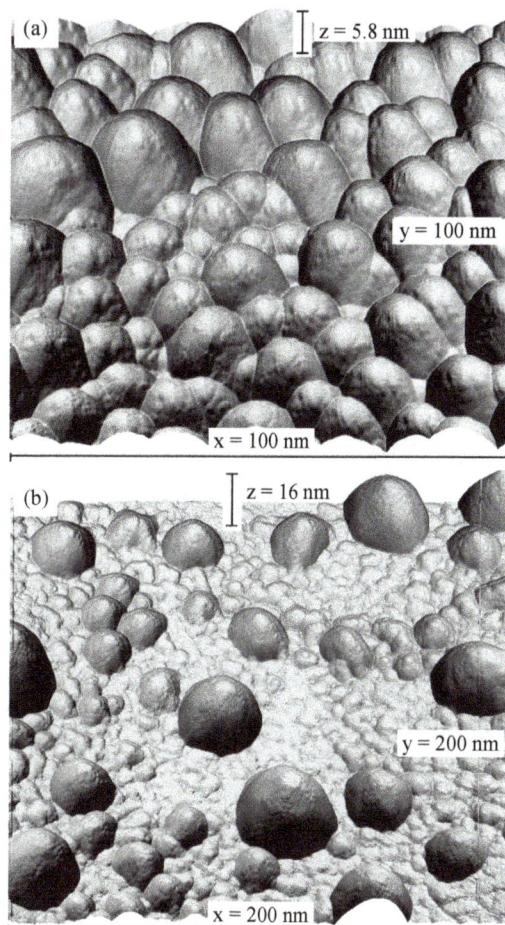

Figure 3. STM images of the oxidized Si(111) surface (a) before and after Ge deposition at the amount of (b) 1, (c) 2 and (d) 3 BL at 590°C. The images have the same lateral scale.

than the rate of reaction (1) in the low temperature range up to about 590°C. So, the reaction (1) acts only at the initial deposition stage until bare Si areas appear. Ge adatoms then start to preferably attach to these areas and form Ge islands by means of reaction (2), whereas reaction (1) is suppressed. As a result, Ge islands laterally grow over the residuals of the SiO$_2$ film. The rate of reaction (1) increases stronger with the increasing temperature due to the greater activation energy and becomes comparable to the rate of reaction (2) at temperatures above 590°C at which the SiO$_2$ decomposition occurs simultaneously with the island growth. This leads to the formation of larger areas of direct contacts between the Si substrate and the growing Ge islands. At the same time, the Ge islands attain a larger lateral size and

smaller height. The continuous action of reaction (1) results in complete decomposition of the SiO$_2$ film.

4. Ge growth at High Temperatures

After Ge island formation at the initial growth stage, the further Ge deposition or the sample annealing lead to the formation of large islands, when the temperature is in the range from 590°C to 650°C (**Figure 4**). The flat areas between the islands exhibit the 5 × 5 reconstruction which is normally observed for the 2 BL Ge wetting layers grown on the bare Si(111) substrates [29-31]. The (111) planes on top of the islands possess the 7 × 7 reconstruction that is similar for that known for Ge islands grown on the bare Si(111) surfaces [32-34]. The traces of the presence of threading dislocations in the islands were not found. The driving force for such surface morphology transformation is the lattice strain.

The Ge growth on the bare Si(111) surfaces has been studied in details [6-18,29-33]. For comparison with the above results the STM data for the Ge growth on the bare

Si(111) surfaces at temperatures of 530°C are shown in **Figure 5**. Deposited Ge forms large flat islands which are surrounded by trenches several nanometers deep [**Figure 5(a)**] [17,18]. The top plane of the islands has the 7×7 reconstruction [**Figure 5(b)**] [32,33]. The reconstructed surface is slightly undulating and that reflects the presence of the dislocation network lying in Ge near the Ge/Si interface. The Ge islands also contain the

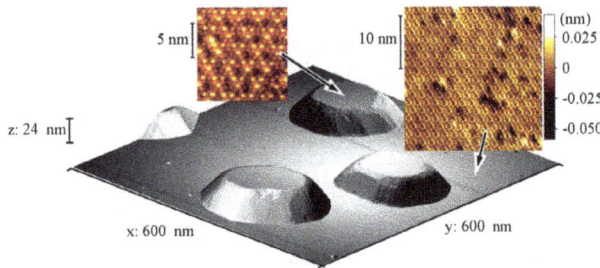

Figure 4. STM data for Ge islands obtained after 12 BL Ge deposition on the oxidized Si(111) surfaces at 630°C. Insets show STM images of an 11 nm² area on the island top plane with the 7×7 reconstruction and of a 26 nm² area between the islands with the 5×5 reconstruction. The images were obtained at the bias voltages of −2.4 and 2.4 V, respectively, and the current of 20 pA.

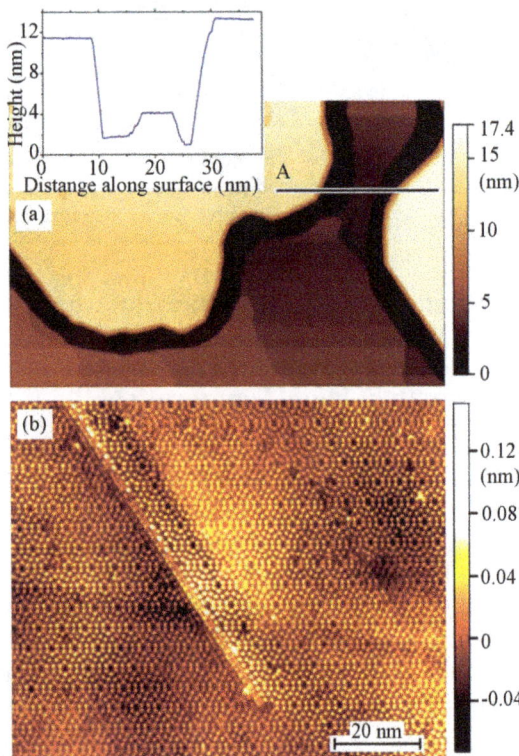

Figure 5. STM images obtained after 10 BL Ge deposition on the bare Si(111) surface at 530°C. (a) Area with the fragments of two flat Ge islands surrounded by trenches. (b) Surface structure of a Ge island obtained at a bias voltage of −2.4 eV. Inset represents the height profile along the line marked by A in (a).

threading dislocations, as shown in **Figure 5(b)** [11-13]. When the islands start to coalesce at the further Ge deposition, the presence of the deep trenches around the islands gives rise to the formation of deep pits in the Ge layer. The important thing is that the dislocations appear in the Ge layer even if the surfactants are used to obtain continuous Ge layers on Si(111).

The surface morphology of Ge layers grown on bare and oxidized Si(111) surfaces at high temperatures is essentially different. The Ge growth on the bare Si(111) surfaces is accompanied by a significant intermixing of Ge and Si atoms that occurs to reduce the lattice strain [17,18,34]. During the Ge wetting layer formation, events of the intermixing happen preferably at the moment of embedding of deposited Ge atoms into the surface. After the appearance of three-dimensional islands, the strongest lattice strain appears along the perimeter of the islands [17,18]. This leads to the formation of trenches around the islands. The absence of the trenches, thus, suggests that the Ge islands are less strained when the oxidized Si(111) surface.

Dark and bright spots in the STM images of the 5×5 reconstructed surface obtained from the areas between the islands (inset in **Figure 4**) shows the presence of structural defects, which are probably traces of the SiO₂ film. The residues of the SiO₂ film can reduce the strain between the Ge wetting layer and the Si substrate. If so, there would be no sufficiently strong driving forces also for Ge-Si intermixing during the formation the Ge wetting layer.

The Ge deposition on the oxidized Si(111) surface at the temperature of 650°C and higher results in the appearance of relatively large and small islands. The large islands are surrounded by rather deep trenches and have a high aspect ratio of 0.2 (**Figure 6**) whose value may reflect the content of Si in the Ge islands [35]. Such characteristics of the surface morphology indicate the appearance of a significant strain between the islands and the substrate due to the reduction of residuals of the SiO₂ film at the Ge/Si interface.

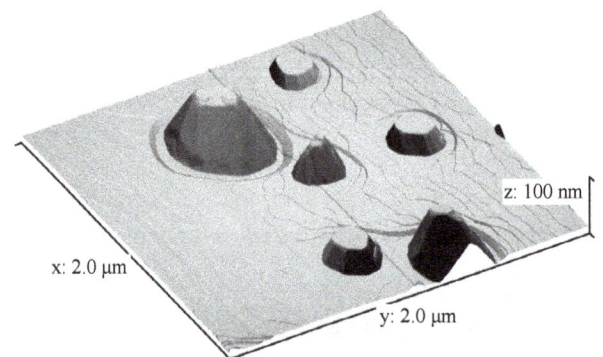

Figure 6. STM image obtained after 4 nm Ge deposition on the oxidized Si(111) surface at 650°C.

5. Shape of Ge Islands Grown at High Temperatures

Figure 7 shows the data that were used to identify the facets orientation at the islands sidewalls appeared after Ge deposition on the oxidized Si(111) surfaces at temperatures in the range of 570°C - 640°C, at which trenches around the islands do not form. The data show that the sidewalls are mainly faceted by {311} planes and groups of facets with orientations close to {110} planes. Other facets occupy a much smaller part of the sidewalls and serve to smoothly join these main facets between themselves and also with the (111) plane on the flat top of the islands. The {311} facets are known to be the major stable planes for unstrained Ge and Si islands [36-40].

Among the facets with orientations near {110}, the largest areas on the sidewalls occupy planes inclined from (111) by angle θ of 27° - 29° and having angular divergence $\Delta\varphi$ = 20.5° - 22.5° in the map of stereographic projections shown in **Figure 7**. The observation of

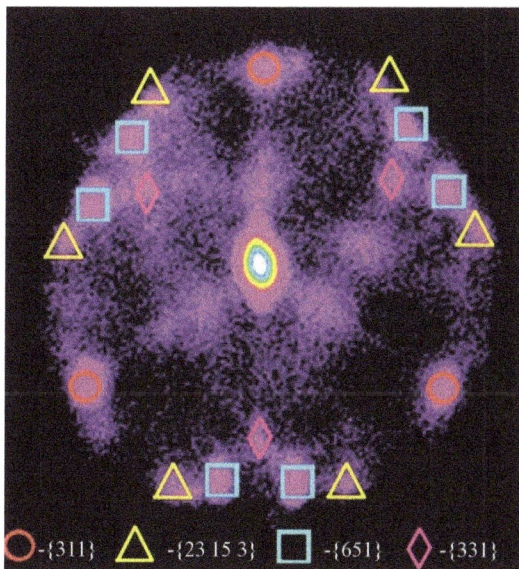

(a)

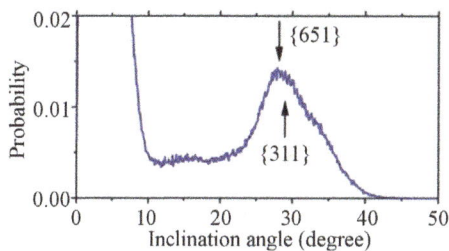

(b)

Figure 7. (a) The map of stereographic projections for surface areas and (b) portion of the areas as a function of their inclination angle with respect to the sample surface obtained from the STM image presented in Figure 4. The brightness in (a) is proportional to the size of the corresponding areas on the surface.

stable Ge facets with such an orientation has not been reported on. For Si surfaces the facets that have close orientations are {23 4 20} (θ = 30° and $\Delta\varphi$ = 16.9°), which were classified as MAJOR, and {651} (θ = 28.4° and $\Delta\varphi$ = 21.5°) (MINOR) [41]. The last facets have the orientations that are in good agreement with our experimental data and they are characterized by bright spots in the map of stereographic projections, covering relatively large sidewalls areas of the Ge islands.

Other facets with the orientations around {110} are identified to be {331} and {23 15 3}. These facets were previously classified as MAJOR for Ge [37]. However, they produce low-intense spots in the map shown in **Figure 7(a)**.

The surface morphology of Ge layers grown at relatively high temperatures on the oxidized Si(111) surfaces composed of Ge wetting layer areas and islands which shape is determined by energetically preferable facets. This is essentially different from that observed for the bare Si(111) surfaces, where the surface morphology contains flat islands without well defined facets on the sidewalls, deep trenches around the islands, threading dislocations and traces of the presence of the dislocation network near the Ge-Si interface (**Figure 5**). The difference arises from the difference in the lattice strain.

6. Influence of the Initial Growth Stage on Subsequent Ge Growth

The above results suggest that, in order to prevent the formation of large islands, the growth of Ge layers on the oxidized Si(111) surfaces should be carried out within several stages that differ in temperature. The initial stage at a high temperature in the range of 570°C - 650°C serves for the formation of nanoareas of bare Si surfaces and then that of epitaxial Ge nanoisland arrays with the concentration of the order of 10^{12} cm^{-2}. The temperature of the next growth stage must be reduced to prevent the decomposition of SiO₂ film residuals. The surface morphology obtained when the temperature was lowered to 430°C is shown in **Figure 8**. The data in **Figures 2** and **8** suggest that the preferable Ge coverage in the initial growth stage should be in the range of 1 - 2 BL. A larger coverage leads to a considerable variation of the islands in size and to non-uniformity in their spatial distribution due to the partial islands coalescence (**Figure 2(d)**)

7. Surface Morphology of Ge Layers Grown within Two Stages

As the Ge coverage increases to 30 BL in the two-stage growth, the facets appear on the sidewalls of the islands and the top of the islands becomes flat with (111) orientation. The data in **Figure 9** show that the largest facets on the sidewalls are {311}. They produce the brightest

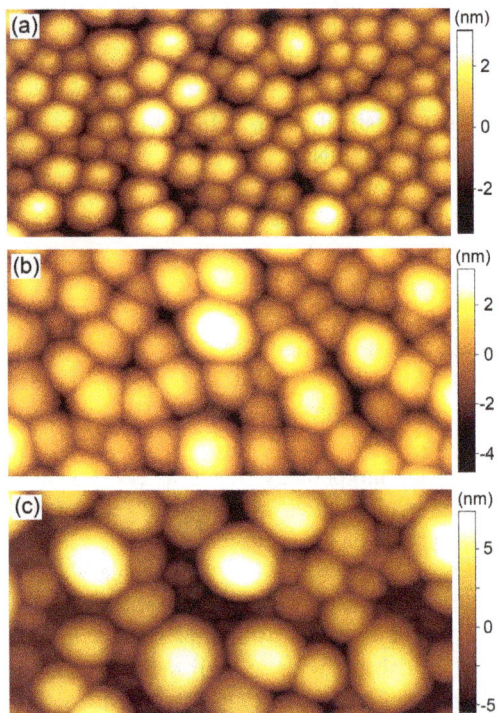

Figure 8. STM images of the oxidized Si(111) surface after Ge deposition at two stages. Ge was deposited for the coverage of (a) 1, (b) 2 and (d) 3 BL at 610°C at the initial stage and then at 430°C for the total coverage of 10 BL. The size of surface areas in the images is 100 × 50 nm.

spots in the map of stereographic projections (**Figure 9(b)**). The sidewalls also contain small-sized facets with orientations around {110}. The large peak in **Figure 9(c)** shows that the facets on the sidewalls preferably incline from (111) by the angles of 28° - 30°.

The coalescence of islands under further Ge deposition leads to a significant increase of the flat (111) areas (**Figure 10**). They undergo c(2 × 4) and c(2 × 8) reconstructions which are typical of the (111) surfaces of bulk Ge [42-44]. The observation of the reconstructions thus indicates that the grown epitaxial Ge layers on Si(111) are unstrained.

The STM data show that the ratio of the top flat areas to the areas of other facets, inclined from (111), increases with decreasing the growth temperature from 550 to 360°C. This occurs simultaneously with the decrease of the depth of depressions between the flat areas. This tendency does not extend to a low-temperature range. **Figure 11** shows that the Ge layers grown at 250°C - 300°C indeed exhibit a relatively low surface roughness; however, the surface morphology is mainly composed of small-sized stepped facets such as {433}, {755} and {775}, whereas the flat (111) areas occupy a relatively small portion of the surface. The root-mean-square roughness was estimated from the STM image, shown in **Figure 11**, to be about 2.8 nm for the 140 nm thick Ge

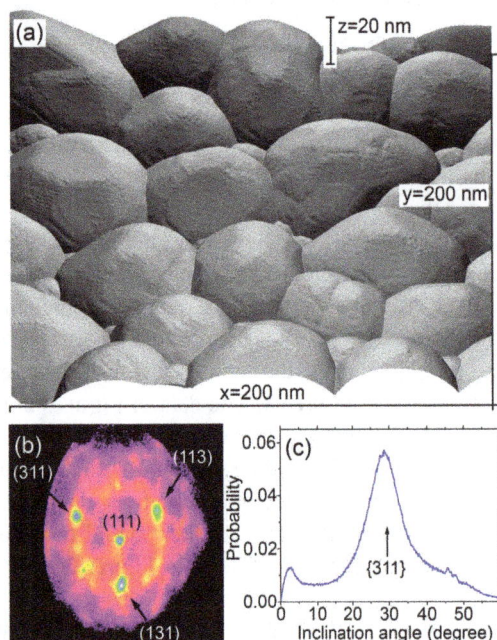

Figure 9. (a) STM image of the oxidized Si(111) surface after Ge deposition in two stages. Ge was deposited for a coverage of 2 BL at 610°C and then at 500°C for 30 BL. (b) The map of stereographic projections of surface areas and (c) portion of the areas as a function of their inclination angle from the sample surface obtained from the STM image.

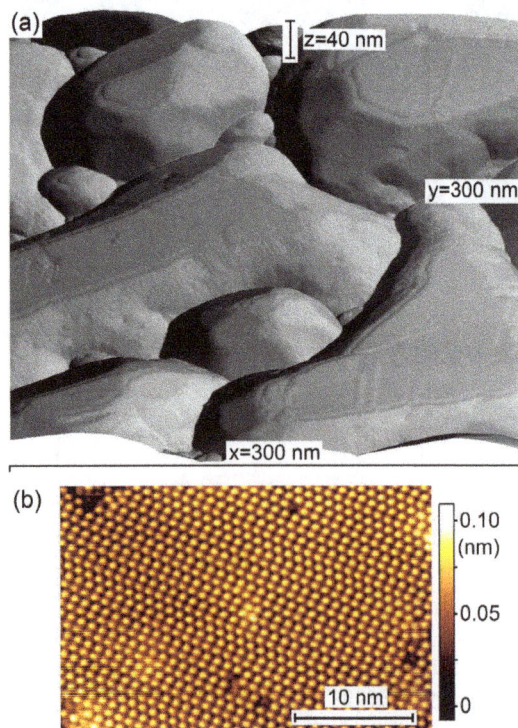

Figure 10. (a) STM image of the oxidized Si(111) surface after Ge deposition in two stages. Ge was deposited for a coverage of 1 BL at 620°C and then at 360°C for 85 BL. (b) STM image of the c(2 × 4) and c(2 × 8) reconstructions observed on flat areas of the Ge layer.

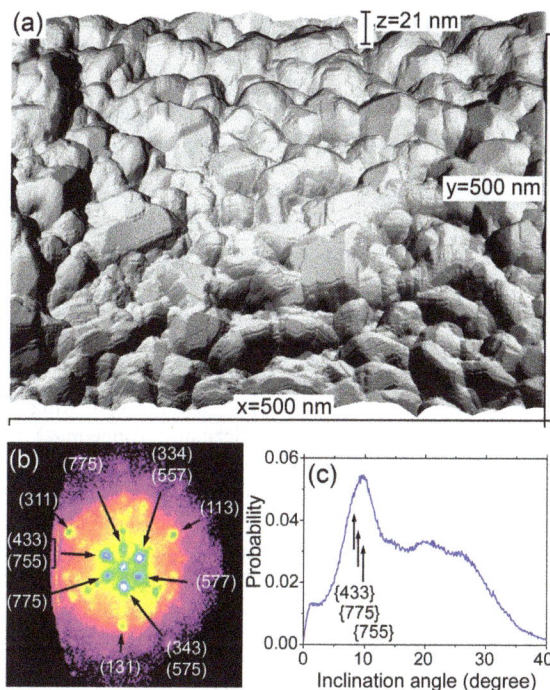

Figure 11. (a) STM image of the oxidized Si(111) surface after Ge deposition at two stages. Ge was deposited for the coverage of 2 BL at 590°C and then at temperatures in the range between 250 and 300°C for the total coverage of 140 nm. (b) The map of stereographic projections of surface areas and (c) portion of the areas as a function of their inclination angle from the sample surface obtained from the STM image.

layer. The most prevalent {433} and {7$\overline{5}$5} facets consist of atomic steps with edges towards [2$\overline{1}$1], which are the sort of stepped surfaces that were observed on the Ge surfaces cleaved in vacuum [45] and found to be stable on the vicinal Ge surfaces inclined from (111) [46].

8. Discussion

The idea of nanocontact epitaxy is to create the conditions for reducing a strain between the growing layer and the substrate by decreasing the area of their direct contact [20,21]. The Ge deposition of the oxidized Si surface produces bare Si areas in the SiO$_2$ film, which are a few nanometers in size and separated by the distance of 7 - 10 nm [22,47]. Transmission electron microscopy data reveal shifts of the Ge lattice with respect to the Si lattice at the Si/Ge interface that occurs in places where Ge is separated from Si by the residuals of the SiO$_2$ film [20]. In order to compensate the 4.2% lattice mismatch between Ge and Si, the shift should occur at the period of ~26 atoms, i.e., over the distance of 10 nm for the interface with (111) orientation. This distance is close to the average separation between the areas of bare Si that appear under Ge deposition on the oxidized Si(111) surfaces in the high-temperature range [22,23]. As shown

here, the Ge layers with a significantly reduced stain can be epitaxially grown on Si(111) using this technique.

After the formation of epitaxial Ge nanoislands, the further Ge deposition at low temperatures leads to the growth and coalescence of the islands. In the places of the coalescence the stacking faults appear. Their concentration essentially depends on the growth temperature. The higher temperature provides the formation of higher crystalline quality Ge layers. At the same time, the temperature influences the surface morphology of the growing Ge layer. At relatively high temperatures, the surface morphology is composed of well-defined facets of energetically favorable planes; this does not, however, lead to a flat surface with the (111) orientation. The roughness of the surface can be decreased by decreasing the growth temperature. Thus, the temperature acts in different ways: i.e., the use of high temperatures improves the crystalline quality of Ge layers, but it induces the formation of a rather large roughness, whereas the use of low temperatures flattens the surface, but introduces crystal defects. This feature is an obstacle in obtaining atomically flat, about 100 nm thick, Ge layers with a high crystalline quality on Si(111) substrates.

The Ge layers grown on bare and oxidized Si(111) surfaces are characterized by the substantially different surface morphology caused by a difference in the lattice strain. The strong Si-Ge intermixing under the strain leads to the formation of deep trenches in Si substrates around large flat Ge islands [17,18] and the network of threading dislocations [11-3]. Even the use of surfactants does not allow one avoiding the dislocation formation. In case of oxidized Si surfaces, the role of strain is significantly reduced. As a result, the surface morphology has no sharp corners and steep facets on the sidewalls. Instead, it includes some additional faceting planes, such as {775}, {755} and {761}, which are not observed for Ge layers grown on the bare Si surfaces.

The use of the oxidized Si surface solves the problem of the strong Ge/Si lattice stain and significantly reduces the Ge-Si intermixing, but it does not provide obtaining thin Ge layers with atomically flat surfaces. It is suggested that the smoothing of the surface of growing layers can be achieved with the help of surfactants, as it has been experimentally and theoretically shown [48-50]. Surfactants also facilitate the surface diffusion of deposited atoms thereby providing the improvement of the crystalline quality of layers growing at low temperatures.

9. Conclusion

The initial Ge deposition on the oxidized Si(111) surfaces at temperatures in the range of 570°C - 650°C results in the formation of nanoareas of bare Si and Ge nanoislands arrays. Further Ge deposition leads to the decomposition of the SiO$_2$ film residuals and to the formation of the Ge wetting layer and large islands. A comparison of Ge deposition on bare and oxidized Si(111) surfaces shows that the Ge/Si lattice stain is substantially reduced in the latter case. To obtain nanoareas of bare Si and homogeneous arrays of epitaxial Ge nanoislands, the preferable Ge coverage in the initial stage is found to be 1 - 2 BL. The temperature must then be decreased to 500°C or less to stop the SiO_2 film decomposition. After lowering the temperature, the shape of the growing islands evolves from rounded to faceted with a flat (111) plane on top. At the coverage of 30 - 140 nm, in addition to the (111) flat areas, the surface morphology is composed of {311} facets and facets lying around {110} and (111). The flat (111) areas exhibit the c(2 × 8) reconstruction that is typical of unstrained bulk Ge. The use of oxidized Si(111) surfaces allows one to obtain thin continuous epitaxial Ge layers on Si(111); however, the suppression of three-dimensional growth is still required to make the layers atomically flat.

10. Acknowledgements

We are grateful for the financial support by the Russian Foundation for Basic Research (Grant 11-07-00475-a), the Program of the Presidium of the Russian Academy of Sciences (project 24.21), and the Ministry of Education and Science of the Russian Federation (contract No. 16.518.11.7091).

REFERENCES

[1] K. Volz, A. Beyer, W. Witte, J. Ohlmann, I. Németh, B. Kunert and W. Stolz, "GaP-Nucleation on Exact Si(001) Substrates for III/V Device Integration," *Journal of Crystal Growth*, Vol. 315, No. 1, 2011, pp. 37-47.

[2] S. G. Ghalamestani, M. Berg, K. A. Dick and L.-E. Wernersson, "High Quality InAs and GaSb Thin Layers Grown on Si(111)," *Journal of Crystal Growth*, Vol. 332, No. 1, 2011, pp. 12-16.

[3] Yu. B. Bolkhovityanov and O. P. Pchelyakov, "GaAs Epitaxy on Si Substrates: Modern Status of Research and Engineering," *Physics-Uspekhi*, Vol. 51, No. 5, 2008, pp. 437-456.

[4] J. G. Cederberg, D. Leonhardt, J. J. Sheng, Q. Li, M. S. Carroll and S. M. Han, "GaAs/Si Epitaxial Integration Utilizing a Two-Step, Selectively Grown Ge Intermediate Layer," *Journal of Crystal Growth*, Vol. 312, No. 8, 2010, pp. 1291-1296.

[5] V. Destefanis, J.M. Hartmann, A. Abbadie, A. M. Papon and T. Billon, "Growth and Structural Properties of SiGe Virtual Substrates on Si(100), (110) and (111)," *Journal of Crystal Growth*, Vol. 311, No. 4, 2009, pp. 1070-1079.

[6] P. M. J. Maré, K. Nakagawa, F. M. Mulders, J. F. Van der Veen and K. L. Kavanagh, "Thin Epitaxial Ge-Si(111) Films: Study and Control of Morphology," *Surface Science*, Vol. 191, No. 3, 1987, pp. 305-328.

[7] U. Köhler, O. Jusko, G. Pietsch, B. Müller and M. Henzler, "Strained-Layer Growth and Islanding of Germanium on Si (111)-(7 × 7) Studied with STM," *Surface Science*, Vol. 248, No. 3, 1991, pp. 321-331.

[8] A. A. Shklyaev, M. Shibata and M. Ichikawa, "Instability of Two-Dimensional Layers in the Stranski-Krastanov Growth Mode of Ge on Si(111)," *Physical Review B*, Vol. 58, No. 23, 1998, pp. 15647-15651.

[9] B. Voigtländer and A. Zinner, "Simultaneous Molecular Beam Epitaxy Growth and Scanning Tunneling Microscopy Imaging during Ge/Si Epitaxy," *Applied Physics Letters*, Vol. 63, No. 2, 1993, pp. 3055-3057.

[10] A. A. Shklyaev, M. Shibata and M. Ichikawa, "Ge Islands on Si(111) at Coverages near the Transition from Two-Dimensional to Three-Dimensional Growth," *Surface Science*, Vol. 416, No. 1, 1998, pp. 192-199.

[11] S. Y. Shiryaev, F. Jensen, J. L. Hansen, J. W. Petersen and A. N. Larsen, "Nanoscale Structuring by Misfit Dislocations in $Si_{1-x}Ge_x$/Si Epitaxial Systems," *Physical Review Letters*, Vol. 78, No. 3, 1997, pp. 503-506.

[12] B. Voigtländer and N. Theuerkauf, "Ordered Growth of Ge Islands above a Misfit Dislocation Network in a Ge Layer on Si(111)," *Surface Science*, Vol. 461, No. 1-3, 2000, pp. L575-L580.

[13] S. A. Teys, "Features of Atomic Processes at the Formation of a Wetting Layer and Nucleation of Three-Dimensional Ge Islands on Si(111) and Si(100) Surfaces," *JETP Letters*, Vol. 96, No. 12, 2013, pp. 794-802.

[14] R. Gunnella, P. Castrucci, N. Pinto, I. Diavoli, D. Sébilleau and M. De Crescenzi, "X-Ray Photoelectron-Diffraction Study of Intermixing and Morphology at the Ge/Si(001) and Ge/Sb/Si(001) Interface," *Physical Review B*, Vol. 54, No. 12, 1996, pp. 8882-8891.

[15] X. R. Qin, B. S. Swartzentruber and M. G. Lagally, "Scanning Tunneling Microscopy Identification of Atomic-Scale Intermixing on Si(100) at Submonolayer Ge Coverages," *Physical Review Letters*, Vol. 85, No. 17, 2000, pp. 3660-3663.

[16] F. Ratto, F. Rosei, A. Locatelli, S. Cherifi, S. Fontana, S. Heun, P.-D. Szkutnik, A. Sgarlata, M. De Crescenzi and N. Motta, "Composition of Ge(Si) Islands in the Growth of Ge on Si(111) by x-Ray Spectromicroscopy," *Journal of Applied Physics*, Vol. 97, No. 4, 2005, pp. 043516-1-043516-8.

[17] T. I. Kamins, E. C. Carr, R. S. Williams and S. J. Rosner, "Deposition of Three-Dimensional Ge Islands on Si(001) by Chemical Vapor Deposition at Atmospheric and Reduced Pressures," *Journal of Applied Physics*, Vol. 81,

No. 1, 1997, pp. 211-219.

[18] F. Boscherini, G. Capellini, L. Di Gaspare, M. De Seta, F. Rosei, A. Sgarlata, N. Motta and S. Mobilio, "Ge-Si Intermixing in Ge Quantum Dots on Si," *Thin Solid Films*, Vol. 380, No. 1-2, 2000, pp. 173-175.

[19] M. Valvo, C. Bongiorno, F. Giannazzo and A. Terrasi, "Localized Si Enrichment in Coherent Self-Assembled Ge Islands Grown by Molecular Beam Epitaxy on (001) Si Single Crystal," *Journal of Applied Phys*ics, Vol. 113, No. 3, 2013, pp. 033513-1-033513-17.

[20] Y. Nakamura, A. Murayama and M. Ichikawa, "Epitaxial Growth of High Quality Ge Films on Si(001) Substrates by Nanocontact Epitaxy," *Crystal Growth & Design*, Vol. 11, No. 7, 2011, pp. 3301-3305.

[21] Y. Nakamura, T. Miwa and M. Ichikawa, "Nanocontact Heteroepitaxy of thin GaSb and AlGaSb Films on Si Substrates Using Ultrahigh-Density Nanodot Seeds," *Nanotechnology*, Vol. 22, No. 26, 2011, pp. 265301-1-265301-7.

[22] A. A. Shklyaev, M. Shibata and M. Ichikawa, "High-Density Ultrasmall Epitaxial Ge Islands on Si(111) Surfaces with a SiO$_2$ Coverage," *Physical Review B*, Vol. 62, No. 3, 2000, pp. 1540-1543.

[23] A. A. Shklyaev and M. Ichikawa, "Extremely Dense Arrays of Germanium and Silicon Nanostructures," *Physics-Uspekhi*, Vol. 51, No. 2, 2008, pp. 133-161.

[24] S. Ghosh, D. Leonhardt and S. M. Han, "Experimental and Theoretical Investigation of Thermal Stress Relief during Epitaxial Growth of Ge on Si Using Air-Gapped SiO$_2$ Nanotemplates," *Applied Physics Letters*, Vol. 99, No. 18, 2011, pp. 181911-1-181911-3.

[25] V. Kuryliuk, O. Korotchenkov and A. Cantarero, "Carrier Confinement in Ge/Si Quantum Dots Grown with an Intermediate Ultrathin Oxide Layer," *Physical Review B*, Vol. 85, No. 7, 2012, pp. 075406-1-075406-11.

[26] N. Miyata, H. Watanabe and M. Ichikawa, "Thermal Decomposition of an Ultrathin Si Oxide Layer around a Si(001)-(2 × 1) Window," *Physical Review Letters*, Vol. 84, No. 5, 2000, pp. 1043-1046.

[27] A. A. Shklyaev, M. Aono and T. Suzuki, "Influence of Growth Conditions on Subsequent Submonolayer Oxide Decomposition on Si(111)," *Physical Review B*, Vol. 54, No. 15, 1996, 10890-10895.

[28] A. A. Shklyaev and S. M. Repinsky, "Investigation of Ge Surface Self-Diffusion by Determination of Changes in the Reflection Intensity Profiles of Low-Energy Electron Diffraction," *Soviet Physics Semiconductors*, Vol. 14, No. 7, 1980, pp. 767-772.

[29] B. Voigtländer, "Fundamental Processes in Si/Si and Ge/Si Epitaxy Studied by Scanning Tunneling Microscopy during Growth," *Surface Science Reports*, Vol. 43, No. 5-8, 2001, pp. 127-254.

[30] N. Motta, A. Sgarlata, R. Calarco, Q. Nguyen, J. Castro Cal, F. Patella, A. Balzarotti and M. De Crescenzi, "Growth of Ge-Si(111) Epitaxial Layers: Intermixing, Strain Relaxation and Island Formation," *Surface Science*, Vol. 406, No. 1-3, 1998, pp. 254-263.

[31] M. Stoffel, Y. Fagot-Révurat, A. Tejeda, B. Kierren, A. Nicolaou, P. Le Fèvre, F. Bertran, A. Taleb-Ibrahimi and D. Malterre, "Electron-phonon Coupling on Strained Ge/Si(111)-(5 × 5) Surfaces," *Physical Review B*, Vol. 86, No. 19, 2012, pp. 195438-1-195438-7.

[32] H. -J. Gossmann, J. C. Bean, L. C. Feldman, E. G. McRae and I. K. Robinson, "7 × 7 Reconstruction of Ge(111) Surfaces under Compressive Strain," *Physical Review Letters*, Vol. 55, No. 10, 1985, pp. 1106-1109.

[33] U. Köhler, O. Jusko, G. Pietsch, B. Müller and M. Henzler, "Strained-Layer Growth and Islanding of Germanium on Si(111)-(7 × 7) Studied with STM," *Surface Science*, Vol. 248, No. 3, 1991, pp. 321-331.

[34] K. N. Romanyuk, A. A. Shklyaev, B. Z. Olshanetsky and A. V. Latyshev, "Formation of Ge Clusters at a Si(111)-Bi-√3 × √3 Surface," *JETP Letters*, Vol. 93, No. 11, 2011, pp. 661-666.

[35] G. Vastola, V. B. Shenoy, J. Guo and Y.-W. Zhang, "Coupled Evolution of Composition and Morphology in a Faceted Three-Dimensional Quantum Dot," *Physical Review B*, Vol. 84, No. 3, 2011, pp. 035432-1-035432-7.

[36] A. Laracuente, S. C. Erwin and L. J. Whitman, "Structure of Ge(113): Origin and Stability of Surface Self-Interstitials," *Physical Review Letters*, Vol. 81, No. 23, 1998, pp. 5177-5180.

[37] Z. Gai, R. G. Zhao, X. Li and W. S. Yang, "Faceting and Nanoscale Faceting of Ge(hhl) Surfaces around (113)," *Physical Review B*, Vol. 58, No. 8, 1998, pp. 4572-4578.

[38] A. A. Stekolnikov and F. Bechstedt, "Shape of Free and Constrained Group-IV Crystallites: Influence of Surface Energies," *Physical Review B*, Vol. 72, No. 12, 2005, p. 125326.

[39] J. T. Robinson, A. Rastelli, O. Schmidt and O. D. Dubon, "Global Faceting Behavior of Strained Ge Islands on Si," *Nanotechnology*, Vol. 20, No. 8, 2009, Article ID: 085708.

[40] A. A. Shklyaev, K. N. Romanyuk, A. V. Latyshev and A. V. Arzhannikov, "Effect of Dislocations on the Shape of Islands during Silicon Growth on the Oxidized Si(111) Surface," *JETP Letters*, Vol. 93, No. 6, 2011, pp. 442-445.

[41] Z. Gai, R. G. Zhao, W. Li, Y. Fujikawa, T. Sakurai and W. S. Yang, "Major Stable Surface of Silicon: Si(20 4 23)," *Physical Review B*, Vol. 64, No. 12, 2001, Article

ID: 125201.

[42] M. Henzler, "Correlation between Surface Structure and Surface States at the Clean Germanium (111) Surface," *Journal of Applied Phys*ics, Vol. 40, No. 9, 1969, pp. 3758-3765.

[43] B. Z. Olshanetsky, S. M. Repinsky and A. A. Shklyaev, "LEED Investigation of Germanium Surfaces Cleaned by Sulphide Films, Structural Transitions on Clean Ge(110) Surfaces," *Surface Science*, Vol. 64, No. 1, 1977, pp. 224-236.

[44] M. Kuzmin, M. J. P. Punkkinen, P. Laukkanen, J. J. K. Lang, J. Dahl, V. Tuominen, M. Tuominen, R. E. Perälä, T. Balasubramanian, J. Adell, B. Johansson, L. Vitos, K. Kokko and I. J. Väyrynen, "Surface Core-Level Shifts on Ge (111)-c(2 × 8): Experiment and Theory," *Physical Review* B, Vol. 83, No. 24, 2011, p. 245319.

[45] M. Henzler, "The Roughness of Cleaved Semiconductor Surfaces," *Surface Science*, Vol. 36, No. 1, 1973, pp. 109-122.

[46] B. Z. Olshanetsky, S. M. Repinsky and A. A. Shklyaev, "LEED Studies of Vicinal Surfaces of Germanium," *Surface Science*, Vol. 69, No. 1, 1977, pp. 205-217.

[47] A. A. Shklyaev and M. Ichikawa, "Effect of Interfaces on Quantum Confinement in Ge Dots Grown on Si Surfaces with a SiO_2 Coverage," *Surface Science*, Vol. 514, No. 1-3, 2002, pp. 19-26.

[48] M. Copel, M. C. Reuter, M. Horn von Hoegen and R. M. Tromp, "Influence of Surfactants in Ge and Si Epitaxy on Si (001)," *Physical Review B*, Vol. 42, No. 18, 1990, pp. 11682-11689.

[49] B. Voigtlander, A. Zinner, T. Weber and H. P. Bonzel, "Modification of Growth Kinetics in Surfactant-Mediated Epitaxy," *Physical Review B*, Vol. 51, No. 12, 1995, pp. 7583-7591.

[50] D. Kandel and E. Kaxiras, "Surfactant Mediated Crystal Growth of Semiconductors," *Physical Review Letters*, Vol. 75, No. 14, 1995, pp. 2742-2745.

25

Dynamic Impact Absorption Behaviour of Glass Coated with Carbon Nanotubes

Prashant Jindal[1*], Meenakshi Goyal[2], Navin Kumar[3]

[1]University Institute of Engineering & Technology, Panjab University, Chandigarh, India; [2]University Institute of Chemical Engineering & Technology, Panjab University, Chandigarh, India; [3]Indian Institute of Technology, Roopnagar, Punjab, India.

ABSTRACT

Boro-silicate glass samples were coated with chemically treated multi-walled carbon nanotubes (MWCNTs) to study the resistance offered by the coatings under the high strain rate impact. Impact testing of these glass samples was performed on Split Hopkinson Pressure Bar (SHPB), where strain rates were varied from 500/s to 3300/s. However, the comparisons were limited to samples subjected to a strain rate of 2300/s to 3000/s so that the effect of only variable deposits of coatings on the stress-strain behavior of glass can be studied. Variable deposits (0.1 mg to 0.8 mg) of MWCNTs were coated uniformly on glass samples having a disc shape with a fixed surface area (79 mm^2) to observe the effect of the coating on the impact absorption capacity of glass. It was observed that the small thickness of about 25 μm formed due to the fact that 0.2 mg of MWCNTs deposit spread over the surface increased the impact absorption capacity of the glass pieces by nearly 70%. However, beyond this amount when the deposit was increased to 0.4 mg, the coating thickness got doubled to nearly 49 μm and this led to a fall in absorption capacity which remained static till 0.8 mg deposit. However, even this decrease in capacity was able to absorb 30% more impact than offered by pure glass sample.

Keywords: Glass Coatings; Impact Behaviour; Strength; Mechanical Properties

1. Introduction

Over the years, impacting resistant materials has been extensively studied using composites that comprise of light weight base matrix and strong filler materials. These materials are tested under extreme impact and static loading conditions so that they can be used for various applications like bullet-proof shields, jackets, resistant surfaces, shock and impact absorbers etc. [1,2].

Apart from fabricating stress resistant materials in the form of composites, absorber coatings also become important when it comes to preserving the basic equipment and acting as a protective coat. These coatings can be sacrificed to protect the base material also. It becomes imperative that such coatings are their light weight so that their own weight does not affect the overall utility of the basic equipment.

One of the most useful equipments for studying material behavior under impact loading is Split Hopkinson Pressure Bar (SHPB). Stress-strain behavior of the specimen when subjected to impact or dynamic loading is

obtained when the specimen is subjected to a strain rate of 100 to 10,000/s.

The SHPB apparatus consists of two long slender bars namely, an input bar and an output bar that sandwich a short specimen between them. Whenever any load is applied on one end, the sandwiched specimen undergoes very high compression loading. A block diagram of a typical SHPB is shown in **Figure 1**.

The details of working of Split Hopkinson bar set up are widely available in literature [3]. It is basically based upon the measurement of wave signal which is generated by the input and output bars due to high strain rate loading. The waves are a measure of strains which are calibrated to find stress and strain in the specimen and in an earlier work. Impact loading using SHPB on carbon nanotube-polycarbonate composites was also studied [4].

To the best of our knowledge, most of the dynamic and quasi-static strength related work has been done on composite structures [4-9]. Static properties like elastic modulus, indentation pressure and fracture toughness of coatings on glass have been studied by Malzbender *et al.* [10,11]. In these studies, the composition of the coatings has also been varied by silica and alumina composition. Fluid based coatings like methyltrimethoxysilane and Ludox were also used. Coatings of thickness nearly 5 μm to 11 μm have also been studied. Static load in the order of 50 mN to 300 mN was applied and observations were measured on the basis of indentations made on the surface. Indentation pressures were greatly reduced after the initiation of any crack or indentation. However, the results have been used only as guidance on how crack. Delamination and chipping of coatings takes place as applied static load is varied.

Thus no study has appeared in the literature that uses a coating of MWCNTs instead of embedding for dynamical impact study. Since MWCNTs have anisotropic behaviors even for elastic properties, these offer great possibilities as protective fronts to soft targets. The Young's modulus as well as tensile strength is significantly different as compared to their bulk modulus [12,13]. Therefore a study that uses vertically aligned coatings as fronts is expected to behave differently as compared to horizontally aligned coating fronts. Usually it is very difficult to control up till now the alignment of carbon nanotubes, therefore a mixture is expected. For horizontally aligned,

resilience of carbon nanotubes is also going to be useful. With this objective in view, we have planned to undertake the present study which aimed to study the modification of resistance offered by pure glass on exposing carbon nanotubes coated surface to the impact. We have prepared variable thickness of coatings of MWCNTs by varying the quantity of deposit on glass and studied them under the high strain rate impact. We have given an experimental methodology for sample preparation. The coating procedure is defined and these samples are then subjected to impact studies using SHPB. In the end, the work is summarized and concluded.

2. Experimental

MWCNTs having diameter about 10 - 30 nm and length 1 - 10 microns were procured from Nanoshel Intelligent Materials Pvt. Ltd., USA. We characterized them using FTIR spectra as shown in **Figure 2** and the peaks are indicative of the MWCNTs. **Figure 3** shows the SEM image of MWCNTs as provided by the supplier. The image indicates the diameter of the material as per specifications.

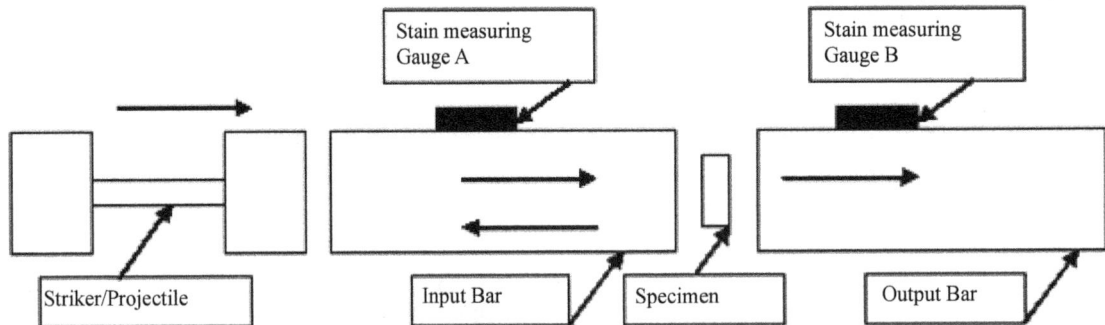

Figure 1. Schematic block diagram of split hopkinson pressure bar.

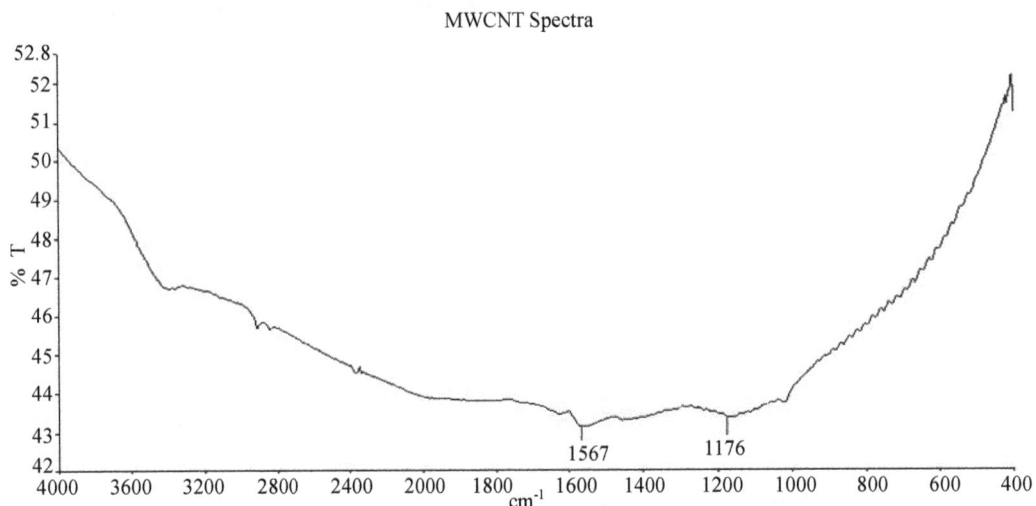

Figure 2. FTIR spectra for MWCNTs purchased from Nanoshel Intelligent Materials Pvt. Ltd.

Coating Procedure

Boro-silicate glass pieces of disc shape having diameter 10 mm and thickness 5 mm were taken as the base material. They were cleaned with ethanol. MWCNTs of variable amounts were mixed with DMF (dimethylformamide) and ultra-sonicated for a few hours to ensure reasonable dispersion. Measured quantities of different concentrations of these MWCNTs solutions were then spread over the glass pieces to form non-covalent bond [14] between the coating and glass surface. The different concentrations of these MWCNTs solutions and amount spread over the glass pieces are given in **Table 1**. On evaporation of the solvent, coatings of varied thickness and quantity of MWCNTs distributed reasonably uniformly as solvent on the surface of glass samples of 79 mm^2 area were obtained.

A simple estimate of a single layer of average thickness D of MWCNTs of bulk density ρ when spread over a surface area A of the glass disc, will have mass as m = ADρ. The average bulk density of our MWNCTs was 100 mg/cm^3, average length = 5 µm, A = 0.79 cm^2 and m = 0.1 mg to 0.8 mg meant that for our samples the thickness was from 10 to 100 µm. It also meant that our samples were coated with about 5 to 20 layers. This way we can control the MWCNT layers to about 50 by varying the deposit of MWCNTs even if the MWCNTs stand vertically. The data of estimated number of layers is also presented in **Table 1**. It may be noted, that the number of layers is based upon the assumption that MWCNTs are vertically aligned, however in reality MWCNTs can be a combination of various alignments. Hence, the number of layers given is a lower estimate.

These different glass coated samples were then used for dynamic impact strength studies and their dynamic impact strengths were compared at high strain rates using SHPB. The variation parameter here was only the amount of coating deposited not the geometry or orientation of the inner structure of specimen.

The setup for SHPB comprised of two high strength maraging steel with yield strength ~ 1750 MPa, diameter 20 mm and length 2000 mm. The projectile diameter was 20 mm and length was 300 mm. Strain gauges of 120 Ω,

900 tee rosette precision stain gauges designated as EA-06-125TM-120) were used.

Projectile of length 300 mm was hit on samples of different deposits one by one which were sandwiched between the two bars.

The projectile was shot at by a pressure gun producing stress-strain curves for different strain rates. Strain rates varied in the range from 500/s to 3300/s.

3. Results and Discussion

The data collected by strain gauges for incident, reflected and transmitted signals leads to evaluation of stress-strain data. Though stress-strain data was obtained for a wide range of strain rates (500/s to 3300/s) for all samples but samples which were limited to a strain rate of 2300/s to 3000/s were compared so that the effect of only variable deposits of coatings on the stress-strain behavior of glass could be studied. This strain rate is a useful range in normal shock conditions, encountered during aviation and defense requirements [15]. Compressive stress-strain behavior for glass pieces coated with MWCNTs of different amount at strain rates of about 2500/s are shown in **Figure 4**.

It is observed from **Figure 4** that a plastic deformation pattern is formed for all samples.

Maximum stress absorbed by each of these samples shows that till a particular deposit of coating, there is a substantial increase in the stress absorbed but after that it starts decreasing. Maximum stress absorbed for pure glass is nearly 389 MPa. When this piece is non-covalently bonded with 0.1 mg of MWCNT coating then this

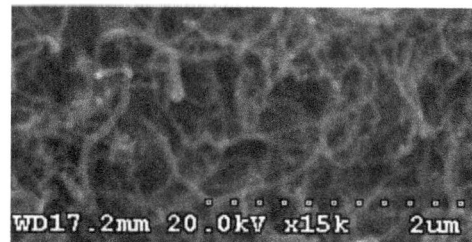

Figure 3. SEM Image for MWCNTs as provided by Nanoshel Intelligent Materials Pvt. Ltd.

Table 1. Samples of various concentrations of MWCNTs solution on glass, thickness of coat, rough estimate of number of layers and quantity of solution that was spread on glass surface.

Sample No.	Concentration of coating on glass (mg/µL)	Quantity of solution poured (µL)	Coating thickness (µm)	Estimated no. of layers
1	10/1000	10	12	3
2	18.6/930	10	25	5
3	20/520	10	49	10
4	25/400	10	80	16
5	15/200	10	95	19

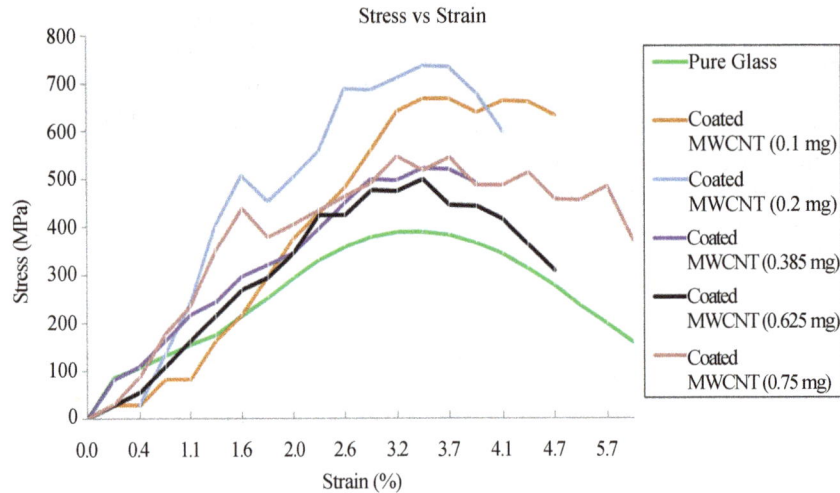

Figure 4. Variation of stress strain for different amounts of coated glass pieces with MWCNTs subjected to strain rates from 2300/s to 3000/s.

maximum limit reaches 667 MPa at nearly the same strain. Similarly, for 0.2 mg coating the stress value is about 736 MPa. But beyond this, for coatings of 0.385 mg, 0.625 mg and 0.75 mg this maximum stress value remains nearly same 500 MPa which is still much higher than pure glass.

So, in comparison to pure glass, the samples which were coated with a very small amount of 0.1 mg and 0.2 mg MWCNTs had about 50% to 70% increased stress absorption capacity. This also implies that a coating thickness of MWCNTs of about 12 μm to 25 μm is sufficient to enhance the stress absorption by almost 2 times.

However, the improved degradation at higher concentration is most likely to be a result of slipping of the layers among themselves as contact with glass gets lost because coatings of nearly 0.4 to 0.8 mg means that thickness of coatings reaches nearly 40 μm to 100 μm. So, the number of layers on the glass pieces increases accordingly.

The effect of variation in deposit of MWCNT coatings on maximum impact stress within the strain rates at about 2500/s as explained above is further depicted in **Figure 5**.

4. Summary and Conclusion

Base materials which have attractive properties like light weight, mould ability, transparency etc. but are vulnerable to impact or shock loads need to be improved in terms of their dynamic strength by either embedding or coating with other stronger materials. In this paper we studied the dynamic impact absorption using SHPB of pure boro-silicate glass as the base material and the same glass coated with variable amounts of MWCNTs.

Boro-silicate glass in the form of a disc 10 mm diameter and 5 mm thickness was used as the base material.

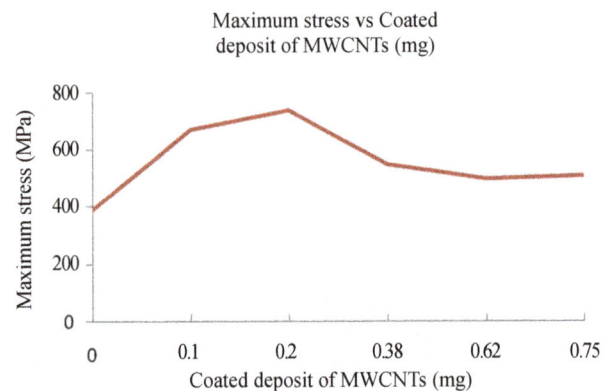

Figure 5. Maximum stress variation with different coated MWCNTs-glass samples subjected to strain rates from 2300/s to 3000/s.

Coated samples were prepared using non-covalent chemical binding techniques. The coated amount of MWCNTs was varied from 0.1 mg to 0.8 mg and accordingly thickness of the coating was also estimated. For smaller concentrations, the thickness of 12 μm to 25 μm meant that the number of layers on the glass surface was nearly 5. But for the higher amount of coatings as the thickness of coating increased to about 100 μm, the layers also reach about 20.

Dynamic impact was applied to these samples and interesting observations were made. Samples which had coatings of about 0.1 mg and 0.2 mg showed significant increase in the maximum stress absorption in comparison to pure glass. The increase was about 50% to 70%. Maximum stress for 0.1 mg and 0.2 mg coating sample was nearly 689 MPa and 736 MPa respectively while pure glass maximum stress was 389 MPa. However, coatings of nearly 0.4 mg, 0.6 mg and 0.8 mg did not show a further increase. The maximum stress absorbed by these samples was nearly 500 MPa, which was still about 30% higher than pure glass but much less than 0.1 mg and 0.2 mg. The reason for this reduction can be the increased thickness of coating that comprises of multiple layers of MWCNTs.

As layers of coatings increase, there is slipping of these layers from the glass surface and amongst the layers themselves. As a result, the coatings slip away from the base glass surface and fail to offer higher resistance to impact.

On the basis of the results obtained in this work, it seems safe to conclude that coating by small concentrations of MWCNT improves the dynamic impact strength of glass.

It not only helps modify glass strength, but also is a useful impact stress sensor. In fact, a stacking of multiple coated glass samples can be used to absorb desired impact as well as sensing unit for such impacts. As the glass piece was covered with minor amounts of MWCNTs, the transparency loss was not significant.

5. Acknowledgements

Prashant Jindal gratefully acknowledges financial support from the Defence Research Organization (DRDO) for a research project (No. ARMREB/DSW/2011/129). He also acknowledges the Director, TBRL and the whole team of Gun Group for extending their lab facilities. Guidance provided by Biomoluecular Electronics and Nanotechnology Division (BEND), at Central Scientific Instruments Organisation (CSIO), Chandigarh is also acknowledged. He is also grateful to Mr. Hitesh Sharma from Accurate Optics, Chandigarh for assistance in providing base material. Dr. Rajesh Kumar, UIET, Panjab University, Chandigarh assistance is also acknowledged.

REFERENCES

[1]   J. N. Coleman, U. Khan, W. J. Blau and Y. K. Gun'ko, "Small but Strong: A Review of the Mechanical Properties of Carbon Nanotube-Polymer Composites," *Carbon*, Vol. 44, No. 9, 2006, pp. 1624-1652.

[2]   P. Raju Mantena, Alexander H. D. Cheng, Ahmed Al-Ostaz and A. M. Rajendran, "Blast and Impact Resistant Composite Structures for Navy Ships Composite Structures and Nano-Engineering Research" Proceedings of the 2008 ONR Solid Mechanics Program—Marine Composites and Sandwich Structures, University of Maryland, Adelphi, 2009, pp. 417-426.

[3]   H. Kolsky, "An Investigation of the Mechanical Properties of Materials at Very High Rates of Loading," *Proceedings of the Royal Society of London*, London, B62, 1949, pp. 676-700

[4]   P. Jindal, S. Pande, P. Sharma, V. Mangla, A. Chaudhury,

Deepak Patel, B. P. Singh, R. B. Mathur and M. Goyal, "High Strain Rate Behavior of Multi-Walled Carbon Nanotubes-Polycarbonate Composites," *Composites Part B: Engineering*, Vol. 45, No. 1, 2013, pp. 417-422.

[5]   N. K. Naik and Y. Perla, "Mechanical Behaviour of Acrylic under High Strain Rate Tensile Loading," *Polymer Testing*, Vol. 27, No. 4, 2008, pp. 504-512.

[6]   A. Jadhav, E. Woldesenbet and S.-S. Pang, "High Strain Rate Properties of Balanced Angle-Ply Graphite/Epoxy Composites," *Composites: Part B*, Vol. 34, No. 4, 2003, pp. 339-346.

[7]   W. Chen, F. Lu and M. Cheng, "Tension and Compression Tests of Two Polymers under Quasistatic and Dynamic Loading," *Polymer Testing*, Vol. 21, No. 2, 2002, pp. 113-121.

[8]   R. B. Mathur, S. Pande, B. P. Singh and T. L. Dhami, "Electrical and Mechanical Properties of Multi-Walled Carbon Nanotubes Reinforced PMMA and PS Composites," *Polymer Composites*, Vol. 29, No. 7, 2008, pp. 717-727.

[9]   Y. H. Xu, Q. F. Li, D. Sun, W. N. Zhang and G.-X. Chen, "A Strategy to Functionalize the Carbon Nanotubes and the Nanocomposites Based on Poly(L-lactide)," *Industrial & Engineering Chemistry Research*, Vol. 51, No. 42, 2012, pp. 13648-13654.

[10] J. Malzbender, G. de With and J. M. J. den Toonder, "Elastic Modulus, Indentation Pressure and Fracture Toughness of Hybrid Coatings on Glass," *Thin Solid Films*, Vol. 366, No. 1-2, 2000, pp. 139-149.

[11] J. Malzbender and G. de With, "Energy Dissipation, Fracture Toughness and the Indentation Load-Displacement Curve for Coated Materials," *Surface and Coatings Technology*, Vol. 135, No. 1, 2000, pp. 60-68.

[12] R. S. Ruoff, D. Qian and W. K. Liu, "Mechanical Properties of Carbon Nanotubes: Theoretical Predictions and Experimental Measurements," *C. R. Physique*, Vol. 4, No. 9, 2003, pp. 993-1008.

[13] A. Sears and R. C. Batra, "Macroscopic Properties of Carbon Nanotubes from Molecular-Mechanics Simulations," *Physical Review B*, Vol. 69, 2004, Article ID: 235406.

[14] Ma, A. Lu, J. Yang, S. H. Ka and M. Ng, "Quantitative Non-Covalent Functionalization of Carbon Nanotubes," *Journal of Cluster Science*, Vol. 17, No. 4, 2006, pp. 599-608.

[15] Mostafa Shazly, David Nathenson and Vikas Prakash, "Modelling of High Strain Rate Deformation, Fracture and Impact Behavior of Advances Gas Turbine Engine Materials at Low and Elevated Temperatures," 2003, NASA CR-2003-212194.

26

Effect of Carbon Content on Ti Inclusion Precipitated in Tire Cord Steel

Yuedong Jiang[1,2], Jialiu Lei[1], Jing Zhang[1], Rui Xiong[1], Feng Zou[1], Zhengliang Xue[1]

[1]Key Laboratory for Ferrous Metallurgy and Resources Utilization of Ministry of Education, Wuhan University of Science and Technology, Wuhan, China; [2]Research and Development Center of Wuhan Iron & Steel Group, Wuhan, China.

ABSTRACT

The precipitation of TiN inclusion during solidification of different carbon content of 0.72%, 0.82% and 0.95% in tire cord steel is thermodynamically studied respectively. The results show that the carbon content has obvious effect on TiN inclusion precipitated in tire cord steel of different strength levels. With the carbon content of tire cord steel increasing, the temperature before solidifying reduced gradually and the required activity product of titanium and nitrogen for TiN inclusion precipitation also declined gradually. With the same condition of initial Ti and N content in liquid steel, the size of TiN inclusion precipitated in tire cord steel of higher carbon content is bigger than that of lower carbon content. In order to control the harmful effects on processability of TiN inclusion precipitated in hypereutectoid tire cord steel of the ultra high strength level, the measures of smelting process must be taken to further reduce the titanium and nitrogen content in liquid steel.

Keywords: Hypereutectoid Tire Cord Steel; TiN Inclusion; Carbon Content; Thermodynamics

1. Introduction

The tire cord steel is a kind of high carbon steel, which is used for the production of car tyres meridian steel wire. With the car lightening and considering the safety of the car, the strength level of car tyre cord steel wire is improving. Before the 1990s, the mainstream brands of the tyre steel wire are SWRH62A and SWRH67A of ordinary strength grade 1750 MPa. Since the 1990s, the mainstream brands of the tyre steel wire are SWRH72A of high strength grade 1870 MPa. Entering the 21st century, the mainstream of the tire steel wire product is the hypereutectoid tire cord steel SWRH82A of the ultra high strength level and brands of a higher strength level in which carbon content reached 0.92% - 0.95%. In order to control the harmful effects of titanium inclusion in hypereutectoid tire cord steel on the drawing performance of wire rod, steelmakers should take tough measures of steelmaking and refining technology to remove the content of Ti and N in the molten steel to the limit as much as possible. In spite of this, in the condition of current steel production technology, the precipitation of titanium inclusion is still inevitable before solidifying [1]. In order to provide theoretical guidance to control TiN inclusion precipitated in the hypereutectoid tire cord steel of the ultra high strength level, the effect of carbon content on Ti inclusion precipitated in tire cord steel of different intensity levels is studied on the basis of thermodynamics in this paper.

2. Temperature Changes in Solidifying Front of Tire Cord Steel

The temperature in the front of solidifying (T_{s-l}) of liquid steel is the solidification interface temperature, this temperature value is between the solidus temperature (T_s) and the liquidus temperature (T_l), It related with the solidification rate (g) of liquid-solid two phase region. The temperature in the front of solidifying is calculated according to formula (1) [2]:

$$T_{s-l} = T_0 - \frac{T_0 - T_l}{1 - g\frac{T_l - T_s}{T_0 - T_s}} \tag{1}$$

In formula (1) T_0 is the melting point of Fe 1538°C, the liquidus temperature (T_l) and the solidus temperature (T_s) are calculated according to formula (2) and formula (3):

$$T_l = 1538 - 65w[C] - 8w[Si] - 5w[Mn]$$
$$\quad - 20w[Ti] - 30w[P] - 25w[S] \qquad (2)\,[3]$$
$$\quad - 1300w[H] - 90w[N] - 80w[O]$$

$$T_S = 1538 - 175w[C] - 20w[Si]$$
$$\quad - 30w[Mn] - 40w[Ti] - 280w[P] \qquad (3)\,[3]$$
$$\quad - 575w[S] - 160w[O]$$

The components of tire cord steel used for calculation are shown in **Table 1**.

According to the components of tire cord steel in **Table 1** the liquidus temperature (T_l) and solidus temperature (T_s) can be calculated, as shown in **Table 2**. With the increase of carbon content, the temperature difference of liquidus and solidus become large.

The relationship between the temperature in the front of solidifying (T_{s-l}) and solidification rate (g), as shown in **Figure 1**.

Figure 1 shows that the temperature in the front of solidifying (T_{s-l}) decreased along with the development of the solidification process. When the solidification rate is the same, the higher carbon content in the tire cord steel, the lower the temperature in the front of solidifying.

3. Effect of Carbon Content on Ti Inclusion Precipitated in Tire Cord Steel

Due to the solidification segregation of the solute elements during solidification process, Ti and N are enriched in the solidifying front continuously, when the activity product of Ti and N is greater than the balanced activity product of TiN precipitation, TiN inclusion will precipitate according to the following formula:

$$[Ti] + [N] = TiN_{(S)}$$
$$\Delta G^0 = -291000 + 107.97T \qquad (4)$$

$$\lg K_{TiN} = 5.64 - 15220/T \qquad (5)$$

In formula (5), K_{TiN} is the equilibrium activity product of TiN precipitation, which related with the temperature T_{s-l} in the front of solidifying. With the solidification rate (g) of liquid-solid two phase region increase, temperature in the front of solidifying dropped and the value of K_{TiN} calculated according to formula (5) declined correspondingly.

The actual activity product of Ti and N in the solidifying front is:

$$Q_{TiN} = f_{Ti} \cdot f_N \cdot w[Ti] \cdot w[N] \qquad (6)$$

In formula (6), f_{Ti} and f_N are the activity coefficients of Ti and N at the temperature of solidifying front, which can be calculated according to formula (7) and formula (8):

Table 1. Chemical composition of high carbon tire cord steel.

Steel grade	C	P	S	Si	Mn	N	Ti
72A	0.72	0.01	0.008	0.2	0.5	0.004	0.0006
82A	0.82	0.01	0.008	0.2	0.5	0.004	0.0006
95A	0.95	0.01	0.008	0.2	0.5	0.004	0.0006

Table 2. The value of T_l and T_s for tire cord steel of different carbon content.

Steel grade	T_l	T_s	T_l - T_s
72A	1486	1386	100
82A	1480	1368	112
95A	1472	1345	127

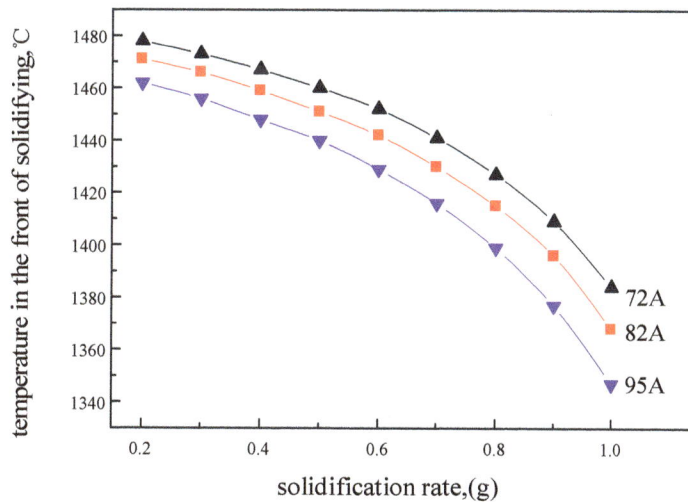

Figure 1. Relationship between the temperature in the front of solidifying and solidification rate at different carbon content of tire cord steel.

$$\lg f_{Ti} = \left(2557/T_{s-l} - 0.365\right)\lg f_{Ti(1873)} \quad (7)\,[4]$$

$$\lg f_{N} = \left(3280/T_{s-l} - 0.75\right)\lg f_{N(1873)} \quad (8)\,[4]$$

In formula (7) and formula (8), $f_{Ti(1873)}$ and $f_{N(1873)}$ are the activity coefficients of elements Ti and N at 1873 K respectively:

$$\lg f_{Ti(1873)} = \sum e_{Ti}^{j} w[j] \quad (9)$$

$$\lg f_{N(1873)} = \sum e_{N}^{j} w[j] \quad (10)$$

The interaction coefficients e_{Ti}^{j} and e_{N}^{j} at 1873 K can be obtained by literature [5,6].

Since the solute elements in the solidifying front enriched during solidification process continuously, so $w[j]$ can be calculated according to formula (9) and formula (10) [7]. It can be seen that $w[j]$ is the function of solidification rate (g).

$$w[j] = w[j]_0 \left(1-g\right)^{k-1} \quad (11)$$

$$w[j] = \frac{w[j]_0}{g(k-1)+1} \quad (12)$$

Formula (11) is used to calculate the quality percentage content of Ti, Si, Mn in the solidifying front and formula (12) is used to calculate the quality percentage content of C, N, P, S in the solidifying front; $w[j]_0$ is the initial percentage of solute element j in liquid steel. k is the equilibrium partition coefficients between liquid steel and γ-Fe which can be obtained by literature [8,9].

The actual activity product Q_{TiN} of Ti and N in the solidifying front of liquid steel can be calculated by formula (6) - formula (12), When the actual activity product Q_{TiN} of Ti and N in the solidifying front is greater than that of the equilibrium activity product K_{TiN}, TiN inclusion will precipitate in the solidifying front. If the initial

Ti, N content of tire cord steel are 0.0006% and 0.004% respectively, the temperature in the front of solidifying and solidification rate for 72A, 82A, 95A, when TiN inclusion precipitated in the solidifying front are shown in **Figure 2**.

Figure 2 shows that with the carbon content in tire cord steel increase, the temperature in the front of solidifying dropped when TiN inclusion precipitated during solidification and the required equilibrium activity product is also declined (**Figure 2(a)**). In addition, with the increase of carbon content in the tire cord steel, TiN inclusion is earlier to precipitate in the solidifying front (**Figure 2(b)**), Therefore, TiN inclusion is also more easily to grow up.

In conclusion, the carbon content has obvious effect on TiN inclusion precipitated in tire cord steel of different strength levels. The higher the carbon content in molten steel, the easier the TiN inclusion will precipitate and grow up during solidification process.

Therefore, in order to control the size of TiN inclusion precipitated in hypereutectoid tire cord steel of ultra high strength level, more strict measures of steelmaking and continuous casting process must be taken to further reduce the initial titanium and nitrogen content in liquid steel. Because the effect of N on the grown up of TiN inclusion is bigger than that of Ti [1], Therefore, it is particularly important to further reduce the N content in hypereutectoid tire cord steel.

The Ti content in the high carbon tire cord steel in molten steel can be reduced effectively through the following measures in the production practice; (1) reducing the TiO_2 content in blast furnace burden and silicon content in hot metal; (2) control of carbon conent and temperature in the bof(basic oxygen furnace) endpoint; (3) control of the amount of converter slag; (4) material control of LF(ladle furnace) refining slag and the TiO_2 content

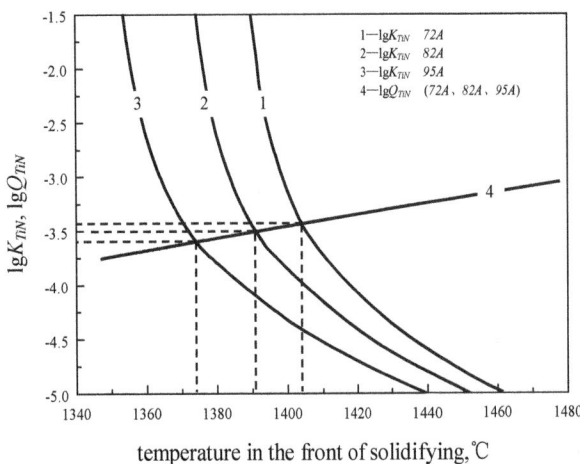

(a)

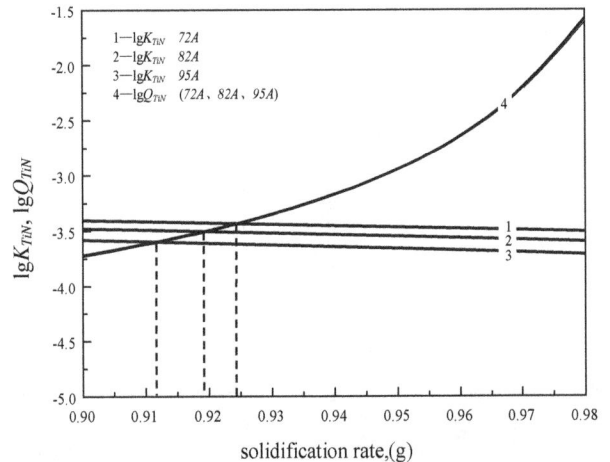

(b)

Figure 2. The effect of carbon content on TiN inclusion precipitated in tire cord steel.

in the tundish flux. The control of nitrogen in the tire cord steel are mainly from the following several aspects: (1) reducing the reblowing operation in endpoint; (2) using pre-melted slag to cover the liquid steel and reducing the inspiration during converter tapping; (3) using low nitrogen carburant; (4) control of LF refining slag and argon stirring; (5) further reducing the content of oxygen and nitrogen in liquid steel by vacuum treatment.

4. Conclusion

(1) The temperature before solidifying (T_{s-l}) decreased along with the development of the solidification process. When the solidification rate is the same, the higher the carbon content in the tire cord steel is, the lower the temperature before solidifying is.

(2) The carbon content has an obvious effect on TiN inclusion precipitated in tire cord steel of different strength levels. The higher the carbon content is in the molten steel, the easier the TiN inclusion will precipitate and grow up during the solidification process.

(3) In order to control the size of TiN inclusion precipitated in hypereutectoid tire cord steel of the ultra high strength level, more strict measures of steelmaking and the continuous casting process must be taken to further reduce the initial titanium and nitrogen content in liquid steel, especially the N content.

5. Acknowledgements

The author gratefully acknowledges the financial support for this work from the Science research plan (No. 201210321098) of Wuhan Science and Technology Bureau, China.

REFERENCES

[1] J. L. Lei, Z. L. Xue and Y. D. Jiang, "Study on TiN Precipitation during Solidification of Hypereutectoid Tire Cord Steel," *Metalurgia International*, Vol. 17, No. 9, 2012, pp. 10-15.

[2] I. Ohnaka, "Mathematical Analysis of Solute Redistribution during Solidification with Diffusion in Solid Phase," *Transaction ISIJ*, Vol. 26, No. 12, 1986, pp. 1045-1051.

[3] E. P. Chen, "Calculation Method and Empirical Formula for Melting Point of Fe-Based, Ni-Based and Co-Based Alloy," *Special Steel*, Vol. 13, No. 2, 1992, pp. 25-30.

[4] J. Fu, J. Zhu and L. Di, "Research on Precipitation Regularity of TiN in Micro Alloy Steel," *Acta Metallurgica Sinica*, Vol. 36, No. 8, 2000, pp. 801-804.

[5] L. K. Liang, "Metallurgical Thermodynamics and Kinetics," Northeast University Press, Shenyang, 1990, pp. 30-31.

[6] Z. T. Ma and J. Dieter, "Characteristic of Oxide Precipitation and Growth during Solidification of Deoxidized Steel," *ISIJ International*, Vol. 1, No. 38, 1988, pp. 46-52.

[7] Y. Qu, "Steelmaking Principles," Metallurgical Industry Press, Beijing, Vol. 8, 1983, pp. 294-313.

[8] Y. Ueshima, S. Mizoguchi and T. Matsumiya, "Analysis of Solute Distribution in Dendrites of Carbon Steel with δ/γ Transformation Solidification," *Metallurgical Transaction*, Vol. 17B, No. 4, 1986, pp. 845-859.

[9] P. A. Manohar, D. P. Dunne and T. Chandra, "Grain Growth Predictions in Microalloyed Steels," *ISIJ International*, Vol. 36, No. 2, 1996, p. 194.

27

Characterization of Pectin Nanocoatings at Polystyrene and Titanium Surfaces

Katarzyna Gurzawska[1,2*], Kai Dirscherl[3], Yu Yihua[4], Inge Byg[5], Bodil Jørgensen[5], Rikke Svava[5], Martin W. Nielsen[6], Niklas R. Jørgensen[1], Klaus Gotfredsen[2]

[1]Research Center for Ageing and Osteoporosis, Departments of Medicine and Diagnostics, Copenhagen University Hospital Glostrup, Glostrup, Denmark; [2]Institute of Odontology, Faculty of Health and Medical Sciences, University of Copenhagen, Copenhagen, Denmark; [3]Dansk Fundamental Metrologi A/S, Lyngby, Denmark; [4]Microtechnology and Surface Analysis, Danish Technological Institute, Taastrup, Denmark; [5]Department of Plant and Environment Sciences, Faculty of Science, University of Copenhagen, Frederiksberg, Denmark; [6]Department of Systems Biology, Technical University of Denmark, Lyngby, Denmark.

ABSTRACT

The titanium implant surface plays a crucial role for implant incorporation into bone. A new strategy to improve implant integration in a bone is to develop surface nanocoatings with plant-derived polysaccharides able to increase adhesion of bone cells to the implant surface. The aim of the present study was to physically characterize and compare polystyrene and titanium surfaces nanocoated with different Rhamnogalacturonan-Is (RG-I) and to visualize RG-I nanocoatings. RG-Is from potato and apple were coated on aminated surfaces of polystyrene, titanium discs and titanium implants. To characterize, compare and visualize the surface nanocoatings measurements of contact angle measurements and surface roughness with atomic force microscopy, scanning electron microscopy, and confocal microscopy was performed. We found that, both unmodified and enzymatic modified RG-Is influenced surface wettability, without any major effect on surface roughness (Sa, Sdr). Furthermore, we demonstrated that it is possible to visualize the pectin RG-Is molecules and even the nanocoatings on titanium surfaces, which have not been presented before. The comparison between polystyrene and titanium surface showed that the used material affected the physical properties of non-coated and coated surfaces. RG-Is should be considered as a candidate for new materials as organic nanocoatings for biomaterials in order to improve bone healing.

Keywords: Surface Properties; Titanium; Polystyrene; Rhamnogalacturonan-I; Osseointegration

1. Introduction

The implant surface plays a crucial role for implant incorporation into the bone and implant surface modifications which are continuously developed in attempts to enhance and accelerate bone formation at the implant surface [1-5]. The development has been approached by chemically and physically modifications of the surface [1,3-8]. The first concept focuses on incorporating inorganic and/or organic molecules at the surface whereas the second focuses on changing surface properties including the surface topography [3,9]. The chemical and physical surface modification can be performed at different levels [1,3,6]. From a biological point of view, the osseointegration process takes place at the cellular level,

and therefore especially micro and nanoscale investigations have great importance for developing new surfaces [4,6]. It has been demonstrated that nanoscale modification of titanium implants affects surface properties, such as hydrophilicity, biochemical bonding capacity and roughness, which influence cell behaviour on the surface such as adhesion, proliferation and differentiation of cells as well as the mineralization of the extracellular matrix at the implant surfaces [2,4-6,9-12].

The inorganic and organic nanocoatings are continuously developed and tested *in vitro* and *in vivo*. For *in vitro* examination, Tissue Culture Polystyrene Surfaces (TCPS) or titanium discs (Grade 2 or 4) are most frequently used [9], whereas for *in vivo* experiments titanium implant surfaces of Grade 4 titanium are the most frequently used [1,12]. To obtain the best conformity be-

---

*Corresponding author.

tween *in vitro* and *in vivo* studies and thereby the best prerequisite for interpretation of *in vitro* results, studies characterizing and comparing how nanocoatings influence the surfaces are important.

It has been shown that both polystyrene and titanium surfaces coated with Rhamnogalacturonan-I (RG-I) affect osteoblast cell responses. By enhancing osteoblast attachment, proliferation and mineralization compared to uncoated surfaces [9-11,13-17]. The biological mechanism when nanocoating RG-Is onto polystyrene and titanium surfaces is however still not fully understood. The positive biological effect might be connected to RG-Is' structure, but also to the change of surface properties caused by RG-Is nanocoatings.

The aim of the presented study was to physically characterize and compare polystyrene and titanium surfaces nanocoated with different Rhamnogalacturonan-Is (RG-Is) and to visualize the RG-I nanocoatings.

2. Materials and Methods

In order to characterize the RG-Is pectin nanocoating, three different types of material surfaces were used: 1) Tissue Culture Polystyrene Plates (TCPS), (Nunc, Roskilde, Denmark) with a diameter of 60 mm; 2) titanium discs (Ti discs) with a diameter of 13 mm (Astra Tech, Mölndal, Sweden); 3) titanium implants (Ti implants) with a diameter of 3.5 mm and a length of 8 mm (ANKYLOS, Dentsply, Konstanz, Germany). We included 7 different surfaces. Five were coated with RG-Is and two were non-coated (**Table 1**).

2.1. Coating Procedure

To obtain a covalent bonding between the surfaces and

Table 1. Surfaces and nanocoatings used in the study.

Surface		Material of the surface		
		Polystyrene	Titanium	
		TCPS	Ti discs	Ti implants
Non-coated	Untreated	+	+	+
	Aminated	+	+	+
Coated	PU	+	+	+
	PA	+	+	+
	PG	+	+	−
	PAG	+	+	−
	AU	+	+	+

+ used in study; − not used in the study; TCPS: Tissue Culture Polystyrene Surface Plates; Ti discs: Titanium discs, Ti implants: Titanium implants; PU: Potato unmodified RG-I, PA: Potato dearabinanated RG-I; PG: Potato degalactanated RG-I; PAG: Potato dearabinanated degalactanated RG-I; AU: Apple unmodified RG-I, RG-I: Rhamnogalacturonan-I.

the RG-I coatings all coated surfaces were aminated. Surface amination of Tissue Culture Polystyrene Plates (TCPS), titanium (Ti) discs and implants was performed by plasma polymerization of allylamine following the procedure described by Morra *et al.* [16] The RG-Is were covalently coupled via reaction between the carboxyl groups present in GalA of the RG-I backbone and the primary amino groups on the surface [16].

2.2. Physical Characterization

Physical characterization of TCPS and titanium surfaces was performed using contact angle measurements for wettability, scanning electron microscopy (SEM) for visualization of the surface texture and atomic force microscope (AFM) for surface roughness measurements and visualization of the nanocoating. The measurements were performed on 3 samples (n = 3) of the TCPS, the Ti discs and the Ti implants and at four different areas (m = 4) for each sample. The measurement areas of TCPS and Ti discs were selected randomly on the surface and at the Ti implants the measurements were performed at the top, valley and flanks according to recommendation by Wennerberg (2010) [18].

1) Contact angles (sessile angle) were measured using a KRUSS Drop Shape Analysis System, DSA10-Mk2 (Kruss GmbH, Hamburg, Germany). A water droplet (2.5 µL on TCPS plates and Ti discs surface and 1 µL on crest of Ti implants surface) was dropped on the surface and recorded photographically. Contact angles were measured in the recorded image by using drop shape analysis software, Scientific Drop Shape Analysis Software, DSA 1, Version 1.70 (Kruss GmbH, Hamburg, Germany) and a curve fitting method (Tangent Method-1).

2) Scanning Electron Microscopy (SEM) was performed with a ZEISS Ultra 55 scanning electron microscope (Carl Zeiss NTS GmbH, Oberkochen, Germany) operating at 3 kV and 20 kV in the secondary electron imaging mode. Images were collected at 1 K, 5 K, 10 K, 20 K, 30 K, 35 K, 40 K or 50 K.

3) Surface roughness was measured with a metrologic Atomic Force Microscope (AFM) DIM3100m (Bruker AXS Inc., Fitchburg, WI, USA). The AFM had a scan volume of 70 µm × 70 µm × 7 µm and intermittent scan mode (tapping mode) was used to minimize interaction force between tip and sample. The applied tip had a specified tip radius of 10 nm. The samples were scanned with a scan rate of 0.1 Hz to minimize scan artefacts in the image profiles. For the analysis of the roughness measurements, the scanned images were pre-processed with a first order lateral plane fit. This corrects for the residual sample tilt and no further filtering was applied. The parameters selected for analysis of surface roughness were S_a and S_{dr} using an area of 20 µm × 20 µm. [3] The same conditions and equipment were used for visualiza-

tion of RG-I nanocoatings on titanium implant surfaces, but the measurements were performed with 1 × 1 μm scan area.

2.3. Visualization of RG-Is Nanocoatings

1) For visualization of RG-Is structures Atomic Force Microscopy (AFM) imaging was performed using a multimode AFM (Bruker AXS Inc., Fitchburg, WI, USA) with a Nano V controller. RG-Is from potato unmodified (PU), potato dearabinanated (PA), potato degalactanated (PG), potato dearabinanated degalactanated (PAG) and apple unmodified (AU) were dispersed into mili-Q water to a concentration of 1 μg·mL^{-1}. Aliquots (10 mL^{-1}) of the diluted RG-Is samples were deposited onto a freshly-cleaved molecularly smooth surface [19-21] (Mica surfaces, Sa ~ 0 nm, Tedpella, Redding, CA, USA) and allowed to dry under ambient conditions before imaging by AFM in air [22]. The samples were scanned in intermittent "tapping" mode with commercial tips "SSS-NCH" from NanoSensors with a tip radius of 2 nm. Scannings was performed in ambient conditions. The scan rate was set to 0.5 Hz to limit the scan speed to 1 μm/s for a typical sized image of 1 μm × 1 μm. Using the image processing software SPIP (Image Metrology, Hørsholm, Denmark), the data was line-wise tilt-corrected with a first order fit restricted to the data points of the Mica surface only. This allows accurate height measurements relative to the flat Mica reference surface with measurement uncertainties below 1 nm.

2) Immunofluorescence labeling and confocal microscopy was performed on four implants, one non-coated aminated Ti implant and three coated Ti implants with (PU, PA, AU). The implants were placed in polystyrene 24-well plate (Nunc, Roskilde, Denmark) in separated wells and blocked for 15 min with 1 ml/well of 5% skimmed milk from Applichem (Darmstadt, Germany) (5% solution of fat-free milk powder in phosphate buffered saline (PBS), pH 7.2). Skimmed milk was removed from the well and 1 ml/well of anti—(1 → 4)-β-galactan LM5 (IgG2c) (PlantProbes, Leeds, UK) diluted 1:10 in 5% skimmed milk was added and placed on a shaker for 2 h. LM5 was removed from all wells and all implants were washed with 5% skimmed milk 3 times (after adding the milk to the wells, the plate was placed on a shaker for 5 min). Secondary antibody, goat anti-rat IgG for LM5 linked to FITC (fluorescein isothiocyanate) from Sigma-Aldrich (Brøndby, Denmark) was diluted 1:200 in 5% of skimmed milk and applied 1ml/well. The plate was covered by aluminum foil and placed on shaker for 2 hours. Subsequently. implants were washed three times. 1 ml of PBS was added to each well in order to store the implants until examination by confocal microscopy at 4 degrees. Confocal images were done with a Leica TCS-SP5 II confocal laser scanning microscope (Leica Mi-

crosystems, Exton, PA, USA) with PL Fluotar 10/x0.30 DRY objective with the same setting and conditions as described in our previous work [23].

2.4. Osteoblast Cell Culture

SaOS-2 osteoblast-like cells were grown on Ti implants under the same conditions as described in our previous studies [23]. The cell morphology observations were done with a Leica TCS-SP5 II confocal laser scanning microscope (Leica Microsystems, Exton, PA, USA).

2.5. Statistical Analysis

Descriptive statistics were calculated as mean values and standard errors of the mean. Results of surface analysis experiments were analysed using ANOVA tests and Bonferroni corrections for multiple comparisons using SPSS 11.5 software. A significance level of 0.05 was used throughout the study. More sensitive statistics between uncoated (untreated and aminated) samples and coated (PU, PA, PG, PAG, AU) samples, as well as between different type of surfaces (TCPS, Ti discs and Ti implants) was applied.

3. Results

3.1. Physical Characterization

3.1.1. Contact Angle
When the contact angles of the 3 surfaces (TCPS, Ti discs, Ti implants) were compared only significant differences were found between TCPS and Ti disc surfaces (p = 0.004). The highest SEM values were found for the titanium implants (**Figure 1**).

When all 3 different surfaces were compiled the con-

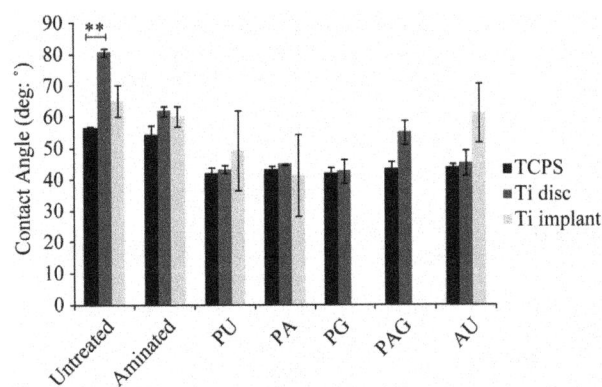

Figure 1. The contact angle (sessile angle) measurements results represent mean contact angle values and standard error of the mean (mean ± SEM). Uncoated surfaces: untreated, aminated and coated surfaces: PU (potato unmodified), PA (potato dearabinanated), PG (potato degalactanated), PAG (potato dearabinanated and degalactanated) and AU (apple unmodified) of TCPS, Ti discs and Ti implants surfaces. Ti: titanium.

tact angle measurements demonstrated that nanocoating with RG-Is (PU, PA, PG, PAG, AU), gave significantly (p = 0.006) lower contact angles compared to the non-coated control surfaces (untreated and aminated) (**Figure 1**). RG-I surfaces coated with PU (p = 0.02), PA (p = 0.01) and PG (p = 0.02) had significantly lower contact angles compare to the untreated surfaces.

When analyzing differences between non-coated and coated TCPS surfaces significant differences were found for all coatings (p < 0.001). For the Ti discs significant differences were found for all coatings compared to controls, except for PAG compared to aminated surfaces. For the titanium implants no significant differences between coated and non-coated surfaces (**Figure 1**).

3.1.2. Surface Roughness

The results for surface roughness (Sa, Sdr) measurements are shown in **Figures 2(a)** and **(b)**. In general the Ti implants were significantly more rough than the Ti discs, which were significantly more rough than the TCPS plates.

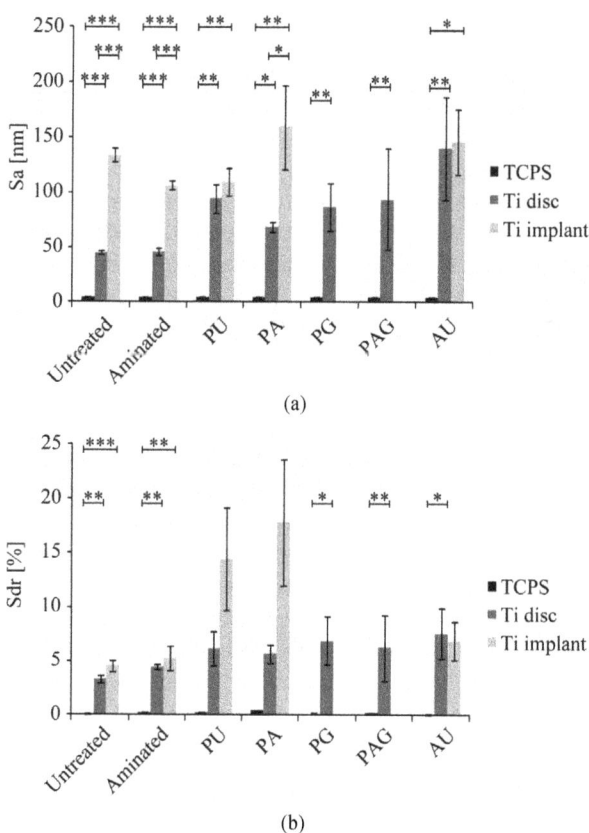

(a)

(b)

Figure 2. Surface roughness measured with AFM and represented by amplitude parameter (Sa) (a) and hybrid parameter (Sdr); (b) (means ± SEM). Untreated, aminated, PU (potato unmodified), PA (potato dearabinanated), PG (potato degalactanated), PAG (potato dearabinanated and degalactanated) and AU (apple unmodified) of TCPS, Ti discs and Ti implants surfaces. Ti: titanium.

When the 3 different surfaces were compiled no significant differences in Sa (p = 0.97) and Sdr (p = 0.86) values were found between coated and non-coated surfaces.

When analyzing differences for the TCPS surfaces, significant differences in Sa value were found for nanocating with PU (p = 0.02), PA (p = 0.003), PG (p = 0.001), PAG (p = 0.001) and AU (p = 0.001) compared to untreated TCPS surface, but not to the aminated surfaces. When analyzing differences in Sa value of non-coated and coated surfaces of Ti discs the only significant difference (p = 0.011) was found between AU coated and untreated Ti discs (p = 0.016) and aminated Ti discs (p = 0.018). When analyzing differences in Sdr values of uncoated and coated surfaces the only significant difference was found on TCPS surfaces (p = 0.013), between PU coated TCPS and non-coated TCPS surfaces (p = 0.026). No significant differences in Sa and Sdr values were found between coated and non-coated Ti implants.

3.1.3. Scanning Electron Microscopy (SEM)

The SEM images of TCPS, Ti disc and Ti implants (**Figure 3**) showed differences in surface texture. On the surface of titanium discs and implants, the texture pattern (machined surface) from the manufacturing process could be observed on the SEM images.

3.2. Visualization of Pectin Nanocoatings

3.2.1. RG-Is Structure Observed with AFM

The structure of potato RG-Is (scan area of 1 μm × 1 μm) is demonstrated at **Figure 4**. The enzymatic modification reduces the size of the pectin molecule. More than 70 measurements were conducted for each type for various individual pectins. The height of the potato RG-Is are shown in **Figure 5**. The results are sorted in decreasing order for better visualization of the height differences. After enzymatic modification of the arabinan and galactan side chains, the height of PU decreased significantly by approximately 3 nm on average (**Figure 5**).

3.2.2. RG-Is Nanocoating on Titanium Implant Surfaces Visualized with AFM

A representative 3D image of PA RG-I nanocoating is shown in **Figure 6** with linear structures and a heterogenous distribution of the RG-I molecules on the surface of the Ti implants.

3.2.3. Immunofluorescence Labeling and Confocal Microscopy

The confocal images showed presence of RG-Is nanocoating on the coated titanium implant surface compared to control aminated titanium implant surface (**Figure 7**).

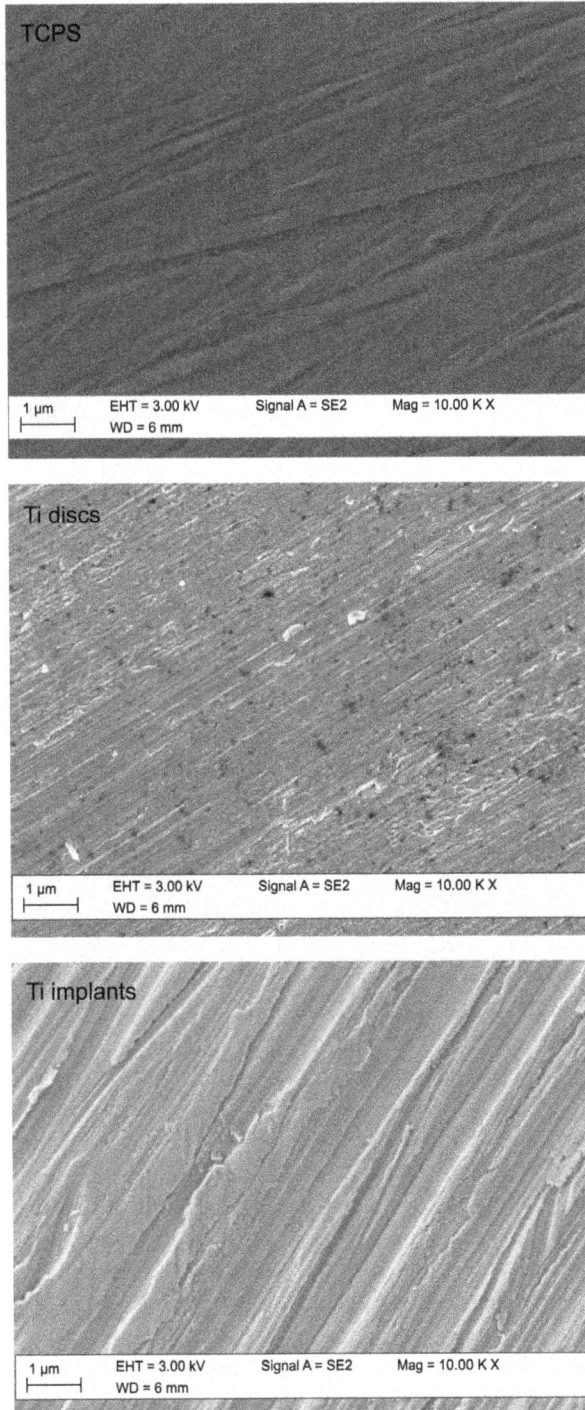

Figure 3. Representative images of untreated control surface of TCPS, Ti disc and Ti implant performed with SEM at 3 kV. TCPS: Tissue Polystyrene Plate, Ti: titanium, WD: working distance, EHT: the high voltage, SE2: type of detector.

3.3. Osteoblast Cell Culture

The confocal images from the Ti implants cultured with SaO2 cells (**Figure 8**) showed spread morphology of the osteoblast-like cells on RG-Is nanocoated titanium im-

plants as well as on the control aminated Ti implant.

4. Discussion

In this work, we assessed the effect on physical properties of nanocoatings with potato and apple RG-Is to polystyrene and titanium surfaces. We found that native (PU, AU) and modified RG-Is (PA, PG, PAG) influenced surface wettability, without any major influence on surface roughness (Sa, Sdr). Furthermore, we demonstrated that it is possible to visualize the pectin molecules and even the nanocoatings on titanium surfaces, which have not been presented before. The comparison between polystyrene and titanium surfaces showed that both materials became more hydrophilic after nanocoating and that the RG-I molecules did not have any major effect on the surface roughness.

In accordance with the present study, a number of other studies have demonstrated that RG-Is nanocoatings influenced the physical properties of polystyrene [11,13, 16,17] and titanium surfaces [10,14,15,23]. According to findings by Morra *et al.* (2004), the difference in wettability between RG-I coated and non-coated polystyrene surfaces is caused by changes in the chemical composition after coating with RG-Is [16]. The same results were presented in a work by Kokonnen *et al.* (2006), where the authors proposed that the RG-Is' side chains produce a hydrated gel-like surface [11]. Our previous work also demonstrated that by changing the chemical composition of the surface with RG-Is nanocoatings the wettability of the polystyrene surface was affected and the surface became more hydrophilic compared to non-coated control surfaces [14]. The same results were observed on titanium surfaces showing that coating with RG-Is gave rise to smaller contact angles compared to the controls *i.e.* more hydrophilic surfaces [10,14,23]. In our study the comparison of different surface TCPS, Ti discs and Ti implants coated with RG-Is confirmed these findings. However at the titanium implants there was a limited access to measurements, which may explain the lack of significant difference in the contact angle between coated and non-coated surfaces. The increased hydrophilicity obtained by nanocoating with RG-Is seemed to be similar at the TCPS, Ti discs and titanium implants. The fact that RG-I coatings created a more hydrophilic surface can have a positive impact on osseointegration, as hydrophilic surfaces are more suitable for interaction with biological fluids, cells and tissues than a more hydrophobic surface [24]. In our study, a significant decrease in the contact angle on the surface coated with PU, PA and PG was shown, compared to the untreated control surface. Therefore, these RG-Is should be considered for further *in vitro* and *in vivo* studies.

The change in wettability of the surface has been reported to be related not only to chemical modification but

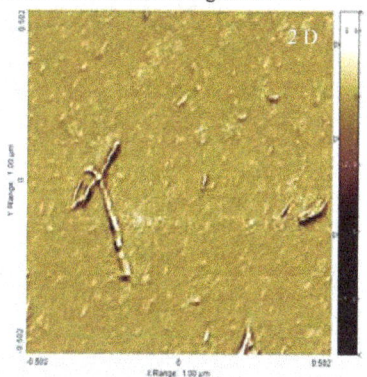

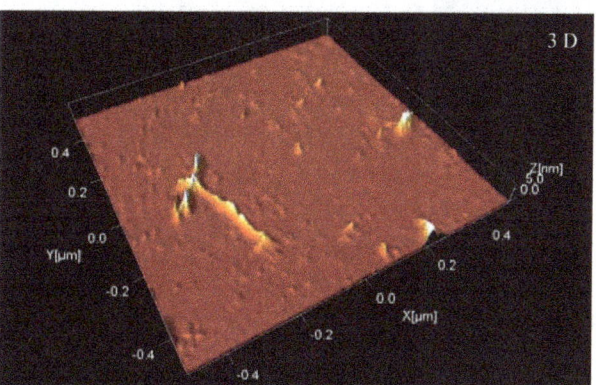

Figure 4. Representative images (2D and 3D) of RG-Is structure: potato unmodified (PU), potato dearabinanated (PA), potato degalactanated (PG), potato dearabinanated degalactanated (PAG) on mica surface measured with a Multimode AFM 1 μm × 1 μm.

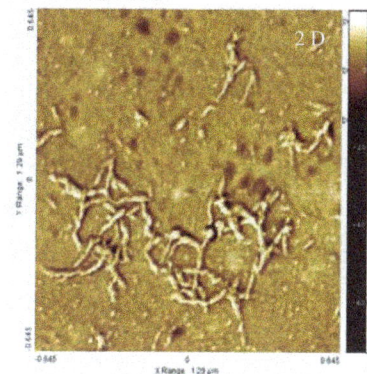

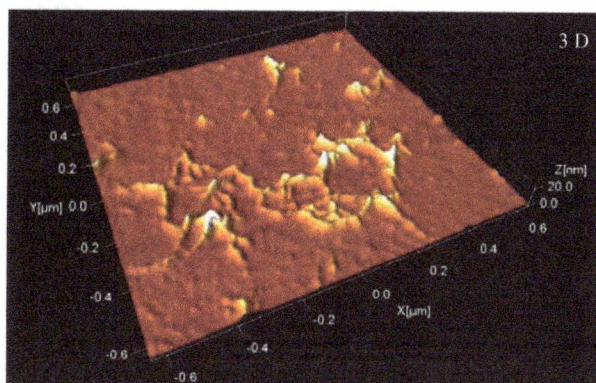

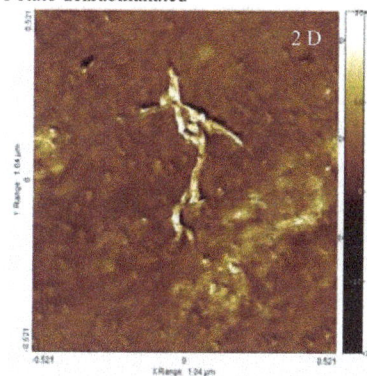

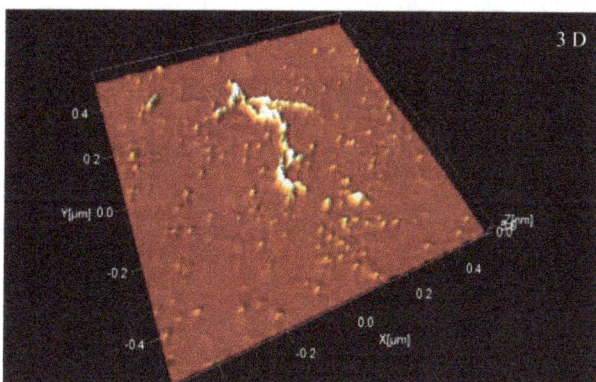

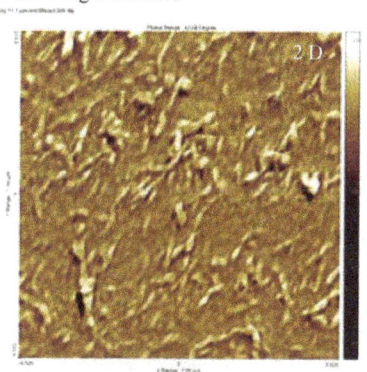

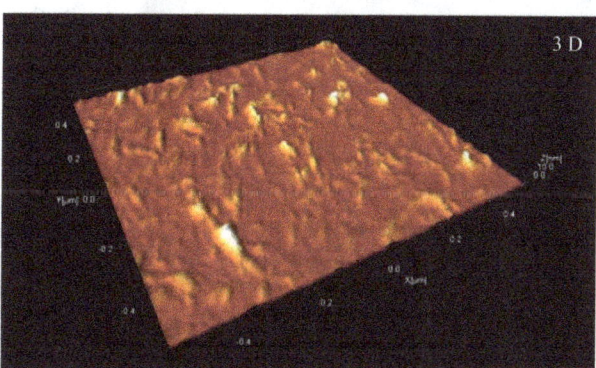

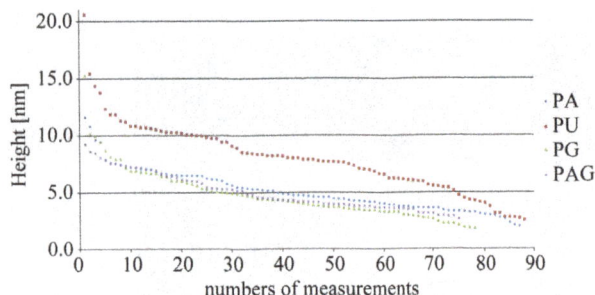

Figure 5. Distribution plot of height measurements in nm of RG-Is structure of PU, PA, PG, PAG performed with AFM on Mica surface with 1 μm × 1 μm magnification. Potato unmodified (PU), potato dearabinanated (PA) and potato degalactanated (PG) and potato dearabinanated degalactanated (PAG).

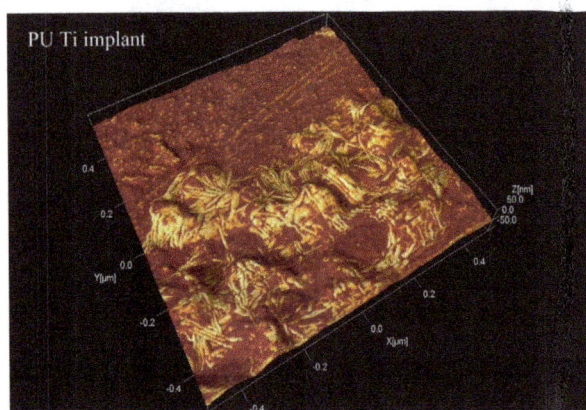

Figure 8. Representative confocal images of SaOS-2 cells stained with Vybrant Cell-Labeling Solutions and cultured on Ti implants surface nanocoated with RG-Is. Control: aminated Ti implant, Coated: PU Ti implant; Ti: titanium, PU: potato unmodified.

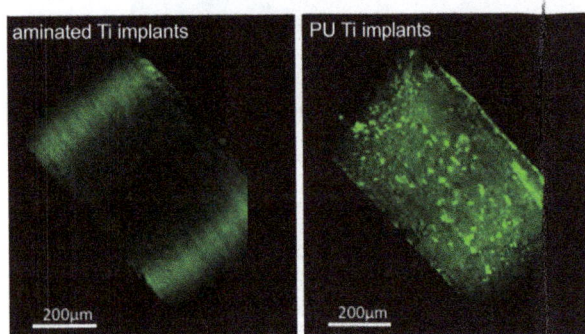

Figure 6. Representative image (3D) of RG-I nanocoating on aminated titanium implant surface coated with potato dearabinanated (PA aminated Ti implant), magnification 1 × 1 μm, measured using AFM with intermittent "Tapping mode".

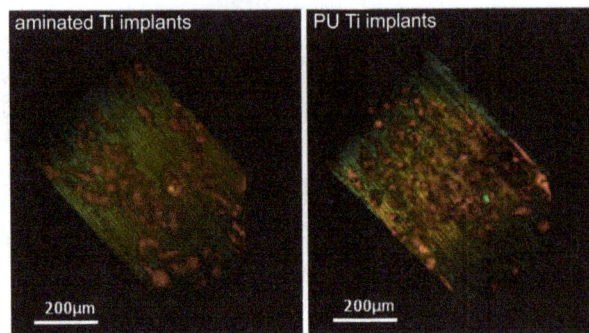

Figure 7. Representative confocal images of RG-Is nanocoating visualised with immunofluorescence labeling by primary antibody, anti –(1 → 4)-β-galactan LM5 (IgG2c) and secondary antibody, goat anti-rat IgG for LM5 linked to FITC. Control: aminated Ti implant, Coated: PU Ti implant; Ti: titanium, PU: potato unmodified.

also to the topographical changes of the surface [25,26]. Our surface roughness results showed that the RG-Is used for nanocoating in general did not significantly affect the roughness of the examined surfaces, which is in-

accordance with findings from Morra *et al.* (2004) [16]. The reason for different results of surface roughness (Sa, Sdr) on polystyrene and titanium surface, when each of the surface groups was compared, can be explained by differences in surface texture, illustrated by the SEM images. It has also to be noticed that the surface roughness was measured by AFM on a 20 μm × 20 μm square area as recommended for a non quantitative overview of the nanotopography [27]. Higher magnification would probably show difference between nanocoated and non-coated surfaces as the RG-I molecules used for coating are around 100 nm in size. Thus, our measurements performed with AFM on a 1 μm × 1 μm area clearly visualized the RG-Is nanocoating also on the titanium implant surface. Nanocoatings of that size have not previously been demonstrated. The visualization of the nanocoating may be important for characterizing the physical properties of the nanocoated surface. The structure, as well as the distribution and the topography of the nanocoating, may play an important role in cell adhesion, as the nanocoating can mimic the extracellular matrix (ECM) [28]. In the present study, we also visualized RG-Is nanocoating on titanium implants by immunofluorescence staining using the primary LM5 antibody, which specifically binds to galactan side chains. By using the AFM technique with atomically smooth surfaces, we were able to visualize and analyze the height and length of RG-Is from potato. The length of the individual molecule was in the range of 100 nm, which is in agreement with other studies [22]. The height measurements showed a decrease in the height of modified RG-Is (PA, PG, PAG) compared to unmodified RG-Is (PU), which showed that enzymatic modification changed the RG-Is structure. This corresponds to our previous findings, demonstrating that enzymatic treatment of RG-Is decreases the amount of galactan and arabinan in modified RG-Is compared to unmodified RG-Is (PU). [14] Our height measurements of modified RG-Is have not allowed us to distinguish between dearabinanated, degalactananated and debranched structures and therefore more detailed investigations with AFM in "liquid cell" (AFM imaging in liquids) are necessary [29].

The osteblast-like cells (SaOS2) grown on Ti implants were spread on the surface, however the morphology, cell viability and proliferation studies remain to be performed to examine the osteoblast behavior on titanium implant surface coated with RG-Is. On the other hand, our previous studies showed a significant increase in cell viability and matrix mineralization of the same type of cells (SaOS2) on polystyrene and titanium discs surfaces coated with RG-Is containing higher amounts of galactose compared to controls [14]. Recent *in vitro* studies [10,11,14-16] showed that the nanocoating of RG-I with high amounts of galactose enhanced osteoblast spreading and growth, in contrast to the nanocoating of RG-I with high amounts of arabinose, which leads to aggregation and decreased proliferation [11,13,14,16,17]. This finding suggests that linear 1.4-linked galactans are important for osteoblast adhesion, and that high content of arabinose can shield the galactans, and thus prohibiting their interaction with osteoblasts [30]. It has previously been shown that Galectin-3 binds specifically to galactose residues [31]. As the osteoblastic cells contain Galectin-3, it could therefore be speculated that osteoblast interaction with the RG-I galactans is mediated through Galectin-3. In addition, titanium surfaces coated with RG-Is have been shown to positively influence cell adhesion, morphology, proliferation and mineralization [10,15,23]. The positive cell response on plant-derived molecules, especially RG-I with high amount of galactose, opens new direction in the development of organic nanocoatings for biomaterials in order to improve bone healing.

5. Acknowledgements

The authors thank Prof. Erik Fink Eriksen from Aarhus Amtssygehus in Denmark for the gift of the SaOS-2 cell line and Marco Morra for the aminated TCPS plates, Ti discs and Ti implants.

REFERENCES

[1]   T. Albrektsson and A. Wennerberg, "Oral Implant Surfaces: Part 1—Review Focusing on Topographic and Chemical Properties of Different Surfaces and *in Vivo* Responses to Them," *The International Journal of Prosthodontics*, Vol. 17, No. 5, 2004, pp. 536-543.

[2]   T. Albrektsson and A. Wennerberg, "Oral Implant Surfaces: Part 2—Review Focusing on Clinical Knowledge of Different Surfaces," *The International Journal of Prosthodontics*, Vol. 17, No. 5, 2004, pp. 544-564.

[3]   D. M. Dohan Ehrenfest, P. G. Coelho, B. S. Kang, Y. T. Sul and T. Albrektsson, "Classification of Osseointegrated Implant Surfaces: Materials, Chemistry and To-

pography," *Trends in Biotechnology*, Vol. 28, No. 4, 2010, pp. 198-206.

[4]   G. Mendonca, D. B. Mendonca, F. J. Aragao and L. F. Cooper, "Advancing Dental Implant Surface Technology—From Micron- to Nanotopography," *Biomaterials*, Vol. 29, No. 28, 2008, pp. 3822-3835.

[5]   M. Morra, "Biochemical Modification of Titanium Surfaces: Peptides and ECM Proteins," *European Cells and Materials Journal*, Vol. 12, 2006, pp. 1-15.

[6]   L. T. de Jonge, S. C. Leeuwenburgh, J. G. Wolke and J. A. Jansen, "Organic-Inorganic Surface Modifications for Titanium Implant Surfaces," *Pharmaceutical Research*, Vol. 25, No. 10, 2008, pp. 2357-2369.

[7]   G. L. Le, A. Soueidan, P. Layrolle and Y. Amouriq, "Surface Treatments of Titanium Dental Implants for Rapid Osseointegration," *Dental Materials*, Vol. 23, No. 7, 2007, pp. 844-854.

[8]   R. Junker, A. Dimakis, M. Thoneick and J. A. Jansen, "Effects of Implant Surface Coatings and Composition on Bone Integration: A Systematic Review," *Clinical Oral Implants Research*, Vol. 20, Suppl. 4, 2009, pp. 185-206.

[9]   K. Gurzawska, R. Svava, N. R. Jørgensen and K. Gotfredsen, "Nanocoating of Titanium Implant Surfaces with Organic Molecules. Polysaccharides Including Glycosaminoglycans," *Journal of Biomedical Nanotechnology*, Vol. 8, No. 6, 2012, pp. 1012-1024.

[10] H. Kokkonen, C. Cassinelli, R. Verhoef, M. Morra, H. A. Schols and J. Tuukkanen, "Differentiation of Osteoblasts on Pectin-Coated Titanium," *Biomacromolecules*, Vol. 9, No. 9, 2008, pp. 2369-2376.

[11] H. E. Kokkonen, J. M. Ilvesaro, M. Morra, H. A. Schols and J. Tuukkanen, "Effect of Modified Pectin Molecules on the Growth of Bone Cells," *Biomacromolecules*, Vol. 8, No. 2, 2007, pp. 509-515.

[12] A. Wennerberg and T. Albrektsson, "On Implant Surfaces: A Review of Current Knowledge and Opinions," *The International Journal of Oral & Maxillofacial Implants*, Vol. 25, No. 1, 2010, pp. 63-74.

[13] C. Bussy, R. Verhoef, A. Haeger, M. Morra, J. L. Duval, P. Vigneron, *et al.*, "Modulating *in Vitro* Bone Cell and Macrophage Behavior by Immobilized Enzymatically Tailored Pectins," *Journal of Biomedical Materials Research Part A*, Vol. 86A, No. 3, 2008, pp. 597-606.

[14] K. Gurzawska, R. Svava, S. Syberg, Y. Yihua, K. B. Haugshoj, I. Damager, *et al.*, "Effect of Nanocoating with Rhamnogalacturonan-I on Surface Properties and Osteoblasts Response," *Journal of Biomedical Materials Research Part A*, Vol. 100, No. 3, 2012, pp. 654-664.

[15] H. Kokkonen, H. Niiranen, H. A. Schols, M. Morra, F.

Stenback and J. Tuukkanen, "Pectin-Coated Titanium Implants Are Well-Tolerated *in Vivo*," *Journal of Biomedical Materials Research Part A*, Vol. 93, No. 4, 2010, pp. 1404-1409.

[16] M. Morra, C. Cassinelli, G. Cascardo, M. D. Nagel, C. Della Volpe, S. Siboni, *et al.*, "Effects on Interfacial Properties and Cell Adhesion of Surface Modification by Pectic Hairy Regions," *Biomacromolecules*, Vol. 5, No. 6, 2004, pp. 2094-2104.

[17] M. D. Nagel, R. Verhoef, H. Schols, M. Morra, J. P. Knox, G. Ceccone, *et al.*, "Enzymatically-Tailored Pectins Differentially Influence the Morphology, Adhesion, Cell Cycle Progression and Survival of Fibroblasts," *Biochimica et Biophysica Acta*, Vol. 1780, No. 7, 2008, pp. 995-1003.

[18] A. Wennerberg and T. Albrektsson, "Suggested Guide-Lines for the Topographic Evaluation of Implant Surfaces," *The International Journal of Oral & Maxillofacial Implants*, Vol. 15, No. 3, 2000, pp. 331-344.

[19] L. Cheng, P. Fenter, K. L. Nagy, M. L. Schlegel and N. C. Sturchio, "Molecular-Scale Density Oscillations in Water Adjacent to a Mica Surface," *Physical Review Letters*, Vol. 87, No. 15, 2001, Article ID: 156103.

[20] T. Matsuura, H. Tanaka, T. Matsumoto and T. Kawai, "Atomic Force Microscopic Observation of *Escherichia coli* Ribosomes in Solution," *Bioscience, Biotechnology, and Biochemistry*, Vol. 70, No. 1, 2006, pp. 300-302.

[21] C. H. Lui, L. Liu, K. F. Mak, G. W. Flynn, T. F. Heinz, "Ultraflat Graphene," *Nature*, Vol. 462, No. 7271, 2009, pp. 339-341.

[22] K. T. Inngjerdingen, T. R. Patel, X. Chen, L. Kenne, S. Allen, G. A. Morris, *et al.*, "Immunological and Structural Properties of a Pectic Polymer from *Glinus oppositifolius*," *Glycobiology*, Vol. 17, No. 12, 2007, pp. 1299-1310.

[23] K. A. Gurzawska, R. Svava, Y. Yihau Dr., K. B. Haugshøj, K. Dirscherl, S. B. Levery, I. Byg, I. Damager, B. Jørgensen, N. R. Jørgensen and K. Gotfredsen, "Osteoblastic Response to Pectin Nanocoating of Titanium Surface," Submitted, under Review.

[24] S. Tajima, J. S. Chu, S. Li and K. Komvopoulos, "Dif-

ferential Regulation of Endothelial Cell Adhesion, Spreading, and Cytoskeleton on Low-Density Polyethylene by Nanotopography and Surface Chemistry Modification Induced by Argon Plasma Treatment," *Journal of Biomedical Materials Research Part A*, Vol. 84, No. 3, 2008, pp. 828-836.

[25] F. Rupp, L. Scheideler, N. Olshanska, M. de Wild, M. Wieland and J. Geis-Gerstorfer, "Enhancing Surface Free Energy and Hydrophilicity through Chemical Modification of Microstructured Titanium Implant Surfaces," *Journal of Biomedical Materials Research Part A*, Vol. 76A, No. 2, 2006, pp. 323-334.

[26] S. Tosatti, M. Textor and N. D. Spencer, "Self-Assembled Monolayer of Dodecyl and Hydroxy-Dodecyl Phosphate at Smooth and Rough Titanium and Titanium Oxide Surfaces," *Langmuir*, Vol. 18, No. 9, 2002, pp 3537-3548.

[27] A. Bagno and B. C. Di, "Surface Treatments and Roughness Properties of Ti-Based Biomaterials," *Journal of Materials Science: Materials in Medicine*, Vol. 15, No. 9, 2004, pp. 935-949.

[28] F. Munarin, S. G. Guerreiro, M. A. Grellier, M. C. Tanzi, M. A. Barbosa, P. Petrini, *et al.*, "Pectin-Based Injectable Biomaterials for Bone Tissue Engineering," *Biomacromolecules*, Vol. 12, No. 3, 2011, pp. 568-577.

[29] V. J. Morris, A. Gromer, A. R. Kirby, R. J. M. Bongaerts and A. Patrick Gunning, "Using AFM and Force Spectroscopy to Determine Pectin Structure and (bio) Functionality," *Food Hydrocolloids*, Vol. 25, No. 2, 2011, pp. 230-237.

[30] H. Kokkonen, R. Verhoef, K. Kauppinen, V. Muhonen, B. Jorgensen, I. Damager, *et al.*, "Affecting Osteoblastic Responses with *in Vivo* Engineered Potato Pectin Fragments," *Journal of Biomedical Materials Research Part A*, Vol. 100A, No. 1, 2012, pp. 111-119.

[31] A. P. Gunning, R. J. Bongaerts and V. J. Morris, "Recognition of Galactan Components of Pectin by Galectin-3," *The FASEB Journal*, Vol. 23, No. 2, 2009, pp. 415-424.

Permissions

The contributors of this book come from diverse backgrounds, making this book a truly international effort. This book will bring forth new frontiers with its revolutionizing research information and detailed analysis of the nascent developments around the world.

We would like to thank all the contributing authors for lending their expertise to make the book truly unique. They have played a crucial role in the development of this book. Without their invaluable contributions this book wouldn't have been possible. They have made vital efforts to compile up to date information on the varied aspects of this subject to make this book a valuable addition to the collection of many professionals and students.

This book was conceptualized with the vision of imparting up-to-date information and advanced data in this field. To ensure the same, a matchless editorial board was set up. Every individual on the board went through rigorous rounds of assessment to prove their worth. After which they invested a large part of their time researching and compiling the most relevant data for our readers. Conferences and sessions were held from time to time between the editorial board and the contributing authors to present the data in the most comprehensible form. The editorial team has worked tirelessly to provide valuable and valid information to help people across the globe.

Every chapter published in this book has been scrutinized by our experts. Their significance has been extensively debated. The topics covered herein carry significant findings which will fuel the growth of the discipline. They may even be implemented as practical applications or may be referred to as a beginning point for another development. Chapters in this book were first published by Scientific Research Publishing Inc.; hereby published with permission under the Creative Commons Attribution License or equivalent.

The editorial board has been involved in producing this book since its inception. They have spent rigorous hours researching and exploring the diverse topics which have resulted in the successful publishing of this book. They have passed on their knowledge of decades through this book. To expedite this challenging task, the publisher supported the team at every step. A small team of assistant editors was also appointed to further simplify the editing procedure and attain best results for the readers.

Our editorial team has been hand-picked from every corner of the world. Their multi-ethnicity adds dynamic inputs to the discussions which result in innovative outcomes. These outcomes are then further discussed with the researchers and contributors who give their valuable feedback and opinion regarding the same. The feedback is then collaborated with the researches and they are edited in a comprehensive manner to aid the understanding of the subject.

Apart from the editorial board, the designing team has also invested a significant amount of their time in understanding the subject and creating the most relevant covers. They scrutinized every image to scout for the most suitable representation of the subject and create an appropriate cover for the book.

The publishing team has been involved in this book since its early stages. They were actively engaged in every process, be it collecting the data, connecting with the contributors or procuring relevant information. The team has been an ardent support to the editorial, designing and production team. Their endless efforts to recruit the best for this project, has resulted in the accomplishment of this book. They are a veteran in the field of academics and their pool of knowledge is as vast as their experience in printing. Their expertise and guidance has proved useful at every step. Their uncompromising quality standards have made this book an exceptional effort. Their encouragement from time to time has been an inspiration for everyone.

The publisher and the editorial board hope that this book will prove to be a valuable piece of knowledge for researchers, students, practitioners and scholars across the globe.

List of Contributors

Jian-Rui Liu, Yi-Na Guo, Wei-Dong Huang
State Key Laboratory of Solidification Processing, Northwestern Polytechnical University, Xi'an, P. R. China

Nambi Muthukrishnan
Department of Mechanical Engineering, Sri Venkateswara College of Engineering, Sriperumbudur, India

Paulo Davim
Department of Mechanical Engineering, Campus Santiago, University of Aveiro, Averio, Portugal

Yoann Joliff, Lénaïk Belec and Jean-François Chailan
MAPIEM, EA 4323, Institut des Sciences de l'Ingénieur de Toulon et du Var, Cedex, France

Umeshkumar P. Khairnar
Department of Physics, S.S.V.P.S. ACS College, Shindkheda, Dhule, India

Sulakshana S. Behere and Panjabrao H. Pawar
Thin Film Laboratory, Department of Physics, Zulal Bhilajirao Patil College, Dhule, India

Bhujang Mutt Girish, Bhujang Mutt Satish and Hanyalu Ramegowda Vitala
Research and Development Center, Department of Mechanical Engineering, East Point College of Engineering and Technology, Bangalore, Karnataka, India

Zhifeng Liu, Liuxian Zhao, Jun Zhong, Xinyu Li and Huanbo Cheng
School of Mechanical and Auto Engineering, Hefei University of Technology, Hefei, China

Nasr-Eddine Belkhouche and Nacera Benyahia
Laboratory of Separation and Purification Technologies, Department of Chemistry-Faculty of Sciences, Tlemcen University, Algeria

Hanène Bedis
UMAO, Faculté des Sciences de Tunis, Campus Universitaire, Tunis, Tunisia; 2ITODYS, 15 Rue Jean-Antoine de Baïf, Paris, France

Vikas Patil, Shailesh Pawar, Manik Chougule, Prasad Godse, Ratnakar Sakhare and Pradeep Joshi
Materials Research Laboratory, School of Physical Sciences, Solapur University, Solapur, India

Shashwati Sen
Crystal Technology Section, Bhabha atomic Research Centre, Mumbai, India

Seung H. Yoon, Won T. Jeong and Kyung C. Kim
Mechanical Engineering, Pusan National University, Pusan, Korea

Kyung J. Kim and Min C. Oh
Electrical Engineering, Pusan National University, Pusan, Korea

Sang M. Lee
Pusan National University, Pusan, Korea

N. A. El Mahallawy
The Design and Production Engineering Department, Faculty of Engineering, Ain Shams University, Cairo, Egypt;

M. A. Shoeib
Surface Coating Department, Central Metallurgical Research & Development Institute, Helwan, Cairo, Egypt

M. H. Abouelenain
Petroleum Marine Service, Cairo, Egypt

Yves Gensterblum
Institute of Geology and Geochemistry of Petroleum and Coal, RWTH Aachen University, Lochnerstr Aachen, Germany

Md. Aminul Islam and Zoheir Farhat
Department of Process Engineering and Applied Science, Dalhousie University, Halifax, Canada

Jonathon Bonnell
Vector Aerospace, Engine Services- Atlantic, Summerside, Canada

Ahmed Ibrahim
Department of Mechanical Engineering, Farmingdale State College, Farmingdale, New York, USA

Abdel Salam Hamdy
Max Planck Institute of Colloids and Interfaces, Am Mühlenberg, Germany

Magdi F. Morks, Ivan Cole and Penny Corrigan
CSIRO Division of Materials Science and Engineering, Clayton South, Victoria, Australia

Akira Kobayashi
Joining and Welding Research Institute, Osaka University, Osaka, Japan

Abodol Rasoul Sohouli, Ali Maozemi Goudarzi and Reza Akbari Alashti
Department of Mechanical Engineering, Babol Noshirvani University of Technology, Babol

I. V. S. Yashwanth
M. V. S. R. Engineering College, Nadargul, Hyderabad, India

Gurrappa
Defence Metallurgical Research Laboratory, Kanchanbagh PO, Hyderabad, India

I. H. Murakami
National Institute for Materials Science, Ibaraki, Japan

Parthasarathy Sampathkumaran and Subramanyam Seetharamu
Materials Technology Division, Central Power Research Institute, Bangalore, India

Chikkakuntappa Ranganathaiah and Jaya Madhu Raj
Department of Studies in Physics, University of Mysore, Mysore, India

Pradeep Kumar Pujari, Priya Maheshwari and Debashish Dutta
Radiochemistry Division, Bhabha Atomic Research Centre, Mumbai, India

Kishore
Department of Materials Engineering, Indian Institute of Science, Bangalore, India

Soheyl Soleymani, Amir Abdollah-zadeh and Sima Ahmad Alidokht
Department of Materials Engineering, Tarbiat Modares University, Tehran, Iran

I. Gurrappa and A. K. Gogia
Defence Metallurgical Research Laboratory, Kanchanbagh PO, Hydereabad

I. V. S. Yashwanth
M.V.S.R. Engineering College, Nadargul, Hyderabad, India

Dongyan Tang, Jie Liu and Fan Yang
Department of Chemistry, Harbin Institute of Technology, Harbin, China

Weiwei Cui
College of Materials Science and Engineering, Harbin University of Science and Technology, Harbin, China

Narayan Chandra Debnath
Department of Physics, Institute of Chemical Technology (ICT), Mumbai, India

Alexander A. Shklyaev, Konstantin N. Romanyuk and Alexander V. Latyshev
A. V. Rzhanov Institute of Semiconductor Physics of SB RAS, Novosibirsk, Russia
Novosibirsk State University, Novosibirsk, Russia

Prashant Jindal
University Institute of Engineering & Technology, Panjab University, Chandigarh, India

Meenakshi Goyal
University Institute of Chemical Engineering & Technology, Panjab University, Chandigarh, India

Navin Kumar
Indian Institute of Technology, Roopnagar, Punjab, India

Yuedong Jiang
Key Laboratory for Ferrous Metallurgy and Resources Utilization of Ministry of Education, Wuhan University
of Science and Technology, Wuhan, China
Research and Development Center of Wuhan Iron & Steel Group, Wuhan, China

Jialiu Lei, Jing Zhang, Rui Xiong, Feng Zou and Zhengliang Xue
Key Laboratory for Ferrous Metallurgy and Resources Utilization of Ministry of Education, Wuhan University
of Science and Technology, Wuhan, China

Niklas R. Jørgensen
Research Center for Ageing and Osteoporosis, Departments of Medicine and Diagnostics, Copenhagen University
Hospital Glostrup, Glostrup, Denmark

Klaus Gotfredsen
Institute of Odontology, Faculty of Health and Medical Sciences, University of Copenhagen, Copenhagen, Denmark

Kai Dirscherl
Dansk Fundamental Metrologi A/S, Lyngby, Denmark

Yu Yihua
Microtechnology and Surface Analysis, Danish Technological Institute, Taastrup, Denmark

Inge Byg, Bodil Jørgensen and Rikke Svava
Department of Plant and Environment Sciences, Faculty of Science, University of Copenhagen, Frederiksberg,
Denmark

Martin W. Nielsen
Department of Systems Biology, Technical University of Denmark, Lyngby, Denmark

Katarzyna Gurzawska
Research Center for Ageing and Osteoporosis, Departments of Medicine and Diagnostics, Copenhagen University
Hospital Glostrup, Glostrup, Denmark
Institute of Odontology, Faculty of Health and Medical Sciences, University of Copenhagen, Copenhagen, Denmark

www.ingramcontent.com/pod-product-compliance
Lightning Source LLC
Chambersburg PA
CBHW050453200326
41458CB00014B/5167

9 781632 404756